The Crisis of Religious Symbolism

&

Symbolism & Reality

THE CRISIS OF RELIGIOUS SYMBOLISM

&

SYMBOLISM & REALITY

by
JEAN BORELLA

Translated by
G. John Champoux

Originally published in French as
La Crise du Symbolisme Religieux

Angelico Press /Sophia Perennis 2016

Series Editor: James R. Wetmore

For information, address:
Angelico Press
4709 Briar Knoll Dr.
Kettering, OH 45429
www.angelicopress.com

978-1-62138-191-4 Pb
978-1-62138-192-1 Cloth

Cover design: Michael Schrauzer

CONTENTS

IN ITS ESSENCE

Note on the New French Edition

This edition of *The Crisis of Religious Symbolism* differs from the previous one, published in 1990 by Editions l'Age d'Homme, on a certain number of points. Carefully reviewed with the help of Pierre-Marie Sigaud and Bruno Bérard, the text has been cleared of faults (in spelling and syntax) that marred it. As for the history of doctrines, two or three paragraphs were rewritten after some of the information proved to be untrue. Similarly, the evolution of certain religious institutions has led to a change of opinion.

Some metaphysical notions have been clarified and reformulated. Two new appendices deal briefly with questions hardly mentioned in the main text. Finally, and without any claim to being exhaustive, the bibliography has been enlarged and updated whenever our readings have enabled us to do so. This edition is therefore the definitive version of our work

Translator's Preface

Combined here in one volume are two books, *The Crisis of Religious Symbolism*, and *Symbolism and Reality.* Although published seven years apart, these two works are integral to one another. *Symbolism and Reality* represents a kind of sabbath rest—its subtitle says "reflection"—after the mighty works of *The Crisis of Religious Symbolism*, where the deep structures of three hundred years of Western philosophical and cultural development are brought to the surface, analyzed, and made meaningful in the light of what Jean Borella has termed "the metaphysics of the symbol." Together, these two works represent a cleansing and restoration of a Christian vision of the world. Through Jean Borella's witness to the death and resurrection of religious symbolism presented here, we are given entrance to a world renewed in Christ.

G. John Champoux

General Introduction
The Aim and Plan of the Work

1. For more than three hundred years a certain strain of philosophical thought, intent on completing the mission with which—it believes—science has invested it, has waged war against the religious soul of humanity. The very place and the stakes of this conflict is the field of sacred symbolism, for the only prizes offered by religion are those forms (whether sensible or intellectual) by which it exists and is culturally expressed. When this battle was first joined philosophy simply wished to *purify* human reason, that is, restore it to its natural state by ridding it of all its accumulated impurities of ignorance and superstition. However, to the extent that this vast critique of religious reason developed, an obligation was imposed, not only to combat it, but even to account for its appearance in human history. Having set itself against religion, reason was not long in perceiving that the enemy dwelt within itself, within the depths of human consciousness. It tried to root it out, an undertaking which, in three hundred years, has led this philosophical critique, once stripped of its claim to hegemony, to reject pure reason itself and then to the kind of speculative suicide evident today in post-Structuralism: in dying, the religious soul carries the rational soul along with it.

Because this self-destruction is in reality impossible (neither God or the intellect can 'die'), we have to call into question three hundred years of European philosophy[1] and seek out, beneath an immense network of antireligious safeguards with which rationalist criticism has surrounded itself, the natural orientation of the intellect to the sacred. And, since the sacred only exists under the form of symbols, to save our intelligence from its waywardness would mean to direct it towards symbolism, to induce it to be converted to the symbol, to send *logos* back towards *mythos*. But the intellect obeys only itself, that is to say the evi-

1. All European philosophy is not anti-religious. But all modern philosophy—any philosophy that wishes to be properly philosophical and modern—is.

dence of the true. So, to effect this conversion, would we not have to prove the truth of religious symbolism rationally? But this is an impossible and even contradictory task: if our intelligence could demonstrate the truth of symbols, it would not need their mediation to attain to the Transcendent presentified and made known in them. In other words, faith would be useless and give way to reason. By this we can gauge the importance of our subject, as well as the extent and abundance of the debates to which it has given rise. However, it is this *direct* (or positive) way that Hegelianism, the broadest philosophical endeavor of modern times, has followed: to reconcile knowledge and faith, spirit and the cultural forms it has assumed, by reducing their contingency to the logical need for their appearance, building up in this way a rationalist and totalitarian pseudo-gnosis. The price paid was that of the Transcendent Itself, which disappeared, sinking into the indefiniteness of Its immanent forms, forms linked together by the most systematic and horizontal of necessitarisms.

2. We must therefore reject the direct way, and surely we have never dreamed of proving the truth of religious symbolism deductively. To the contrary, we believe that a humanly insurmountable hiatus must be maintained between the intellect and the symbol (analogous to the one separating the knowing subject from objects known either naturally or by revelation), precisely because, in accepting this distance, the intellect realizes the truth of its nature: intellect is relationship and only has access to its own identity by an assented-to orientation towards the alterity of being; it only 'integrates' that to which it submits itself. What we designate as the conversion of the philosophical understanding to the symbol is the speculative *consent* to this submission, and to bring this about is the task imposed on us. But there is no other solution for this task than the one called the indirect or negative way.

This way consists in showing how the revolt against the symbol, having reached its goal, leads reason to its own destruction. But it is obviously not within the power of reason to annihilate itself: anyone who denies reason rationally is only affirming it. What remains then is to gain entry into an understanding of the symbol so to receive its light. As can be seen, our course is akin to that of Descartes in his *Meditations on First Philosophy*: to establish, by means of a *dubitatio universalis* (the exercise of a universal doubt), the need for an intellectual conversion to symbolism. But there are differences. A difference of object: no longer is consciousness attempting to break with its ideal representations (to *experience* the resistance that its own conscious existence offers to this doubt); this attempt involves its religious representations. A difference of terrain: it is no longer knowledge but culture that is the place for exercising this *dubitatio*, in conformity with the philosophical crisis of our times, which is no longer only, as with Descartes, a crisis of our cognitive situation (tied to the advent of science in the seventeenth century), but one of cultural rootedness (tied to the unsettling of our living conditions wrought by technology and to the collapse of societies in the twentieth century). And finally, a difference of method: it is useless to proceed 'artificially' with, as Descartes says, meditations that are not quite 'natural'; it is enough to follow the work of the symbol's deconstruction as realized in the history of European thought for more than three hundred years and which, today, probably stands complete.

But every work of deconstruction lays bare the components and articulations of what it has deconstructed, the various phases of its execution necessarily corresponding to the various components of the deconstructed entity, and the succession of phases governed by the order of their relationships with one another. Such is the very simple idea that presides over the composition of our work: the crisis of symbolism is governed by the very structure of the symbolic sign and can only unfold according to the logic of its articulations. Therefore we need to call to mind this structure and logic.

3. We have shown, in *History and Theory of the Symbol*, that the apparatus of symbolism was composed of a living relationship—called the 'semantic triangle'—which unites *signifier, meaning* and the particular *referent* under the jurisdiction of a fourth element, called the *metaphysical* (or transcendent) *referent*, in which the first three terms find their unitary principle: the signifier (or 'symbolizer') is usually of a sensory nature; meaning, mental by nature, is identified with the idea that the signifier evokes, either naturally or culturally, in our mind; the particular referent is the non-visible (either accidentally or essentially so) object that the symbol, as a function of its meaning, can designate (the designation of referent, or the completion of meaning, is the task proper to hermeneutics or the science of interpretation); as for the metaphysical referent, always forgotten and yet fundamental, since this is what makes of the sign a true symbol, it is the archetype—or the metacosmic principle—of which signifier, meaning and particular referent are only distinct manifestations. Take the symbolism of water for example: the signifier is the liquid element, the thing designated by this name; its meaning is the idea, evoked by the image of water, of a 'material' able to take on all forms yet retaining none; the particular referent, what the symbol designates, will, according to the case, be concerned with the formation of the world ('the spirit of God moved over the waters'), the

regeneration of the soul (baptismal water), or other matters; finally, the metaphysical referent is universal Possibility, the divine Essence as an infinity of possibilities. Now, for this universal Possibility, the signifier 'water' is Its corporeal image; the meaning 'protoplasmic substance' Its mental form; the referents 'primordial waters' or 'baptismal water' Its modes of manifestation, the former cosmogonic and the latter ritual.[1] Clearly then, each of these elements finds in the metaphysical referent its unique and unifying principle. As a result there is no basic difference, from the point of view of this supreme referent, between symbolizer, meaning, and referent, since they are all modes of the referent-archetype's manifestation, and therefore what is symbolized can become symbolizer in its turn: the primordial or purifying waters are cosmic or ritual symbols of infinite Possibility, just as protoplasmic and ever-virgin substance is Its mental or conceptual symbol. The only basic distinction is situated between the Uncreated, always symbolized, never symbolizing,[2] and the multiple degrees of the created, each one of which, except for the lowest, is at once symbolized by a lower degree and symbolizes a higher degree: a means for the higher to be present in the lower, the symbol therefore symbolizes by *presentification* and not by representation; this is its specific act, its proper mode of signification.

But the symbol does not only effect, from the vantage point of the metaphysical referent, a *vertical* 'distinction/unification' of various degrees of reality; accordingly, it also effects a *horizontal* 'differentiation/mediation' on the level of human existence. As a sign it interposes its mediation between man and the world, awakening us to the differential consciousness of subject and object at the same time that it enables us to enter into relationship with things. In short, the signifier-meaning-referent triangulation is a consequence, within the symbol itself, of

1. We could apply this analysis to other symbols, the cross for example: the signifier is the orthogonal intersection of two straight lines; the meaning is the idea of a conjunction between two different elements or orders; the particular referent can be the sacrifice of Christ, the Holy Trinity, the encounter of the creative ray with a level of existence, either heaven and earth or the divine and the human; the metaphysical referent is the reciprocal implication of absolute Transcendence and total Immanence.

2. There exists, however, an uncreated prototype of symbolism: the Son is the symbol of the Father in the mirror of the Holy Spirit. In this sense the Word, the divine place of the archetypes, the synthesis of all the possibilities of creation, is identical to Being as the principle of existing beings, and must be regarded as the supreme Symbolizer, whose supreme Hermeneut is the Holy Spirit with the Father as supreme Referent. And so we rediscover, transposed into principial mode, the semantic triangle. As for the metaphysical referent, which is obviously not a referent in the proper sense of the term, but which constitutes the identity of the triangle's poles, it corresponds to the Divine Essence or Godhead.

the sign-man-world (either culture-consciousness-nature, or revelation-soul-creation) triangulation which structures the field of human existence.

These two, normally indissociable ways[1] of working can lead, however, to two antagonistic conceptions of the symbolic sign. Seen integrally, according to the rules of traditional societies, the symbol is synthetically defined as the *semantic ray* which, crossing all the degrees of being, unites the corporeal signifier with the metaphysical referent,[2] which is analytically translated, with respect to its reality as *sign*, into the form of a semantic triangle: semantic ray and triangle respectively define the 'symbol' aspect and the 'sign' aspect of a symbolic sign. But, for a rationalistic and scientistic civilization, such a metaphysical referent is quite simply nonexistent—unless (and this is Kant's thesis) its transcendence rightly precludes all cosmic presentification.

This negative point of view is obviously that of the philosophical critique of symbolism, and even its initial premise. We see then that there would have been no crisis of religious symbolism if the symbolic sign had never included this transcendental dimension of the semantic ray, since this alone has given rise to the rationalistic and naturalistic reaction; but, without it, there would be neither symbolism nor religion. If such thinkers stubbornly insist on giving symbolism a purely logical and formal definition, a definition applied, however wrongly, to all entities so labeled (this is even the case with some Scholastic treatises), they forego any understanding of this questioning of sacred forms unfolding beneath our very eyes. Let us realize once and for all: in speaking about the symbol, tradition and modernity are not speaking about the same thing. Everything odd and difficult in the field of symbology stems from this. And the reduction of the symbol to a semantic triangle, or even to the signifier/meaning binomial, is only based on the explicit negation of its metaphysical dimension. Thus we discover that symbolism and metaphysics have severed their ties, and with the self-same movement the dullness of rationalism grinds down that 'outcropping' of the spirit which is the metaphysical vision of things, that mysterious inner dimension residing in symbolic forms—such is the major conclusion of our book.

4. Now we must return to a point that we have only mentioned in pass-

1. The first corresponds rather to the essence of the symbol, the second to its existence as a signifying entity.

2. Seen from above to below, starting from the Principle, the semantic ray is the same as the creative ray.

ing: hermeneutics. This is what makes the symbol 'function', and assigns, as we have stated, a referent to the symbol. But hermeneutics can only speak about the symbol in its own language, that of the rational intellect, and therefore according to the understanding's idea of the real. Every hermeneutic is the function of an either explicit or implicit philosophy of being and beings, and stems from what we will call an onto-cosmology of reference which conditions the choice of referent. In short, hermeneutics, which is the effectuation of meaning, will assign a particular referent to a particular signifier only on the basis of what appears to be cosmologically and ontologically possible. Thus, for the signifier 'water' to symbolize 'prime matter' (cosmological level) or 'creative possibilities' (ontological level), two conditions are required: first, that these ideas correspond to objective realities; and second, that water, in its very substance, can be taken for the physical manifestation or presentification of *materia prima* or creative potentiality. In other words, the hermeneut needs to espouse the doctrine of the hierarchical multiplicity of the degrees of reality (or *scalar ontology*), and the doctrine of the essential unity of these degrees (or the theory of universal correspondence). Ultimately, this amounts to affirming the theophanic function of the cosmos: not only do heaven and earth reveal God under the form of a forever unknowable Cause, they also 'tell His glory'. Creation as a whole, inasmuch as it is 'God visible', is the hermeneut of the invisible God, just as religious symbolism is the hermeneut of cosmic theophany.[1]

This is exactly what the appearance of Galilean science, at the beginning of the seventeenth century, seems to definitively condemn: the cosmological revolution that it effects destroys all possibility of natural theophany. Thus it inaugurates the crisis of religious symbolism in Europe, and this is why we begin with its study.

Certainly we are not claiming that Galileo proposed to lay down a critique of symbolism, even though he broached the question several times. But, albeit only indirectly, the new science could not remain without effect on sacred forms. It constitutes the first moment in the *crisis* of symbolism because it destroyed the onto-cosmological foundation of symbolism. What is more, Galilean physics is only being studied

1. This implies no pantheism. The world only tells of God that which is utterable: His glory (or the irradiation of His creative word), not His absolute essence. And yet this is not simply nothing: thus we know that God has the beauty of the rose, the strength of a lion, the purity of water, the splendor of light, the majesty of a mountain, the immensity of the ocean, the softness of milk, the nobility of the eagle, the wisdom of the elephant, the royalty of the sun, the depth of the night, the perfection of the sky, the rigor of death, the joy of life, the centrality of man, and so forth; but all of that subsists in Him according to a supereminent and ineffable mode.

here under this heading. Now, if one admits that the final act of symbolism is accomplished with the choice of a referent, it will also be understood why the initial or inaugural moment of this crisis may be defined as a suspension of reference: under the influence of the new physics, symbols lose both their metaphysical referent, the presentification of which they cease being, and their particular referent, the objective existence of which is denied. As we see, the first moment of this crisis necessarily concerns the last pole of the symbolic triangle. This is why the first part of our book can, with good reason, be titled: *The Negation of the Referent or the Destruction of the Mythocosm.*

On the whole, the effect of the Galilean revolution on religious symbolism was to transform the symbolic sign into a 'fictive substitute', to confer on the word 'symbol' the unique and debased sense of 'unreal entity'. The symbolic is what explicitly aims at functioning as something real in appearance only: here sign has in some manner absorbed symbol; or again, the symbol, amputated from the semantic ray, finds itself reduced to the horizontal segment of its triangular structure. Moreover, we must point out that the particular referent itself becomes problematic, since the relationship that it maintains with the signifier has lost its objective basis. This relationship, which we have rightly called the meaning of the symbol, when deprived of such a basis, amounts to a subjective production of one's religious consciousness.[1] What is the referent 'creation of the world' or 'purification of the soul' without its being designated by a symbol? Modern philosophy answers: No more than a superstition or a forever unverifiable hypothesis. And what is the signifier 'water' without its symbolic usage? The new physics answers: No more than a corporeal element. It is the human mind and it alone, or rather the imagination, which unites the one to the other; and it is therefore in the functioning of this mind that the explanation for the production of symbols resides.

As we see, the *crisis* of symbolism expressly takes on the form of a *critique* of religion. And this critique necessarily has a bearing on the second pole of the symbol's triangle, that is to say, on the meaning whose genesis is situated, unbeknownst to itself, in the hidden recesses of a religious consciousness. This is the central and rightful place for the cri-

1. The word 'consciousness' only takes on the currently held sense of 'self-knowledge and states of the subject' towards the middle of the eighteenth century. It seems that it may have been Descartes, a century earlier, who had inaugurated this usage (in French): "*ma pensée ou ma conscience*, my thought or my consciousness" (letter of 1/19/1642); for the Latin version, see *Réponses aux IIIe objections*, édition Alquié, 605. Malebranche follows his example: *Recherche de la vérité*, III, II, VII, 4.

tique of the symbolic sign, and therefore the most important moment in the crisis of symbolism as well. To the traditional hermeneutic of sacred symbols, philosophic reason opposes a demystifying hermeneutic of religious consciousness: the meaning of symbols is not what it is thought to be because religious consciousness does not know what it is saying. This is why our second part can, with good reason, be titled: *The Subversion of Meaning or the Neutralization of Religious Consciousness.*

In this way philosophy thought that it had forever cured the soul of its religious folly. However, by concentrating all the efforts of its critique on the subjectivity of an alienated consciousness, philosophy forgot the symbols themselves. Perhaps it took into account the process of symbolization (at least in its own eyes), but it unquestionably left symbols as such, in their contingency and variety, unexplained. And yet it is within them, in the very words of this critique, that man discovers the truth about himself. How could symbols not end up monopolizing the attention of modern thought? To learn out of what unconsciousness our consciousness is made, and how, deceiving itself, it produced the entirety of religious symbolism without knowing it, is enough to awaken a most lively curiosity and even an admiring recognition for those subtle hermeneuts who have known how to foil its stratagems. The moment arrives, however, when with the waning of this kind of interest the question shifts: granted that consciousness deceives and hides itself, why under one particular disguise and not another? Now the first pole of the semantic triangle, the signifier, as such and in its singularity, finally comes to the fore, after referent and meaning—definitively neutralized—have ceased being important in the play of criticism. By this we are led to the third stage of the crisis of symbolism, and hence to the title of our third part: *The Reign of the Signifier or the Elimination of the Symbol*, devoted essentially to contemporary structuralisms.

We can now gauge how these three critiques differ in the treatment that they inflict on the symbol: if the first critique negates the referent, if the second subverts the meaning, the third brings the signifier to the fore by laying on it a crushing burden. Henceforth, they claim, signifying unities will organize the cultural field, and therefore structure reason as well as consciousness. The *logos*, stripped of its royalty, becomes a mere functional effect in the category of signs.

By its own admission, such is the general conclusion reached by contemporary philosophy, a conclusion which then puts the question of sacred symbolism at the very heart of Western speculative debates. This conclusion also leads to the metaphysical principle of every intellectual procedure, and forms the title of our fourth part: *The Semantic Principle or the Primary Evidence of the Logos.* The *logos* must in fact go to the very end of its self-purification to *experience* the suicidal yet impossible character of this process (the need for meaning is absolute) and to recognize the *de facto* indissolubility of its relationship to the symbol. And this is where our interpretation of the famous paradox of Epimenides leads, a paradox to which we accord the value of an initiatic test for entry into the philosophic Way. Thus the essential conjunction of *logos* and *mythos* is established *per absurdum*, and, therefore, the fact of religious symbolism is recognized in its necessity: no one can root out the sacred from the human soul without destroying it. From the side of symbols we can proceed no further: to rationally demonstrate their logical necessity would amount to a denial of *mythos* while deductively basing it on *logos*, and thereby reducing it to *logos*. But, from the side of the intellect, it is possible to transform the relationship of fact which unites it to the symbol into a legitimate coordination. Meditating on the ontological argument using St. Anselm's formulation, the intellect, confronted with the supreme task of thinking the Infinite, discovers its own theophanic nature.

With this we come to our fifth and final part, in which *The Conversion of the Intellect to the Symbol* is at last accomplished. This is truly the fundamental *hermeneutic principle*. If the need for meaning inherent to the intellect prevails absolutely, this can only be accomplished by an (apparent) renunciation of its own light and its submission to the revelation of the symbol. Meditate on the road just travelled and you will see that, with respect to the needs of an authentic philosophy, there is no other way. This also means that the conflict of faith and reason, of the universality of the *logos* confronted with the contingencies of religious cultures, is resolved: here meaning is united with being; the non-formal intellect is united with sacred forms, dead in themselves but resurrected in their transfiguration. To the impossible speculative *suicide* of a reason illusorily demystified corresponds the *sacrifice* of an intellect that finds its fulfillment only in the crucifying mediation of the symbol, as exemplarily taught in the mystery of the paschal Night.

On the Feast of St. Irenaeus of Lyon,
defender of truly-named gnosis,
June 28, 1989

JEAN BORELLA

PART I

The Negation of the Referent or the Destruction of the Mythocosm

1

On Traditional Mythocosmology

The Problematic of Semantic Cosmology

Claude Lévi-Strauss observes, in connection with a shamanic rite performed by a Cuna Indian to assist a woman with childbirth: "That the mythology of the shaman does not correspond to an objective reality does not matter. The sick woman believes in the myth and belongs to a society which believes in it."[1] A remark made in passing and seemingly self-evident. The West readily imagines a 'socio-cultural placebo effect' that acts by means of a collective auto-suggestion.[2] One would have to be a fool to think otherwise. What the modern *episteme*[3] rejects, under the circumstances, is what we propose to call symbolic or semantic causality, that is the efficacy of the sign's intelligible contents. That a particular sign is the cause of health because it signifies it, directly or indirectly, by virtue of an analogy or a kinship *semantic* in nature between the one and the other, is an axiom inseparable from a traditional conception of symbolism. In sacred rites, in astrology, in alchemical operations, it is not so much occult forces that are active as the power proper to intelligible similitudes, to relationships semantic in expression which unite signs to things, like the Christian sacraments which produce what they signify, not magically, but because the symbolic form really participates in the being of its referent, the virtue of

1. *Structural Anthropology* (Garden City, NY: Anchor Books, 1967), 192.

2. Besides, how such a process occurs is no less mysterious than in the case of the shaman.

3. In a sense akin to that of Foucault (*The Order of Things* [London: Tavistock Publications, 1970], xxii), by *episteme* we are designating the concept that a time or a culture has of true knowledge.

which it communicates accordingly. We will return later, in connection with Johannes Kepler, to the notion of semantic causality. But we can already see that to call this into question is to call into question all sacred symbolism and the ontological relationship that it establishes between signs and things. In short, this is to proceed to an onto-cosmological neutralization of the symbol.

Not only atheistic philosophers participate in this neutralization, however. Christian thinkers are doing the same: "Whoever listens to myth's original discourse," writes Antoine Vergote, "in the proper sense of the term, does not believe in it.... The mythic period has had its day.... Nor can we ever situate ourselves in myth's original light."[1] The situation of a Christian thinker is, however, much more delicate than that of an atheistic philosopher, for whom the rejection of semantic causality is literally not a problem. As to the believer, he comes up against the fact that his own religion presents itself to him under a symbolic form and rightly seems to imply faith in the efficacy of signs. Can he keep his religious faith, continue to believe in what Revelation says, without for all that clinging to what is implied by the manner in which it is said? Would it not be right to proceed with a hermeneutic? But is there not then a risk of 'throwing out the baby with the bathwater', as Kepler already feared in connection with those who despised astrology?[2] How are we to discriminate, in revealed data, between what hermeneutics reveals, because merely figurative in nature, and what its proper and direct expression reveals? Does not this boundary shift at the whim of the hermeneuts? As we see, the tangle of questions is nearly inextricable.

We will begin with a very simple question: why do Claude Lévi-Strauss or Antoine Vergote no longer believe (or no longer can believe) in what myth affirms? The answer is clear: by virtue of their own ontology of reference they reject the one that myth seems to imply. But this answer raises two objections: first, is it exact to say that religious discourse implies this unacceptable ontology of reference, and second, is it true that this ontology of myth is scientifically unacceptable?

To this second question I have tried to reply very briefly in *Symbolism and Reality*.[3] I will only say that, even though contemporary science in

1. *Interprétation du langage religieux* (Paris: Edition du Seuil, 1976), 85, n.1.

2. "Advice to some theologians, doctors, and philosophers, and especially to Philip Feselius, not to, by condemning the superstitions of the star-gazers, throw out the baby with the bathwater...." (*Reply to Roslin*, cited by Géard Simon, *Kepler, astronome astrologue*, Paris: Gallimard, 1979, 92).

3. *Infra*, 415–31.

certain respects makes it possible to restore a sense of a symbolic cosmos, Galilean physics is absolutely opposed to it. Now, although today it may no longer be able to lay claim to the title of an adequate picture of material phenomena,[1] this physics continues to prevail in the majority of minds as the general model of reality,[2] a model in such direct contradiction to the cosmos of Judeo-Christian revelation that one can only coexist with the other in the same intellect at the cost of a truly schizoid culture.

There remains the first question, concerning the solidarity of religious discourse with its ontology of reference. Is there not an interest in dissociating them? And would not this dissociation constitute precisely the birth-certificate for a true consciousness of the 'symbol'? To grasp the 'symbolic' nature of a religious statement: is this not in fact to understand that it is saying something other than what it seems to declare, that, according to Ricoeur's expression, its aim is a second meaning through a first meaning? Hence, do we not need to admit that 'symbolic' consciousness is built on the disappearance of mythic unconsciousness? Should not the term 'myth' be reserved to characterize a thought unable to perceive the 'symbolic' nature of scriptural statements, that is, basically, unable to dissociate words from the things they designate? This is surely the conviction of a majority of contemporary exegetes, and especially Bultmann: "Mythological conceptions can be used as symbols or images which are perhaps necessary to the language of religion and therefore also of the Christian faith . . . statements which describe God's action as cultic action, for example, that he offered his Son as a sacrificial victim, are not legitimate, unless they are understood

1. Cf. the remarkable work by Prof. Wolfgang Smith, *The Wisdom of Ancient Cosmology: Contemporary Science in the Light of Tradition* (Oakton, VA: Foundation for Traditional Studies, 2003).

2. "Who would suspect today that the notions of the *trajectory* of a moving body, the *speed* of a body in space, and the *distance traveled* by a rocket are *pre-Galilean* notions to which it is strictly impossible to attach a precise significance, unless we consider reference to an immobile Earth a dogma?"; J.-M. Souriau, "L'évolution des modèles mathématiques en mécanique et en physique," in *Revue de l'Enseignement philosophique*, 22e année, n° 3, février-mars 1972, 8. To speak of *pre-Galilean notions* in connection with the (Newtonian) concept of absolute space and time should not mislead us. In this article the author upholds the thesis that, in reality, Galilean space-time is *relativist*, which, according to him, would have escaped Newton. There does exist in fact a Galilean relativity principle. But it is doubtful that Galileo himself was able to draw out all its consequences, since he did not go so far as to conceive of an absolutely infinite space, and since he harbored traces of Aristotelianism (Galileo, *Dialogo, I, Opere*, vol. VII, 43; cf. Koyré, *Galileo Studies*, trans. John Mepham, Atlantic Highlands, NJ: Humanities Press, 1978).

in a purely symbolic sense."[1] Once this is accomplished, myth can be restored to its true significance: "Against its real intention [mythological thinking] represents the transcendent as distant in space and as only quantitatively superior to human power."[2] And so the true Christian should be grateful to modern science, which compels us to demythologize. This is exactly what a Catholic theologian, and a Thomist at that, asserts: "The full development of cosmology is only possible if heaven is 'demythologized' by ceasing to be the dwelling-place of the gods. Now nothing is more demythifying than the affirmation of monotheism and the recognition of a God who does not inhabit a portion of cosmic space.... In this sense Galileo's work is profoundly Christian...."[3]

The situation of sacred symbolism seems therefore excellent, above all in the West. Freed, thanks to science, from a reifying outlook and from the naïve realism of a mythology ignorant of its own true significance, symbolism should be able to adequately express the truth of religious revelation. However, and without even taking into account the all-too-obvious contradictions inflicted on us by factual reality, the demythologizing approach is far from exhausting the question. For we would still need to be assured of the true intentions of the sacred texts and traditional mythologies. This line of questioning is, then, inconclusive. Who has proven that it is contrary to its true intentions that myth lodges the divine in distant space? Perhaps we would be inclined to admit this if the mythic discourse as a whole developed under the rule of the symbolic unconscious. It could then be entirely converted into 'symbol'. But it clearly seems that this is not so in every case. First it would be necessary, confining ourselves to Western culture, to take into account very ancient disputes about mythological discourse (those of Greek rationalism, as well as those imposed by the Christian faith), as Jean Pépin has shown in his study *Mythe et Allégorie*.[4] The birth of 'symbolic' consciousness would be then very much prior to the advent of Galilean science. The very notion of allegory is there to testify to this. But that is not all. We also must ask ourselves if the sacred texts do not themselves make a distinction between what is mythic and what is historical. For the New Testament this is incontestable (cf. the notion of

1. 'Jesus Christ and Mythology', in Rudolf Bultmann, *Interpreting Faith for the Modern Era*, ed. R.A. Johnson (Minneapolis, MN: Fortress Press, 1987), 318, 319.

2. Rudolf Bultmann, *New Testament and Mythology and Other Basic Writings*, trans. S. M. Ogden (Philadelphia, PA: Fortress Press, 1984), 161.

3. Jean-Michel Maldamé, O.P., 'Cosmologie et théologie. Réflexion théologique sur la science cosmologique moderne', in *Revue thomiste*, janvier-mars 1978, LXXXVIe Année, t. LXXVIII, nº 1, 83.

4. Paris, Éditions Montaigne, 1958.

parable). This is also brilliantly verified by the famous medieval distinction between allegory in word and allegory in word *and* fact, which is taken into account by nearly all commentators.[1]

Now, from the philosophical point of view, this distinction strangely complicates the question, for we needs must ask: where then do we draw the line of demarcation between *allegoria in verbis* and *allegoria in factis*? What then is the criterion that will enable us to trace it? Obviously this criterion is not the same for an Augustine and a Bultmann, who are, however, both Christians. Where then does the difference lie? Nowhere else than in their respective 'ontologies of reference'. Some statements that are factual for Augustine are purely 'symbolic' for Bultmann, because contrary to what is *physically possible* in his eyes.

Will we now demonstrate that it is indeed Galilean physics that comes into play in the question of symbolism? Yes, to a certain extent, but basically what a paltry certainty! It is high time that everyone becomes aware of this. All that remains, some will say, is to draw out its consequences from the vantage point of sacred scripture, that is to transfer what still belonged, with regard to Augustine, under the rule of history to the rule of 'symbolism'. True, there are a certain number of Christians who still cling to the historicity of the Judeo-Christian revelation, without counting the mass of those who prefer not to pose the question. This is not surprising: time is needed to drink so bitter a wine, even if poured out some time ago. In any case, that changes nothing as to the question of 'symbolism' itself, which, to the contrary, will only gain full self-consciousness by it. And so the objection stands: Galileism marks perhaps the disappearance of the 'symbol's' ontological referent; but, far from constituting the first stage of the critique (that is, of the deconstruction) of the 'symbol', it makes possible the true birth of this critique.

There is one point that our line of questioning has left in the background, surely because its possibility is set aside *a priori*, as if unconsciously; but it is nevertheless the essential point. The hermeneutic 'compatibility' of a sacred text is not made in a twofold way, by arranging 'factual' passages in the right column and solely allegorical passages in the left. A third column must also be prepared, concerned with passages at once historical and symbolic, that is to say relating to *sacred facts*. And is this not preeminently the case with the divine deeds related

1. I have explained this distinction in *Histoire et théorie du symbole* (Lausanne: L'Age d'Homme, 2004), 51–58. But it is of course to be found in the Fathers (Ambrose, John Chrysostom, Tertullian, etc.); cf. Henri de Lubac, *Medieval* Exegesis, vol. 2, 86–89, and vol. 4 [French ed.; Eng. ed. of vol. 4 not yet published], 131–149.

in the books of the Old and New Testaments, from Abraham's sacrifice to Christ's resurrection and ascension? This is surely the conviction that animates Jewish, Christian, and Islamic hermeneutics. The historical and symbolic senses are not juxtaposed according to a horizontal bipartition, but superposed according to a vertical hierarchy. This is actually why only the 'middle column' is essential. This means that it is to this one that the other two, which are only its fragmentary aspects, must be related. In other words, *in sacred scripture there is no purely factual or historical sense*, whatever the exegetes, even the most illustrious and the most open to symbolism, may say.[1] And far from volatilizing the reality of historical facts, their relationship to a symbolic significance is the only way to establish and fix this reality. True, they can be considered merely from the vantage point of their spatio-temporal manifestation, but, insofar as they are sacred facts, they remain in and by themselves *potentially* open to a symbolic significance. The hermeneutic that deploys it does not superadd it from without, but completes the reality of fact and history.[2] Similarly, there is not, in sacred scripture, any purely symbolic sense, that is, any sense which possibly implies the actual reality expressed by the signifier. In the last analysis every symbolic expression is based on the ontological relation that the corporeal signifier maintains to the reality expressed.

Now, there are precisely these two types of relationship that Galilean physics renders impossible. The symbol, as we have repeatedly shown,[3]

1. Thus we see Cardinal Daniélou affirm: "The cross is important in Christianity, not primarily because of its symbolic qualities, but because Christ was put to death on a particular kind of wooden gibbet. The historical event came first; it was later that liturgical development, seizing upon the approximately cruciform shape of this instrument, enriched it with all the natural symbolism of the cross, the figure of the four dimensions, or of the four cardinal points.... This is all secondary, in comparison with the historical facts" (*The Lord of History*, trans. N. Abercrombie [Cleveland and New York: Meridian Books, 1968], 125). Such a doctrine is, to our eyes, a pure and simple negation of the incarnation of the Word in Jesus. True, this illustrious scholar seems to express a different opinion when he writes: "For the Jewish Christian then, the Cross is something more than the wood on which Jesus was crucified. It is a spiritual, mysterious, living reality which accompanies the risen Christ" (*The Theology of Jewish Christianity*, trans. J.A. Baker [London: Darton, Longman & Todd, 1964], 270). It is, according to St. Justin, "the great *Symbol* of the strength and power of Christ" (*First Apology*, LV, 2; P.G., t. VI, col. 412).

2. It is symbolism that confers its reality and its understanding on an historical event. We have shown this in particular for the symbolism of Christ's wounds: cf. *The Sense of the Supernatural*, trans. G.J. Champoux (Edinburgh: T&T Clark, 1998), 69–96. We should also ask: what is the true reality of the 'two pieces of wood'? Quantum physics replies: an electron mist. Today the spontaneous and unconscious materialism of modern exegetes no longer has the least basis.

3. *Histoire et théorie du symbole*, page 154 in particular and *passim*.

constitutes a *sui generis* and autonomous order of reality, a place of exchange for the sensible and the intelligible, a universal athanor of their common transformation. It is therefore necessarily correlative to a cosmos compatible with this alchemy. It has need of a corporeal universe in which it is not impossible to introduce something of the spiritual (this is the case of the historical open to the symbolic), just as it has need of an intelligible universe into which something of the corporeal can be introduced (this is the case of the symbolic relative to the naturalness of its signifier). Now the Galilean world rejects both: the corporeal becomes purely spacial, and the intelligible sheer mathematical rationality. With Galileo it is not only the world here below that is transformed and emptied of every qualitative presence, it is also the intelligible heavens that are brought down to human thought alone. Nothing is more significant, in this respect, than the Cartesian dualism of soul and body (with, however, for Descartes the metaphysician, the presence of the divine heavens in the substance of the soul), and the 'Humanist' development of Locke and Hume's philosophy.

Also, confining ourselves to the here and now, the Galilean revolution has not only rendered impossible the producing of certain phenomena (attested to by Holy Scripture) within the general framework of the universe; it has also transformed the general framework itself, or rather has made it disappear as the finite totality of corporeal beings: the possibility of a cosmology seems definitively excluded. But, to know this expressly, we must wait for Kant, the dialectic of pure reason, and the resolution of cosmological antimonies, which Kant calls the *Critical Solution of the Cosmological Conflict of Reason with itself.*

The Cosmology of the Ancients[1]

METAPHYSICAL SYMBOLISM AND SCIENTIFIC REALISM

One could hardly exaggerate the importance of the crisis unleashed by Galilean physics in Europe at the beginning of the seventeenth century.[2] The reason for this is that the advent of the new science *simultaneously* shook the religious world, the 'intellectual' world of the scholars and philosophers, and lastly the everyday world of ordinary people, an upheaval whose repercussions are still being felt three hundred years

1. Cf. Appendix 1, 134: *Very Elementary Notions of Ancient Cosmography.*

2. Arthur Koestler has composed a popular history of the different cosmological systems from Pythagoras to Newton (*The Sleepwalkers*, New York: Grosset & Dunlap, 1963). The chief work on this topic remains Pierre Duhem's *Le système du Monde. Histoire des doctrines cosmologiques de Platon à Copernic*, in ten volumes (Paris: Hermann, 1913–1958,

later. Never, we think, has such a combination of circumstances occurred in the history of humanity (at least historical humanity).

Why was Galilean physics the specific cause of this crisis? The answer to this question is at once immediate and of such complexity as to defy analysis. There is agreement that in all likelihood it above all played the role of catalyst, making possible the crystallization of numerous elements that had 'exercised' medieval thought for a long time. The tensions were close to their breaking point. The introduction of one extra

reprinted 1965). The work of Duhem (d. 1916) is recommended above all for the truly monumental mass of hard-to-find texts collected and translated. Completing his effort is the more recent two-volume work of A.C. Crombie, *Augustine to Galileo* (Cambridge, MA: Harvard University Press, 1961). For an understanding and analysis of the cosmological crisis under discussion, we are chiefly indebted to Alexandre Koyré, through the following works: *The Astronomical Revolution: Copernicus, Kepler, Borelli* (trans. R.E.W. Maddison [Ithaca, New York: Cornell University Press, 1973]); *From the Closed World to the Infinite Universe* (Baltimore: Johns Hopkins Press, 1957); *Galileo Studies* (trans. John Mepham [Atlantic Highlands, NJ: Humanities Press, 1978]); 'Le vide et l'espace infini' (1949), an article reprinted in *Etudes d'histoires de la pensée philosophique* (Paris: Gallimard, 1971). Whoever wishes to be introduced not only to the history but also to the scientific nature of the questions posed, the clearest and most instructive work is Jacques Merleau-Ponty and Bruno Morando's *The Rebirth of Cosmology* (trans. H. Weaver [New York: Knopf, 1976]). On Galileo in particular we have taken as a guide Maurice Clavelin's *The Natural Philosophy of Galileo* (trans. A.J. Pomerans [Cambridge, MA: MIT Press, 1974]), which includes a masterly analysis of the Aristotelian theory of movement (25ff). Concerning the religious aspect of the Galilean crisis, the most comprehensive study is that of Giorgio Santillana, *The Crime of Galileo* (Chicago: University of Chicago Press, 1955). The 'militant' Galilean Emile Namer has published, in the series *Archives*, *L'affaire Galilée* (Paris: Gallimard-Juillard, 1975), which provides the principal documents relevant to the question. The (traditional) Catholic viewpoint is expressed by Pierre de Vrégille in his article 'Galilée' for the *Dictionnaire apologétique de la foi catholique* (Beauchesne, 1924). The joint anthology *Galilée—Aspects de sa vie et de son oeuvre* (Paris: P.U.F., 1968) contains often illuminating studies on particular points. Finally, the studies devoted to the astronomy of the ancients and Copernicus by Lenis Blanche, published in *Revue de l'enseignement philosophique* from 1965 to 1980, are invaluable: nothing better informed or more illuminating has been written in the history of astronomy on the question of geocentrism and geostatism. We have read Pietro Redondi's *Galileo: Heretic* (trans. R. Rosenthal [Princeton, NJ: Princeton University Press, 1987]) too late to take it into account here. Besides, the 'discoveries' that he refers to are not directly concerned with my subject: cf. *infra*, 62, n. 1. In 1988 Editions Arléa published a French translation of an Italian work, *Galilée—Entre le pouvoir et le savoir*, by Franco Lo Chiatto and Sergio Marconi. This work provides some of the trial records. The purely Marxist introduction in no way answers to the demands of historical science, and, although it is the most complete available in French, the documentation shows significant gaps. On Galileo's physics, the best introduction is by Françoise Balibar: *Galilée, Newton lus par Einstein* (Paris: PUF, 1984).

element was enough to burst the cultural universe of ancient and medieval humanity to pieces. This element was the realism of one especially pugnacious physicist who 'proved' Copernicus through the telescope, that is by direct observation.[1] Or rather, Copernican heliocentrism appears, for Galileo, tied to the 'empirical' discoveries made possible by the astronomical telescope, which revealed behind "the great number of fixed stars that up to now men have been able to see by their natural sight ... innumerable others," and which showed the Moon to be not a divine body with a crystalline substance, but "just like the face of the Earth itself ... full of enormous swellings, deep chasms and sinuosities."[2] Hence the shock to the entirety of the European mindset. Not only did the Sun not rise in the east and set in the west, contrary to what everyone saw each day, but even that natural intuition which makes us raise our head to the sky to contemplate what is most noble and most pure is deceived, for matter itself is spread throughout a universe which also has neither high or low. As the following analyses will show, the most violent blows that Galileo dealt to the medieval universe were much less of the astronomical than of the physical order. Aristotelian physics, astronomically confirmed by Ptolemy's remarkable construct, is dual: celestial physics is not the same as earthly physics, in such a way that the hierarchical structure of the universe in space is in agreement with the qualitative difference between supralunar and sublunar substances.[3] Copernicus' own system by no means abolishes this distinction

1. The telescope cannot actually prove Copernicus (and besides, strictly speaking, nothing proves the movement of the earth around the sun, not even Foucault's pendulum). But the telescope showed that the heavenly bodies were composed of matter, like the earth, and the stars, more numerous than previously thought and of variable luminosity, were not fixed to a crystalline sphere, but arranged in tiers at variable distances in depth.

2. Galileo, *Sidereus nuncius* (1610), *Opere*, Edizione Nationale, v. III, Firenze, 59ff.; cited by Koyré, *From the Closed World*, 89.

3. J. Merleau-Ponty, *The Rebirth of Cosmology*, 32–34; also Aristotle, *On the Heavens*, L. III, 1, 298 a–b. We say Ptolemy confirmed Aristotle. This means that Aristotelian physics, accepted by Ptolemy because of its scientific character, was in need of being reconciled with astronomy in a more adequate manner than Aristotle himself had done (Koyré, *The Astronomical Revolution*, 82, n.43). Ptolemy even recognized that "there is perhaps nothing in the celestial phenomena which would count against that hypothesis, *at least from simpler considerations*," that is for "the heavens to remain motionless, and the earth to revolve from west to east about the same axis [as the heavens], making approximately one revolution each day." But it is physics that is opposed to this thesis and its partisans: "From what would occur here on earth and in the air, one can see that such a notion is quite ridiculous" (*Almagest*, book 1, chap. VII, trans. G.J. Toomer [Princeton, NJ: Princeton University Press, 1998], 44–45). As Duhem shows (*Le système du*

and therefore can be in a certain manner considered as a simple way of seeing things, of 'saving the phenomena'.[1] But this is no longer possible with Galileo: Copernican heliocentrism is allied to a physics of material bodies, which marks it as implacably realist.

This was not, however, the first time that such a hypothesis on the 'earthen' nature of the Moon was proposed. We read in a most curious short treatise of Plutarch, *Concerning the Face Which Appears in the Orb of the Moon*: "The Moon . . . is a celestial Earth . . . let us not think it an offense to suppose that she is Earth and that for this which appears to be her face, just as our Earth has certain great gulfs, so that Earth yawns

monde, tome II, 57), the physics of Aristotle dictates that all celestial movements are circular and *homocentric* to the earth's center, which is not verified by observation, hence the necessity to construct (but this was already done by the Platonist Eudoxos of Cnidus) an extraordinarily complex system of eccentric circles and deferents. But it is hard to know what reality Ptolemy attributed to them, since, in the *Almagest*, the physical reality of circles is never positively stated (Koyré, *The Astronomical Revolution*, 83, n. 44). Recall that the Alexandrian Claudius Ptolemy lived until the second century after Christ, and his best-known work, the *Mathematical Synaxis*, to which the Arabs would give the name *Almagest*, that is to say the 'Very Great' (Composition), "from the mathematical point of view is one of the finest and most outstanding works of the human mind" (Koyré, *The Astronomical Revolution*, 23). Cf. Appendix 2, 138.

1. Did Copernicus believe in the reality of his astronomical construct, or did it have only a hypothetical character? The expression 'save the phenomena', which has had such an extraordinary history, is probably of Platonic origin (Robin, *Platon* [Paris: PUF, 1935], 158). But it is open to multiple interpretations, one of which would favor a certain conventionality: to account for appearances, in an intelligible manner, by a geometric construction for example, whatever might be the relationship of this construction to factual reality. Now the chief work of Copernicus, the *De revolutionibus orbium coelestium* (1543), is preceded by three prefaces: one titled "Advice to the Reader," a letter from Cardinal Schönberg urging Copernicus to publish his discoveries, and a dedicatory letter of the author to Pope Paul III. In the "Advice to the Reader," Copernicus states: "For these hypotheses need not be true nor even probable. On the contrary, if they provide a calculus consistent with the observations, that alone is enough." Volumes have been written on this "Advice": is it from the hand of Copernicus, or added by a prudent friend? And what does the word 'hypothesis' mean? From the second half of the nineteenth century, historians have thought that this "Advice" could be neither from Copernicus or a follower, under the pretext that Copernicus has never used the term 'hypothesis' to designate the Earth's movement. But Lenis Blanche has shown that in Book V, chap. 20 of the *De revolutionibus*, he speaks of "our hypothesis of the mobility of the earth" (*Revue de l'enseignement*, 30^e A., n° 6, 5), that this expression has nothing anti-Copernican about it, and that nevertheless certain mathematical errors contained in this "Advice," which Galileo—who believed it to be a forgery—had stressed in his *Considerations on the Copernican Opinion* (1615), reinforce the thesis of those who see here the work of Osiander, a Protestant theologian corresponding with Copernicus (L. Blanche, Revue, 28^e A., n° 3, 34–36).

with great depths and clefts."[1] As we see, this text is quite close to Galileo's. In the same way, Copernican heliocentrism was not a complete novelty. Copernicus himself did not claim this.[2] But we need to go further. Far from being exceptional, the doctrine of the Earth's mobility, either in place, or around a center, or both, has been upheld by numerous philosophers, not only by modest Pythagoreans, but by the divine Plato himself.

We would not dream of restating here all the documentation on this question, since it is to be found among the specialists and especially Lenis Blanche, who has presented it in the different studies indicated. This material is beginning to be known, and yet our minds are far from recognizing the consequences that flow from it. We will address only two points, one relating to the movement of the Earth, the other to heliocentrism.

THE EARTH: MOBILE OR IMMOBILE?

The Rev. D. Dubarle writes in a study on *La méthod scientifique de Galilée*: "Ancient physics was accomplished by a rational idealization of naïve perception.... The spontaneous feeling that the earthly observer has of being, together with the Earth, at rest and at the center of the totality of all things, has been integrated into the geocentric system."[3] Now this thesis, which basically echoes the one explained by Leon Brunschvicg in *Les âges de l'intelligence*,[4] does not correspond to reality. If the traditional cosmologies were a 'rational idealization of naïve perception',

1. *Opera Moralia*, vol. 12, chap. 21, 935b; trans. H. Cherniss and W.C. Helmbold (Cambridge, MA: Harvard University Press, 1957), 141–143.

2. In the dedicatory letter to Pope Paul III, and in chapters 5 and 10 of *De revolutionibus*, Book I, Copernicus explains that he has already encountered the idea of the earth's rotation around the central fire in Nicetas, Heraclides of Pontus, Ekphantos the Pythagorean, and Martianus Capella. But in the manuscript we can also read the crossed-out name of Aristarchus of Samos (third-century BC), who, in the words of Archimedes (*The Sand-Reckoner*, in *The Works of Archimedes*, trans. T.L. Heath [Cambridge: Cambridge University Press, 1897], 222), would have taught the double revolution of the Earth upon itself and around the sun. Now, it must be pointed out that *The Sand-Reckoner* was only published in 1544, and that Copernicus was already in possession of his doctrine thirty years before, as proven by the publishing of the Commentariolus in 1514 at the latest (L. Blanche, *Revue*, 29ᵉ A., nº1, 17), which excludes (Koyré's thesis) that he could have read it in manuscript. Camille Flammarion, in his *Vie de Copernic* (Paris: Didier, 1872) has made a systematic study of all the authors cited by Copernicus and some others (128–154).

3. In *Galilée—Aspects de sa vie et de son oeuvre*, 88.

4. Paris: PUF, 1953. For him there are two views, one that is prisoner of the sensory (dominant until the end of the Middle Ages), and one that breaks away from the sensory and constructs the real scientifically.

they should all be similar, since perception (naïve or not) is identical for everyone. But there is nothing to this, as is proven by the fact that certain models depict the Earth as a flat body and others as a sphere, which can hardly be verified, however, except on the seashore.[1] On the other hand, it is not at all certain that ancient physicists started with the sensation of an immobile Earth under their feet and at the center of the world. The majority of physicists might even be thought of as teaching the contrary, with Aristotle one of the rare ones, if not even the first—at least as far as the rigor of his demonstration—to immobilize the Earth at the center of the world. In any case this is what stands out in a reading of *De Caelo*. If this were not so, it would be hard to explain why Aristotle takes such care to separate himself from his predecessors, and the numerous arguments that he deploys to deny the movement of the Earth.[2] Actually:

• Far from having the feeling that the Earth was an immobile observatory under their feet, the Ancients thought rather that it might fall (and perhaps was falling); this is why the majority sought to support it and to have it carried upon some unshakable foundation.

• Far from *seeing* the Sun traverse space above their heads, the ancients saw it as in rotation upon itself, around a fixed axis, which did not admit of an orbital translatory movement. Besides, this is what Aristotle himself attests. If the stars possessed a rotational movement, he says, "they would then stay where they were, and not change their place ... the only star which appears to possess this movement is the sun, at sunrise or sunset."[3] How can someone *see* the Sun rotating? Lenis Blanche conjectures—and we follow him—that this involves the perceiving of sun-spots, which are in fact observable to the naked eye when a haze makes looking at the Sun on the horizon bearable.[4] Aristotle is alluding then to this observed fact, but only to deny it and correct perception with reason: "This appearance [the rotation in place of

1. On this topic let us recall that the earth's curvature was known since very remote antiquity, and that this concept is sometimes concurrent and contemporaneous with a planar representation. For Cosmas Indicopleustes (sixth century AD), the earth is a rectangle. But for him, this is an attempt to break with the received opinion of its roundness. The first article of St. Thomas Aquinas' *Summa Theologiae* affirms that "*terra est rotunda*." For an informed judgment on medieval science read *La science de la Nature: théories et pratiques*, edited by G.H. Allard and J. Ménard, Cahiers d'Etudes Médiévales 2, Bellarmin-Montréal, Vrin-Paris, 1974.

2. *On the Heavens*, II, 13–14.

3. *On the Heavens*, II, 8, 290a, 10–15 (J.L. Stocks trans., Oxford: Clarendon Press, 1922).

4. L. Blanche, *Revue*, vol. 18, no. 5, 24. This is how one treatise from *Cosmographie élémentaire* (by Paul Blaize, for use in philosophy classes [Magnard, 1949], 17) expresses it: "Sun-spots of 40 to 50,000 km are not rare; they are *visible to the naked eye* as black

the sun] is due not to the Sun itself, but to the distance from which we observe it. The visual ray being excessively prolonged becomes weak and wavering."[1] Notice that we are quite far from an idealization of a naïve perception; to the contrary, this involves a rational construction set up in opposition to the evidence of the senses. Other texts also testify to this immobility of the sun, such as the following, an excerpt from Diogenes Laertius in his life of Pyrrho: "The objects, therefore, appear different to us according to the disposition of the moment; for, even madmen are not in a state contrary to nature. For, why are we to say that of them more than of ourselves? For *we too look at the Sun as if it stood still*."[2] If, then, the Sun is immobile, and since what we observe is a relative movement of the Earth and Sun, we clearly need to endow the Earth either with an axial rotation, or even with an orbital revolution.[3]

It seems that this is clearly the case with Plato, according to Aristotle himself: "Others, again, say that the Earth, which lies at the centre, is 'rolled', and thus in motion, about the axis of the whole heaven. So it stands written in the *Timaeus*."[4] Here Aristotle is alluding to a well-known passage from Plato, the disputed translation of which has caused much ink to flow: "The Earth, which is our nurse, stands firm *by rolling* about the pole which is extended through the universe; *by this mechanism she is the guardian and artificer of night and day*" (40b).[5] The disputed term is the one translated as 'by rolling' (in Greek: *eillomenen*). The ancient commentators (except for Plutarch and Proclus), at the head of whom Aristotle must be listed, interpreted it as meaning 'in rotation', and this is indeed why the Earth is the 'artificer' of days and nights, that is to say, by its working there is succession of day and night. Obviously, from this it must be conceded that Plato did not teach the total immobility of the Earth, nor the real motion of the celestial sphere (or sphere of the fixed stars) around the axis of the universe. But this was

dots when the Sun is near the horizon.... The sighting of the spots makes it apparent that the Sun rotates." In a general way, it is probable that the capacity for observation by the naked eye is underestimated. Thus, the historians of the Chinese sciences today affirm that the 'moons of Jupiter' were located and described by the astronomer Gan De in 364 BC, two thousand years before Galileo announced the discovery of the 'Medicean' planets' (after the Duke de Medici to whom the *Sidereus Nuncius* was dedicated) as "something no one has ever seen."

1. *On the Heavens*, II, 8.
2. *The Lives and Opinions of Eminent Philosophers*, trans. C.D. Yonge (London: H.G. Bohn, 1853).
3. L. Blanche, *Revue*, vol. 18, no. 6, 5.
4. *On the Heavens*, II, 13, 293b, 31–32.
5. B. Jowett translation, *Timaeus* in *Plato, The Collected Dialogues*, ed. E. Hamilton and H. Cairns (Princeton, NJ: Princeton University Press, 1961).

clearly in fact the opinion of the ancients, as reported by Cicero: "Hiretas [or Nicetas] of Syracuse, as Theophrastus tells us, thinks that the Sun, and Moon, and stars, and all the heavenly bodies, in short, stand still; and that nothing in the world moves except the Earth; and, as that turns and revolves on its own axis with the greatest rapidity, he thinks that everything is made to appear by it as if it were the heaven which is moved while the Earth stands still. And, indeed, some people think that Plato, in the Timaeus, asserts this, only rather obscurely."[1] The difficulty with this interpretation comes from what certain commentators have seen, at other points in the Platonic corpus, such as the affirmation of the Earth's absolute immobility and the motion of the sphere of the fixed stars. But an attentive examination of all these passages shows that this is not so and other interpretations should be rejected.[2] This is why Diogenes Laertius is faithfully summarizing the Platonic thesis when he writes, in his account of the doctrine of the master, that "the Earth . . . was formed in order to be the dispenser of night and day; and as she is placed in the center, she is constantly in motion around the center."[3] Lastly, a passage from the *Epinomis* must be cited (a dialogue whose authenticity, thanks to computer science, is fully recognized today) that clearly teaches the rotation of the Earth: "It cannot be that earth and sky, with all the stars . . . , if no soul had been connected with, or perhaps lodged in, each of them should move so accurately, to the year, month, or day."[4] To be included among the proponents of the Earth's rotation are, besides Plato, Heraclides of Pontus, to whom Plato

1. *The Academic Questions*, II, §39, trans. C.D. Yonge (London: George Bell & Sons, 1880), 81.

2. For the earth's immobility, *Phaedo* 99b and *Timaeus* 62d–63 are advanced; for the movement of the sphere of the fixed stars, *The Republic*, X, 616c–617d is advanced. L. Blanche has shown, by taking into account *all* the textual data (op. cit. 14–16), that the circle in motion, in the myth of Er the Pamphylian, is that of the ecliptic, the bearer of the equinoctial points (and not the circle of the celestial equator), and that its revolution represents the precession of the equinoxes and the retrogression of the vernal point. If therefore the circle of the ecliptic revolves, this is because the circle of the fixed stars is immobile; otherwise it would be impossible to estimate the shifting of the vernal point, that is to say the shifting of the point where the circle of the ecliptic intersects the circle of the celestial equator at the Spring equinox.

3. *The Lives and Opinions of Eminent Philosophers*, Plato, XLI. It is even possible that, at the end of his life, Plato may have abandoned geocentrism and sided with the Pythagoreans. According to Plutarch, this is what Theophrastus of Eresa taught (L. Blanche, *Revue*, vol. 18, no. 2, 2).

4. *Epinomis*, 983c; trans. A.E. Taylor, *Plato, The Collected Dialogues* (ed. Hamilton & Cairns), 1525.

entrusted the direction of the Academy at the time of his third voyage to Sicily; Seleucus (second century BC); the 'Pythagoreans', in the words of Aristotle; and, Aristotle adds, "many other philosophers."[1]

It is, then, altogether obvious that the hypothesis of geostasis was not held by everyone, and Aristotle was even, if we are to follow Lenis Blanche, the first to have affirmed and demonstrated it for 'scientific' reasons. Now—and this is how we will conclude—the idea of a universe in which the apparent movements of the Sun and stars are explained by the real movements of the Earth, contrary to the most immediate data of sensory experience, entails no incompatibility with the symbolism of the cosmos. Quite the opposite: the most directly symbolist of the philosophies (the Pythagoreans and the Platonists), the very ones that saw in the sensible a reflection and an image of the intelligible, at the same time favored a non-geostatic cosmology.[2] Conversely, it is the hardly symbolist philosophy of Aristotle, who even reproaches Plato for speaking symbolically, that strives to establish the most rigorous geostatism rationally. But both representations have been deployed in a world where symbolism held sway. And yet it does not seem that Aristotle's cosmology made it any more difficult for this symbolism, even if, as we will see, his cosmology was much less favorable to it. Surely, by very reason of its physicist realism, Aristotelian cosmology entailed the ruin of cosmological symbolism, which had been avoided by the maintenance of Platonism, as proven by the examples of Nicholas of Cusa as precursor, and Johannes Kepler, Henry More, and perhaps even Newton as corroborators. It could be asserted, which is hardly contestable, that the error or sin of geostatism and geocentrism was to seek to assign a geometric center to cosmic space, this is to say, basically, to wish to confer on space a purely spatial meaning, for each spatial center negates the centrality of all other centers, and vice versa. The true centrality, even for space, is necessarily spiritual in nature. But, as a matter of fact and as we shall see,

1. *On the Heavens*, II, 13, 293a, 15–30.

2. This is the moment to stress our disagreement with the concept of Platonism upheld by Koyré, who seems to draw Plato's geometrism in the direction of a conventionalism opposed to Aristotle's physicist realism. There is certainly a good deal of truth in this interpretation. But it should be added that—and this is, moreover, what we propose to show—for Plato, if our understanding of nature remains hypothetical, this is not because of the weakness of our intellect; it is because of the lack of reality of the object to be known. Consequently, the only adequate knowledge for a deficient being is symbolic knowledge, because it first posits its object for what it is, a symbol, but a real symbol, that is to say an image that participates ontologically in its model. In other words, the hypothetical geometrism of Platonic cosmology is equilibrated by its symbolist realism. Platonism is not an idealism.

Aristotle has not ignored this principle altogether. Let us at least keep in mind that, however else it might be, the crisis of cosmological symbolism was not directly tied to the rejection of a world image conformed to the most immediate appearances. Something else was needed.

THE SUN: CENTRAL OR PERIPHERAL?

We come now to the second point mentioned above: geocentrism. Its denial by Copernican heliocentrism is presented, still today, as the major event triggering one of the most important religious and philosophical crises of the West. Here is what one astronomer declares in one of our current encyclopedias: "Nearly all systems of thought, and especially Christian thought in the Middle Ages, was based on the supremacy of man in creation, hence the importance of the Earth in the universe, an importance that naturally sprang from its place at the center of everything. Copernicus destroyed this myth, proposing instead a much simpler system in which the Earth was but an ordinary planet, entirely under the dominion of the sun."[1] Nothing is more widespread than the ideas expressed in this text, nothing more false.

And from the very first it is clear that if Plato or Aristotle are geocentrists, this would not be because of Christianity, and therefore that the place of the Earth at the center of everything might follow from considerations other than those of human supremacy over the created. But we must also examine the significance of this geocentrism.

As for Plato, even the most superficial reader cannot ignore that, in his doctrine, the Sun performs the central function; next, as the *Republic* teaches, it is the 'offspring', the 'child' of the Supreme Good, the visible image of the anhypothetical Principle, of the superontological Absolute.[2] The Earth itself is, here below, an inferior region, the one from which we must depart to turn toward the heights. Platonism is then, although astronomically geocentric, functionally heliocentric. But the same remark is valid for Aristotle, who is careful to distinguish two meanings for the word 'center', the center of an expanse and the center of a thing: "It is better to conceive of the case of the whole heaven as

1. A. Boischot, "Astronomy" article in the *Encyclopedia Universalis*, tome II, 691. We find a kindred but more elaborated concept in Koyré, who writes: "It is only in the Christian tradition that a connection between these two [geocentrism and anthropocentrism] is established and asserted, for in this tradition the earth is the place where the divine-cosmic drama—the Fall, Incarnation, Redemption—unfolds itself, and which alone gives meaning to the creation of the universe" (*The Astronomical Revolution*, 72–73, note 8).

2. Book VI, 508b; likewise in Book VII, 516b: It is the Sun that "presides over all things in the visible region."

analogous to that of animals, in which the center of the animal and that of the body are different. For this reason they [the Pythagoreans] have no need to be so disturbed about the world, or to call in a guard for its center: rather let them look for the center in the other sense [the Sun] and tell us what it is like and where nature has set it. That center [the Sun] will be something primary and precious; but to the mere position [the Earth] we should give the last place rather than the first. For the middle is what is defined, and what defines it is the limit, and that which contains or limits is more precious than that which is limited, seeing that the latter is the matter and the former the essence of the system."[1] There is no doubt then that Aristotle is essentially a heliocentrist (the Sun is the natural center of the universe) but 'locally' a geocentrist. And if the Earth is at the center, this is because, in a spherical universe, *the local center is the lowest point.* Now the Earth, being a heavy body (a solid body), falls until the moment when it reaches its natural place, that is until the lowest point. Having arrived there, it can no longer move about naturally. It remains there by its own weight, and therefore has no need to be 'carried' by something else. This is a perfectly rigorous concept, given the principles of Aristotelian physics. The commentators Simplicius and Theon of Smyrna are not mistaken: the world is like an animal, with the Earth its navel and the Sun its heart. Cicero summarizes this concept in the grandiose tableau of *Scipio's Dream.* He depicts "the Sun—the leader, governor, the prince of the other luminaries; the soul of the world, which it regulates and illumines, being of such vast size that it pervades and gives light to all places." Next, after having situated Venus and Mercury, he comes to the Moon: "Below this, if we except that gift of the gods, the soul, which has been given by the liberality of the gods to the human race, every thing is mortal, and tends to dissolution."[2] This text is actually as much Platonic as Aristotelian: Platonic by mention of the light which spreads throughout the universe, Aristotelian by recognition of the two cosmic regions on either side of the Moon. One of the last Platonists, Proclus Diadochus, expresses himself no differently when he declares that we should be more or less content with everything sublunar, "we who are lodged, as they say, at the bottom-most point of the universe."[3]

True, the case of the Pythagoreans might be raised as an objection.

1. *On the Heavens*, II, 13, 293b, 1–15.

2. Yonge translation, *The Treatises of M.T. Cicero* (London: Henry G. Bohn, 1853), 384. Cited by L. Blanche, *Revue*, vol. 18, no. 5, 6.

3. *In Platonis Timoeum commentaria*, ed. E. Diehl, Lipsiae, tome 1, 353 (*Tim.*, 29c–d); cited by P. Duhem, *Le système*, tome II, 107.

For them the center is not the world's lowest point, but, to the contrary, its most noble. However, as we have seen, Aristotle's criticisms are directed at them when they are reproached for confusing essential center with local center. But, even if they differ in this from Aristotle and Plato (the young Plato), they are in agreement with them on the Earth's inferior situation, since it is the Sun that they put in the noblest place. We therefore must conclude with L. Blanche that "pagan antiquity is unanimous in scorning the earthly globe."[1]

As for the connection joining essential and qualitative geocentrism to anthropocentrism, it is basically weak. It must be observed first that there are two kinds of anthropocentrism: one that intends to reduce everything to human measure—like Protagoras, whom Plato censured—and another that posits man as sovereign of the universe, as the noblest of all beings. Next, it must be remarked that, in this second sense, Aristotle is altogether anthropocentrist, "for man, when perfected, is the best of all the animals."[2] To tell the truth, this superiority stems from his nature, characterized by the intellect's presence within it and aptly reflected in his body by its vertical stance: "Of all living beings with which we are acquainted man alone partakes of the divine, or at any rate partakes of it in a fuller measure than the rest . . . in him alone do the natural parts hold the natural position; his upper part being turned towards that which is upper in the universe. For, of all animals, man alone stands erect."[3] This is why J.M. Leblond can write: "Man is clearly the center of reference and pole of Aristotelian biology . . . he is the most natural of all the animals, the one in whom the intentions of nature are fully realized."[4] Such is Aristotle's doctrine that is more or less shared by all ancient philosophers.

This is also the doctrine of the Christian philosophers. And yet, here too, we are unable to provide every piece of documentation. But it is clear that, for Christians, the Earth was situated at the lowest point of the universe, while the Sun was the very image of Christ, *Sol invictus*, creation's king. In no way, though, does this hinder man from being more perfect and noble than all other earthly beings. Commenting on Aristotle's *De Caelo*, St. Thomas declares: "Containing bodies are the most formal, but contained bodies are more material. This is why in all

1. L. Blanche, *Revue*, vol. 18, no. 5, 8.

2. *Politics*, Book I, chap. I, 1253a; Jowett trans. (Oxford: Clarendon Press, 1885).

3. *On the Parts of Animals*, II, 10, 656b, 7–13; Ogle translation.

4. Introduction to *On the Parts of Animals* (1st bk.) in *Aristote, Philosophe de la vie* (Aubier: 1945), 44.

the universe . . . the Earth, which is contained by all the rest of the universe, since situated in the midst of all other locales, is the most material and ignoble of all the bodies."[1] Conversely, the heavenly bodies, which are composed of aether, are superior to all that is sublunar. Each heavenly body is, moreover, unique to its species, like the separated substances: "We do not find among them several individuals of the same species; there is only a single sun and a single moon." In this way he presents a remarkable analogy with the angels.[2]

With St. Thomas we are dealing with an Aristotelian, at least in natural philosophy.[3] But the breadth of his doctrine should not hide the existence of a Neo-Platonic and Pythagorean current which, in natural philosophy as in metaphysics, pervaded the whole of the Middle Ages. Now this tradition is often heliocentrist. A typical case is that of John the Scot, who drew from sources other than Isidore of Seville or Bede the Venerable. John the Scot had read the Greek Fathers, Pliny the Elder, and Ptolomy. He also had read Martianus Capella, who provided the Middle Ages with the classical manual for the *trivium* and *quadrivium*, that is to say the seven liberal arts;[4] he knew *Scipio's Dream* in the Macrobius version, as well as the commentary on the *Timaeus* by Chalcidius. All of these authors follow Heraclides of Pontus, who asserted that the Sun is at the center of the orbits of Mercury and Venus.[5] Now the system of Heraclides seems to have been quite popular. We have seen that Theon of Smyrna favors it. Before him Vitruvius speaks of it as a generally adopted system.[6] This is also what John the Scot does, who writes: "[T]he body of the sun, since it possesses the middle [region] of the world for, as the philosophers affirm, the distance from the Earth to the Sun is the same as that from the Sun to the stars—is understood to

1. *Sententia de caelo et mundo*, II, 20, in: Joseph de Tonquédec, *Questions de cosmologie et de physique chez Aristote et Saint Thomas* (Vrin: 1950) 15.

2. *De spiritualibus creaturis*, 8, 2; Tonquédec, op. cit., 26.

3. On St. Thomas as precursor to Copernicus, cf. Appendix 3, 138–39.

4. *De nuptiis Philologiae et Mercurii libri duo, de grammatica, de rhetorica, de geometria, de arithmetica, de astronomia, de musica libri septem* (*On the Marriage of Philology and Mercury*).

5. The heliocentrism of Heraclides of Pontus (a direct disciple of Plato, who was entrusted with the direction of the Academy during the travels of the master) is denied by some authors. Besides, his works are lost. But Jacques Flament, studying an eighth- or ninth-century planisphere illustrating an astronomical text, has shown that Mercury and Venus are represented as revolving around the Sun ("Un témoin interéssant de la théorie héliocentrique d'Héraclide du Pont," in *Hommages à Maarten J. Vermaseren*, vol. I [Leiden: E.J. Brill, 1978], 381–391; the planisphere is reproduced on Plate LXXVI).

6. Duhem, tome III, 47–51.

occupy a kind of midway position.... But the planets which revolve about it change their colors in accordance with the qualities of the regions they are traversing, I mean Jupiter and Mars, Venus and Mercury, which always pursue their orbits around the Sun, as Plato teaches in the *Timaeus*."[1] As we see, John the Scot adds the two planets Jupiter and Mars to the doctrine of Heraclides. But do we have to grant, as Koyré contends, that John the Scot is an exception, and that he alone took the system of Heraclides of Pontus seriously?[2] This is certainly not the opinion of Duhem, who, after reviewing a certain number of medieval authors, such as Manegold, Bernardus Sylvestris and the School of Chartres to which he belonged, Pseudo-Bede and William of Conches, concludes: "Thus, thanks to Chalcidius and Martianus Capella, and to Macrobius, the majority of those who, from the ninth to the twelfth century, have written on astronomy and whose books are preserved, have known and acknowledged the planetary theory envisioned by Heraclides of Pontus."[3] Now the least to be said is that these authors, and foremost John the Scot, the greatest of all, are not adverse to symbolism, and to a symbolism based on an ontological foundation, that of a true participation of things below in higher realities.

Similar statements can be made if we turn to different cultural settings, where a certain heliocentrism may coexist with a certain geocentrism. Chinese astronomy, which "practices little planetary astronomy" is a dubious case, lacking a well worked-out geometrical doctrine,[4] which would have enabled it to construct celestial trajectories in space. Even so the Sun is considered Master of the world and "emblem of the leader."[5] India shares with China the concept of an extremely vast cos-

1. *De divisione Naturae*, liber III, *P.L.* CXXII, col. 697–698; trans. I.P. Sheldon-Williams, revised by John O'Meara, *Periphyseon, The Division of Nature* (Montréal: Bellarmin/Washington: Dumbarton Oaks, 1987), 327–328.

2. Koyré, *The Astronomical Revolution*, 40–41.

3. Duhem, t. III, 110.

4. Joseph Needham, *The Grand Titration*, (London: Allen & Unwin, 1969), 85.

5. Marcel Granet, *La pensée chinoise*, (Paris: Albin Michel, 1968), 241. Let us add that, as Needham has shown (op. cit., 45), the concept of cosmic space is much more vast for the Chinese than for the Greeks. On this topic he speaks of "the infinite and empty spaces of the Hun-tian Chinese school or the relativistic Buddhist philosophers." Conversely, and contrary to Needham, it must be said that it is precisely because their space is bounded that the Greeks were able to create a science of it. This is in particular what Charles Mugler shows ("L'infini cosmologique chez les Grecs et chez nous," an article published in *Lettres d'Humanité*, tome VIII, [Paris: Les Belles Lettres, 1949], 43–66). He explains that the illusion of the sphere of the fixed stars—the belief that all the stars are fixed on an immense sphere—made possible the formation of an astronomical science. Otherwise the phenomena would have seemed too chaotic (53).

mic space, which, from the outset, relativizes geocentrism considerably.[1] Here the Sun likewise plays the major role it assumes in all traditional doctrines. Certain astronomers, such as Aryabhata in the sixth century, have even taught the rotation of the Earth upon itself.[2] More than Chinese culture, Indian culture was broadly diffused through Asia as well as the Near East. Its influence on Arab science is certainly undeniable. This is why it is not surprising to encounter in al-Biruni, that great polyglot traveler who had a lengthy stay in India, an assertion (not shared by al-Biruni, moreover) of the Earth's rotation. He actually attests that he saw an astrolabe "called *Zuraqi*" in compliance with "the belief that universal visible motion is produced in fact by the Earth and not by that of the heavens."[3] This belief is not an isolated fact. Al-Biruni himself attributes its origin to a Hindu (Aryabhata) and tells of its numerous adherents in his treatise *al-Qanun al Masudi* (Masud is the name of the sultan to whom the book is dedicated). This is why we can conclude, along with Carra de Vaux, that around the tenth or eleventh century (Biruni's work dates from 1066) the hypothesis of the Earth's rotation "had occurred in rather numerous discussions."[4] Even in those places where the Greek Aristotelian-Ptolemaic tradition predominates most directly, a kind of functional and, in a certain manner, astronomical heliocentrism shows through. This is especially the case for one of the greatest Arab metaphysicians, the Sufi Muhyiddin Ibn 'Arabi. Although not an astronomer like Biruni, what is essential for him is not the objective exactness of a description of the world, but the metaphysical and transcendent truth of the symbolism of appearances: "[T]he terrestrial position of the human being serving as the fixed point to which will be related all the movements of the stars—here symbolizes the central role of man in the cosmic whole, of which man is like the goal and the center of gravity."[5] And yet this anthropocentrism is spiritual in nature and in no way connected to any pride whatsoever. The proof is precisely the status of servant, which is the status of man as

1. Jean Filliozat, "Science de l'Inde ancienne," an article published in *Approches de l'Inde* (Cahiers du Sud, 1949), 251.

2. Jean Filliozat, "La science hindoue," in *Histoire des sciences*, under the direction of René Taton, tome I, *La science antique et médiévale*, PUF, 165.

3. S. Pines, "La théorie de la rotation de la Terre à l'époque d' Al-Bîrûnî," article published in *Journal asiatique*, tome CCXLIV (Paris: Librairie orientaliste Paul Geuthner, Société asiatique, 1956), 301–306.

4. Carra de Vaux, *Les penseurs de l'Islam*, tome II (Paris: Paul Geuthner, 1921), 218.

5. Titus Burckhardt, *Mystical Astrology According to Ibn 'Arabi*, trans. B. Rauf (Cheltenham, UK: Beshara Publications, 1977), 10.

such in Islam: "Man is at once transient and eternal."[1] And yet this spiritual geocentrism is combined with a functional heliocentrism. In fact, as we have pointed out with John the Scot, the Sun (*ash-Shams*) is *in the middle* of the heavens. Ibn 'Arabi even expands the Ptolemaic system, since there are for him seven heavens inferior to the Sun, with seven superior to it going as far as the Divine Throne, the synthesis of the whole (visible and invisible) cosmos, while the Earth is both the inferior conclusion and "center of fixation."[2] At the center of this immense cosmic tableau is to be found the sun, which is called *Qutb*, the 'Pole', and *Qalb al-'âlam*, the 'heart of the world'. It is this place 'in the middle of the heavens' that the Earth, or rather the terrestrial orb, occupies in the Copernican system, while for him the Sun holds the same place that the Earth does among the Ptolemeans. This is why highly regarded specialists have stated that Copernicus was actually a geocentrist, at least functionally.[3]

If we turn lastly to the Kabbalah, the most important text of which, the *Zohar*, 'appears' in the twelfth century, we notice that, even though its general cosmological scheme remains Ptolemaic, traces of Pythagorean-Platonic astronomy are nevertheless thought to be found in it. This was in any case the opinion of A. Franck and Jean de Pauly, who interpret the following text in a Copernican sense: "The whole inhabited Earth turns around as in a circle.... As a result there are, on Earth, certain regions where it is truly night while, in other regions, it is daytime.... There is also a region where daylight is constant and where the night is quite short."[4] We could likewise read another text in this

1. Muhyiddin Ibn 'Arabi, *The Bezels of Wisdom*, trans. R.W.J. Austin (Mahwah, NJ: Paulist Press, 1980), 51.

2. Burckhardt, op. cit., 12. The arrangement given by Ibn 'Arabi is the following: the inferior skies (earth, water, air, aether, sky of the Moon, sky of Mercury, sky of Venus)—sky of the sun—the superior skies (sky of Mars, sky of Jupiter, sky of Saturn, sky of the Fixed Stars, sky without stars, sphere of the Divine Pedestal, sphere of the Divine Throne).

3. "When Copernicus places the Sun at the center of the Universe, he does not place it at the center of celestial motions: neither the center of the earth's sphere nor that of the planetary spheres is placed *in* the sun, but only *near* to it, and the planetary motions are referred not to the sun, but to the center of the earth's sphere—eccentric with respect to the Sun.... As a result, we have the paradox, that in the celestial *mechanics* of Copernicus the Sun plays a very unobtrusive part. It is so unobtrusive that we might say that it plays no part whatsoever" (Koyré, *The Astronomical Revolution*, 59, 65). In fact, as underscored by Koyré, for Copernicus the Sun gives light to the universe, which connects the Polish astronomer to Pythagorean light mysticism.

4. *Zohar*, III, fol. 16 a, trans. J. de Pauly, tome V (Paris: Ernest Leroux, 1906), 29; cited by Duhem, *Le système*, tome V, 143. To the contrary, we read in the translation of the

way, one dealing with the numerous wheels that cause the turning of corporeal aggregates: "These wheels make the Earth turn in a circle and upon itself."[1] Still, although Father Duhem does not think that these passages are truly Copernican, we should not rely on the Jean de Pauly's translation, and nothing of this is to be seen again in the one provided by Charles Mopsik.

We now need to draw a conclusion from this brief historical review. It is categorical: pre-Copernican geocentrism is unconnected to the naïve anthropocentrism of a Christianity that affirms the 'supremacy of man in the universe' and would see its cosmological validation in the Earth's central position. This is not because "two [geo- and anthropocentric views] are established and asserted, for in this tradition the Earth . . . is the place where the divine-cosmic drama—the Fall, Incarnation, Redemption—unfolds itself, and which alone gives meaning to the creation of the universe."[2] At the very least, the meaning of this 'divine-cosmic drama' should not confer on the Earth a privileged situation incompatible with its Copernican decentering. For the Incarnation is also a 'fall', an abasement, a *kenosis*, according to the word of St. Paul, that is to say, in its proper meaning, an annihilation. By coming to earth, God approaches nothingness. He does not establish Himself at the center of the world, but, to the contrary, on its periphery. In the Bible, the Sun, Moon, and the heavenly host (the stars) are not considered to be inferior to the Earth, which symbolizes the 'vanity', the nothingness, the deficiency of created being, its vocation to die: "thou [shalt] return to the earth out of which thou wast taken: for dust thou art, and into dust thou shalt return" (Gen. 3:19).[3] And this is why the Ecclesiast can cry out: "Why is earth, and ashes proud?" (Ecclus. 10:9). And St. Paul commands: "Seek the things that are above. . . . Mind the things that are above, not the things that are upon the earth" (Col. 3:1–2). Without doubt, it has been pointed out insistently how the Bible proscribes what is called 'star worship'. Testimonies are not lacking.[4] How-

Midrach ha Neelam by Bernard Maruani at the end of volume I of the *Zohar* translated by Charles Mopsik (Lagrasse: Verdier, 1981, 578–579) the following text: "The Holy One, blessed be He, has made a fourth space in which He has placed the sun, and the latter turns around the world according to moments and periods for the gain of the world and for mankind" (15c). These translations were obviously not available to Duhem.

1. *Zohar*, II, fol. 235b; P. Duhem, *ibidem*. Tome IV, 261–262.

2. Koyré, *The Astronomical Revolution*, 72–73, note 8. This asserts precisely the thesis that we reject.

3. Also Job 10:9, 34:15, Psalm 103:29, etc.

4. Texts relating to this question are studied by A.M. Dubarle in *La manifestation naturelle de Dieu d'après l'Ecriture* (Paris: Cerf, 1976), 65–67, 132, 134, 144–145, 149.

ever, the clearest text in this regard is somewhat startling. Moses expresses himself in this way: "And beware lest you lift up your eyes to heaven, and when you see the Sun and the Moon and the stars, all the host of heaven, you be drawn away and worship them and serve them, things which YHWH your God has allotted to all the peoples under the whole heaven" (Deut. 4:19).[1] YHWH has therefore set aside a people to whom He has revealed His Name. But, for the other peoples of the universe, He reveals Himself through the heavenly symbols of the Sun and the stars. He Himself therefore establishes and authorizes such a symbolism. The host of heaven forms, moreover, part of the God-King's court: "I saw YHWH sitting on his throne," says the prophet Micheas, "and all the host of heaven standing beside him" (1 Kings 22:19). This is why these cosmic beings can be considered divine in nature, the Sun and Moon as "rulers of the world" (Wisd. 13:2). Is not the Sun, moreover, a symbol of the Messiah?—"But unto you that fear my name, the Sun of justice shall arise" (Mal. 4:2). This announcement by Malachi will be taken up again and ratified by all of Christianity. From St. Ambrose to St. Francis of Assisi, in whom it achieves something like its perfect expression, the 'Canticle of the Sun' is unfolded and repeated: "The Sun begins to arise. Cleanse, now, the eyes of your mind and the inward gaze of your soul, lest any mote of sin dull the keenness of your mind and disturb the aspect of your pure heart.... It is true that it [the sun] is the eye of the world, the joy of the day, the beauty of the heavens, the charm of nature and the most conspicuous object in creation."[2] How can we not relate these Ambrosian images to those strangely consonant ones met with in Hindu tradition: "The sun," says the *Bhavishya Purana*, "is visible divinity, the eye of the world, the maker of the day."[3] But there would be no end to it if we were to cite all those texts that manifest this symbolic heliocentrism; and, as we have shown, the Christian voice is not least in this concert.

We must then face the facts. The explanation currently upheld by historians of science is quite simply false. What is at issue in the Copernicus-Galileo affair is not the destruction of a reassuring biblical universe bursting apart under the blows of a finally discovered real universe. It was much more profoundly—and more or less confusedly perceived—

1. This interpretation is that of St. Justin, *Dialogue with Trypho*, LV, 121; *P.G.*, tome VI, 596; Clement of Alexandria, *Stromata*, VI, 14; *P.G.*, tome IX, col. 333; Origen, *Commentary on St. John*, Book II, 3; *P.G.*, tome XIV, col. 112; etc.

2. St. Ambrose, *Hexameron, Paradise, and Cain and Abel*, trans. J.J. Savage (New York: Fathers of the Church, 1961), 126–27.

3. Cited by Alain Daniélou, *Hindu Polytheism* (New York: Bollingen Foundation, 1964), 93.

the substitution of a universe-machine for a universe-symbol. This is not so much an astronomical revolution as a physical and cosmological revolution. And this is why the astronomical restructuring proposed by Copernicus hardly brought any reaction, while its revival by Galileo the physicist (who, besides, made no major astronomical discovery and even 'ignored' Kepler's works)[1] unleashed a crisis of culture and civilization still in progress.

In summary, we believe we have established the following points:

- The rotation of the Earth about itself or around the Sun is not, by itself, contrary to the cosmological symbolism of the various religious traditions—even though it may contradict, in a certain manner,[2] the universe perceived by the senses—since this rotation has been affirmed by eminently religious and symbolist cultures.

- Geocentrism and geostatism are not tied to the religious anthropocentrism of the Judeo-Christians since, on the one hand, we find an analogous anthropocentrism among non-Christians (Aristotle), and, on the other, the local or astronomical geocentrism of the Judeo-Christians and even the Muslims is accompanied by a functional or axiological heliocentrism, and by a depreciating of the Earth's local situation, residing as it does at the lowest point of the universe.

FROM MYTHOCOSMOLOGY TO PHYSICS

We must now attempt to define the true nature of the Galilean revolution, the very effect of which was the almost total destruction of contemporary humanity's symbolic universe. Everyone agrees that this was a major event that 'proves' with the same blow the true stakes of a metaphysic of the symbol.

Now, if we wish to grasp the most radical significance of Galileism, we have to stop seeing the pre-Galilean cosmology as a whole, and distinguish within it a Pythagorean-Platonic form succeeded by an Aristo-

1. Kepler formulated his famous laws on the elliptical nature of the planetary orbits around the Sun between 1614 and 1616. And yet Galileo, who knew him well, only wanted to envisage circular orbits. As a good Copernican, he upheld this point of view until his death in 1642. On the necessity of circular movement for Galileo, see Clavelin, *The Natural Philosophy of Galileo*, 212–213.

2. We say 'in a certain manner', for, in another manner, it provides an altogether satisfying explanation, one which the people of the Middle Ages had accounted for perfectly. Thus Nicolas Oresme writes: "... if the Earth is moved with a daily movement and not the Heavens, it would seem that the Earth was at rest and the Heavens were in motion." (This is from a commentary on the pseudo-Aristotelian treatise *On the World*, cited by Lenis Blanche, *Revue*, vol. 18, no. 5, 17); cf. likewise Koyré, *The Astronomical Revolution*, 56.

telian form. Up until this point we were not compelled to view them separately, it being given that sacred symbolism, appreciated from our vantage point, seemed to accommodate itself to either one, as history shows. However, as a matter of fact the Galilean revolution was directly opposed to Aristotle much more than to Plato,[1] and it was basically the Aristotelians who combatted it. But this was not due to an accident of history. There was in fact a considerable difference between these two forms of ancient cosmology, which is important to emphasize and which explains, at least in part, the bitter conflicts. The difference is between a basically metaphysical and religious perspective, and a basically rational and physics-oriented perspective.

In reality and properly speaking, there is no autonomous and 'scientific' cosmology for the Pythagoreans and Platonists.[2] What is posited from the outset is the divine and the sacred. The point of departure and the convergence of doctrine is fundamentally bound up with a 'mystical' vision of primal and ultimate Reality. The order of the world, the concept of a universe, is a *consequence* of this Divine Reality. It springs from it as a sensible illustration of what, in itself, is invisible and transcendent. The world is not encumbered *a posteriori* with a more or less adventitious symbolic meaning. From the outset, and in its very substance, it is endowed with an 'iconic' function. The cosmos *is* the manifested image of a Reality and Order both unmanifested and unmanifestable in itself: "for Pythagoras, the sensible universe is spherical because this form is the perfect form; spherical also all the stars and, among them, the Earth. Likewise, if the stars and, among them, the Earth turn around a central fire, this is for a reason of the metaphysical order."[3] Analogous remarks can be made in connection with Plato: "This world," Timaeus informs us at the beginning of his explanation, "must of necessity ... be a copy of something."[4] And that of which it is an image is the most beautiful of models, since it is itself the most beautiful of things in the sensible order,

1. In his treatises on movement (*De Motu*) Galileo attacked Aristotle quite violently, but declared himself for the Ancients (*nos pro antiquis*). And the Ancients? Essentially Archimedes, but also Plato (A. Koyré, *Galileo Studies*, 34).

2. This implied relationship between Pythagorean and Platonic cosmology might be contested. But, apart from the kinship that seems obvious between number with Pythagoras and essence with Plato, it should not be forgotten that Timaeus, the spokesman for Platonic cosmology, is a Pythagorean, according to a tradition undoubtedly going back to Plato himself.

3. Ivan Gobry, *Pythagore ou la naissance de la philosophie* (Paris: Éditions Seghers, 1973), 50.

4. *Timaeus*, 29b.

"God accessible to the senses."[1] Moreover, the dialogue *Timaeus* is as a whole a kind of myth, a symbolic manner of speaking about a subject—being in becoming—which, by its very nature, eludes true knowledge: Platonic cosmology is ever but 'a probable myth' (*ton eikota mython*), as Plato expressly declares.[2]

This symbolic cosmology, as a result, does not intend to provide a realist description of a changing world, nor to enclose the nature of things within the form of scientific concepts. We must make up our minds about it: there is no certain knowledge of an object, the world is found wanting on this point of reality.[3] This is why its study has some chance of being exact only if one connects this object to its principle, which alone possesses a full reality. Here the world is in some manner seen from the vantage point of God, because there is in fact no other vantage point under which it can be perceived in its true nature. Not from a scorn for cosmology does Plato act in this way, but from a care for 'scientific objectivity'. Moreover, we will return to this question and show that only with symbolism is there true cosmology.

Quite otherwise is the vantage point of Aristotelianism: this is basically that of a nature philosophy, a physics. There the world is not viewed first as a moving and ephemeral image of motionless and principial Reality, but directly *in itself*. The revolt of Aristotle against Plato is foremost a revolt against metaphysics for the sake of the physical world's reality. Aristotle does not intend to tell us a 'probable story', but to construct "the theoretical science of being in movement" which is called physics.[4] True, there is a theoretical first science which is not physics and is called theology. But if it differs from the former by its object, being as being, "immobile and separated nature,"[5] and if it is in this sense the highest science, it does not differ from it as to the mode of knowing. *All sciences are, for Aristotle, sciences for the same reason*, and can be set in parallel, a concept directly opposed to that of Plato, who distinguishes a hierarchy of degrees of knowing: sensation, empirical knowledge, discursive reason, and intellectual intuition. These degrees

1. Ibid., 92c.

2. Ibid., 29d.

3. This does not mean that a science of the world is impossible, but that, as such, its certainty should not be apodictic. If Kepler had known the various positions of the earth around the Sun with greater precision, he would not have been able to bring them together through an ellipse! Cf. Ivar Ekeland, *Mathematics and the Unexpected* (Chicago: University of Chicago Press, 1988), 40–46.

4. *Metaphysics*, Book 6, 1, 1025c–1026a.

5. Ibid.

differ not only by their respective objects, but also by their modes. Certainly, the last two degrees pertain equally to science (*episteme*), whereas the first two are a matter of opinion (*doxa*). However, as stated in the *Republic*, the difference separating empirical from rational knowledge (the average person from the expert and the mathematician) is *less great* than the one separating rational knowledge from intellectual intuition (the expert from the 'philosopher'). Metaphysical knowledge, in Platonism, is therefore of another nature than rational science: it is a direct and deifying vision that assimilates us to its transcendent Object, beyond any discursive and mediate knowledge, while, for Aristotle, *first* philosophy is only the upper limit of physical knowledge.

This point of view implies that the universe, having been 'posited in itself', must find its principle of intelligibility within itself. At the same time, there must be a conceptual consistency to scientific discourse, which becomes as realist as its object has become real, to correspond to the ontological consistency of cosmic substance. To the science of things corresponds a science of discourse on things, hence the invention of 'logic'. For the first time a style of thinking comes into the world that exhibits its own formal coherence as the norm and guarantee of its truth. What a difference between the Platonic language, continually renewing itself and adjusting, fundamentally open to the real—for this reason always aware of its inadequacies—with its recourse to symbols, and the language of Aristotle, a precise and meticulous clockwork, a technical language because philosophy is conceived of as a technique of the *logos*, in short, the wherewithal to satisfy scientific needs, that is to say the needs of a conceptual possession of reality, that great imprisonment of things!

True, for Aristotle there is an equilibrium between the science of the world and the science of discourse, because there is a profound agreement between the two, the forms of things informing the cognitive intelligence. However, this equilibrium, like every non-hierarchic equilibrium, is fragile. It is the equilibrium of a scale beam, not that of a building with its segments rising in tiers hierarchically.[1] Seen from without, the Aristotelian construct as a whole shows itself much more

1. For a St. Thomas Aquinas, by what only seems a paradox, this horizontal equilibrium is in some manner protected by a vertical religious one. Brought into relation with the transcendence of revelation, the diverging tensions at work in the horizontal equilibrium of Aristotelianism are neutralized: the *nominalist* need for a formality of discourse, and the *realist* need for the openness of the intellect to things, are thoroughly relativized by another need, that of eternal salvation. When this loses its urgency the latent conflict becomes apparent.

strongly structured and unified than the Platonic effort. But, just as with buildings made of concrete, an earthquake occurs and it collapses into a single heap.

However that may be, the essential point at present is the one indicated just a moment ago: the intelligibility of things is to be found in things themselves. The ordering of the world is sheer immanence to the beings of the world. For want of conceiving of the *transcendence* of the intelligible and its being *participated in* by the sensible,[1] Aristotle rendered it inseparable from things. By this very fact, he conferred on the sensible world the coherence, permanence, and necessity of the intelligible. Hence the eternity of the world and its order. In a certain manner, there is no longer any appearance. Everything is true. The figure of the universe is integrally justified. Nothing is more significant, in this respect, than Aristotle's philosophy of motion, analyzed so remarkably by Clavelin. For Plato, motion, which is only one form of change, that is to say of the non-permanent, is necessarily bound up with the ephemeral nature of a reality subject to becoming. The image of immutable Reality, being image and not model, is inevitably mobile and changing. The mobility of the image does not contradict the immutability of the Model; to the contrary, it expresses it, is a participation in it. Movement is thus made intelligible to the very extent that the world does not have its raison-d'être in itself, and is as if the *very expression* of this 'hetero-sufficiency'. Aristotle has forbidden *a priori* such a metaphysical explanation of movement. It is in the world itself that movement's meaning should be found. But movement, being a process, has in a certain manner no proper essence. Quite curiously, there is with Plato an essence for becoming, and for the most inferior of realities, 'hair, mud, and dirt', as an aging Parmenides taught the youthful Socrates—while, for Aristotle, movement can find its meaning only in its term. In itself it is nothing

1. That a genius as extraordinary as Aristotle was unable to understand this Platonic doctrine, and misinterpret it to the extent of producing an absurd caricature (in particular by reifying the *eidos*), is one of the history of philosophy's great mysteries. This mystery teaches us that a human intellect can display the most universal genius without for all that exceeding the limits of its own mental horizon. This mental horizon is basically determined by what we will call the 'sense of the real'. What is 'instinctively' real for Aristotle is a particular tree, cat, or man, in short, an individual substance. For him, nothing exists except as an individual substance, or in a substance (for an accidental being). He has no 'way of imagining' that certain modes of the real can be altogether different from the one conferred by the experience of the senses, which is precisely the case with the Platonic essence: in order to be real it has no need to exist in the manner of a thing.

but *realization* of this term: there is no *ousia* of movement, says Aristotle.[1] Movement is therefore the actualization of a nature which, in order to be what it ought to be, has need of becoming. The meaning of movement, its intelligibility, is to be a realization of this nature.[2] Hence, since the meaning of movement is in the term to which it leads, which once attained implies the cessation of movement, that is to say rest, it follows that every rest from a movement is a sign that the movement has reached the term to which it should attain *by nature*. The (unforced) rest of a body naturally defines its *natural place*, the place (or state) to which its nature destines it. If then a body falls, this is because it desires to realize its nature, for example when it has been violently impeded, as when a stone is thrown into the air. As a consequence, cosmic space is found to be provided with an 'ordered structure'[3] formed by the qualitative difference of all natural places able to be known by observing the sectors of the cosmos where the different bodies find their rest.

Basically, everything transpires as if Aristotelianism were a translation of Platonism into cosmology and physics. Metaphysics becomes astronomy, and the distinction between intelligible Reality and sensible reality becomes that between the supralunar and sublunar worlds. The ordered structure with which the cosmos is provided determines its different qualitative regions that account for, that is to say assure, the intelligibility of all physical phenomena. As a result there are two series of consequences:

- Intelligibility and ordered structure of cosmic space are absolutely integral, the first being but the reflection of the second, which is itself

1. *Physics*, V, 225b. For Plato, becoming exists before the creation of the world: "being and space [the cosmic receptacle] and generation [becoming], these three, existed in their three ways before the heaven" (*Timaeus*, 52d; Jowett trans.). Clearly, then, there is in a certain manner an archetype of becoming. But for Aristotle, movement cannot have any essence to the very extent that there is no world other than the spatio-temporal universe, that is to the extent that the intelligible world is denied. Motion, being in this case an actualization of essence (of what is moving), cannot itself be an essence; and there is no other essence than that realized in the sensible world. Likewise, the Aristotelian analyses of movement and rest in his *Physics* could be compared with the *metaphysical* analyses of them given by Plato in the *Sophist* (247–249).

2. Obviously, Aristotle is right to make of movement the physical realization of the essence; but he is wrong to reduce it to that. Moreover, we are describing here a 'common' and 'global' Aristotelianism, because it does not seem to be without responsibility in the Galilean crisis, but we are not intending to restore the doctrine in its profound complexity. Aristotelianism is a cosmology *rather than* a metaphysics, his point of view being that of *samkhya* rather than *vedanta*; but clearly neither is there a true cosmology—which is Aristotle's philosophy—without metaphysics.

3. Clavelin, *The Natural Philosophy of Galileo*, 14.

nothing but the perceived structure of the universe when we construct a rational justification for it.

• No longer is there any symbolism, since there is no longer any cosmos-symbol.

True, such a cosmos is, by itself, not opposed to a possible symbolic treatment. The historical example of its adoption by Christianity proves it. However, a non-incompatibility rather than a true accord is involved here. More precisely, although by its hierarchical *structure* this cosmos lends itself to the symbolic usage that religion must make of it, *ontologically* it is opposed to it by the nature of its substantiality.

Religious Symbolism and Physical Realism in the Middle Ages

THE SCHOLASTIC MISUNDERSTANDING

As we have shown, all human culture is witness to the fact that the sensible world, the world of perception, is open to being treated symbolically. Such a symbolism is connatural with religion: the divine speaks to man the language of creation. This theme is likewise present in Christianity. The heavens and the earth tell the glory of God; the Invisible manifests its *Being* and *Power* in the visible universe.[1] For more than a thousand years Christianity had found in Platonism and Neoplatonism a metaphysics especially appropriate to such a belief, a metaphysics that provides an altogether adequate mode of expression. Undoubtedly this was a Platonism 'rectified' by St. Augustine in which the intelligible forms are situated in God Himself, in the divine Word. But it is finally from Platonism and by Platonism that Christian thought was intellectually nourished, particularly in reading the *Phaedo* and the *Timaeus* with the commentary by Chalcidius. To this we must add, which is too often forgotten, the Latin poets, Virgil and Ovid above all, who supplied Christian symbolism with many of its elements.[2] Christianity lived in this way until the middle of the twelfth century, until the advent of the Aristotelian 'freshet'. And we know that Aristotle did not conquer with-

1. Psalm 18:2, Rom. 1:20: "For the invisible things of him from the creation of the world are clearly seen, being understood by the things that are made. His *eternal power* also and *divinity*." It is striking that these two terms (Divinity and Power) correspond to the Hindu couple of *Brahma* and his *Shakti*, which Coomaraswamy calls the 'divine Biunity'.

2. Among others, cf. M.M. Davy, *Initiation à la symbolique romane* (Flamarion, 1964), 147; and J. Pépin, *Dante et la tradition de l'allégorie* (Vrin, 1970), 101–118, which

out resistance, nor was his triumph definitive. Quite significantly, the author most cited by St. Thomas Aquinas, thanks to whom the philosophy of the Stagirite became, however, the 'official' speculative vehicle for Catholic theology, is not Aristotle, but St. Dionysius the Areopagite. Besides, an intellectual genius of St. Thomas' breadth should not be reduced to the role of a Christian Aristotle. And yet it is incontestable that the conceptual form of the theological discourse becomes with him (and first with St. Albert the Great) basically Peripatetic.

The reason for this success was brought to light by several studies, especially the celebrated work of Father Chenu: *La théologie comme science au XIIIe siècle.*[1] This reason is that Aristotle supplied the model for what is scientific, and that the rigor of this model seemed to relegate the supple perspectives of Augustinian Platonism to the level of poetic approximations. As a consequence, the Platonic distinction of the modes of knowledge was eclipsed when confronted with the unity of speculative science, thus verifying what we previously stated about the Aristotelian *episteme* which brought what Plato had so clearly distinguished down to a single level. This happened because its author *no longer understood* what the metaphysical intuition of the Intelligibles truly is, and, beyond the Intelligibles, the superessential and superontological Good.[2] True, St. Thomas Aquinas can do this in the name of the transcendence of the revealed light, with respect to which noetic intuition and the dianoetic approach seem indistinguishable. For all that, the cost of such a reduction seems exorbitant: on the one hand, in its most essential act faith is deprived of the light of the intellect, which, outside the Thomist synthesis, cannot but lead to fideism; on the other hand, and as a conse-

cites abundant testimonies to an allegorical interpretation by Christianity of pagan culture. Dante provides the key to this interpretation: "Plato called them [the pure Intellects] 'ideas', which is as much as to say universal forms and natures. The pagans call them Gods and Goddesses, although they did not think of them in a philosophical sense as did Plato, and they venerated images of them and built great temples to them, as, for example, to Juno whom they called goddess of power, to Pallas or Minerva whom they called goddess of wisdom...." *The Banquet*, II, 4, 5–6, Lansing translation. There is then an analogy between the Ideas of Plato, the pagan gods, and also the angels of the Abrahamic tradition.

1. Vrin, 1957.

2. Characteristic in this respect is a question raised and decided in the affirmative by Aristotle: "Should our study [first philosophy] be concerned only with substances, or also with the essential attributes of substances?" (*Metaphysics*, Book *Beta*, I, 995b, 20). The unicity of philosophy, says Aristotle, which distinguishes only a first and a second philosophy, is similar to that of mathematics. There is the same relationship, as to science, between metaphysics and physics as between arithmetic and geometry (*Metaphysics*, Book *Gamma*, 2, 1004–05).

quence, theology becomes, in its most immediate act, a purely rational way of knowing—at the extreme, an atheist could do theology if he worked along the same axiomatic as the believer—which will lead directly to rationalism.[1] "The temporal and spiritual," declares the Thomist Chenu, "are no longer, with respect to revelation and its light . . . disparate objects, known by different processes: God and creature, mystery hidden in the Father, or biblical history, Eternal Word or Word made flesh, contemplative speculation or rules for the moral life . . . all plainly stem from the same principle of knowledge . . . these ways of knowing are encamped in the same field of intelligibility."[2]

Here, then, we have laid the groundwork for speaking of a misunderstanding. It was the strong faith of a St. Thomas, and the power of his spiritual gaze, that made him, if not insensible, at least indulgent toward Aristotle's basically non-religious naturalism. As Chenu says: "with respect to revelation and its light. . . ." As an inevitable result, the speculative order is no longer *intrinsically* religious or sacred, and its 'soul' is somehow added to it from without. What is true for theological discourse is also true for cosmology. The Aristotelian world is no longer, substantially, an epiphany of God. Its character as theophanic revelation is conferred upon it from without, by virtue of the power native to the symbolist instinct of a whole civilization. Such a situation of cosmological symbolism is clearly reflected in aesthetic forms, especially in the transition from Romanesque to Gothic style. The first is characterized by an immanence pregnant with the divine that transfigures the heaviest masses and lightens them from within without needing to reduce the thickness of their volumes. The second, to the contrary, seems to want to defy gravity and assert itself as the most visible and most vertical sign of transcendence in a lace-work of stone, as if the material was ashamed

1. When we say that Thomism (not St. Thomas) leads to fideism and rationalism, this can at least be granted as an observation of fact, for, two centuries after St. Thomas, both arose with Luther and Galileo—while a thousand years of Christian Platonism had known nothing of the sort. We acknowledge the sophism of a *post hoc ergo propter hoc* conclusion, and it is clear that from within the Thomist synthesis such a deviation is excluded. But it is *a priori* impossible for any synthesis whatsoever to remain at once living and unchanged. It is also certain that Augustinian Platonism can lead to gnosticism and illuminism, as well as to a mutilating dualism. It seems, however, that Platonism, by very virtue of the inward and inspiring nature of its light, more easily avoids the degradations of history than the more exterior formality of the Aristotelian system, and that, by the same token, it more efficaciously assures its function for speculative orthodoxy. In a certain manner, and by simplifying overmuch, Plato is a spirit to be rediscovered and Aristotle a letter to be preserved. Lastly, we will add that there is in any case a great difference between Thomism and St. Thomas Aquinas.

2. *La théologie comme science*, 97–98.

of its weight. Again we find the misunderstanding that we emphasized in connection with knowledge: the universe, which is thus enveloped in the light of biblical symbolism, is a universe that possesses both its own consistency and intelligibility independent of all symbolism. Such a universe, as we have said, is not opposed to symbolism; but, in its very being, does not require it.

This neutrality of the Aristotelian cosmos, that allows it to be treated symbolically in the light of biblical revelation, is tied to two of its essential characteristics, as noted above. We will say on the one hand that, in this universe, intelligibility and structure are completely interdependent, and, on the other, that this structure was nothing but the structure of the universe as perceived by us.[1] Now, as we have likewise made clear, the symbolic cosmos of a revelation is that of perception. Religion is addressed to man as he is: it is founded on the symbolism of appearances, appearances that are the object of a quite real (on its own level), coherent and therefore legitimate experience. In this sense the Aristotelian world is not opposed to the biblical one. On the other hand, by linking the visible structure of the world with intelligibility, by forming a *topology*, a veritable logic for the regional structure of cosmic space, Aristotelianism brought, in some manner and despite itself, the philosophic guarantee of rational knowledge to the religious meanings of symbolism. That there is no heaven other than the visible one, as Aristotle asserts,[2] would not be conceded by the religiously-minded.[3] But, for the physical heaven to be the natural place for the noblest bodies,

1. No idealization of the sensible perception of the world is involved here. We have discarded this overly superficial thesis. More profoundly, this involves a 'pre-existing' accord between perception and world, an accord which is itself only a particular mode of that accord which all knowledge in general realizes as the common act of the knower and the known. And sensation is already knowledge, the "act in common of the one sensing and the sensible," according to the Greek commentators (L. Robin, *Aristote* [Paris: PUF, 1944], 182). In such an act, the sensible 'informs' the one who senses. Hence the objectivity and realism of sense knowledge. For Aristotle, what is first is the *act*. The realism of sensible qualities is not a more or less naïve idealization of an ignorant 'science', but the rigorous consequence of a philosophical thesis.

2. *On the Heavens*, I, 9, 278b, 5ff.

3. Thus St. Thomas Aquinas (*Summa theologiae* I, q.68, a.4) set aside Aristotle to justify the biblical mention of 'heavens'. And St. Albert the Great is likewise altogether capable of setting aside Aristotle, not only for religious reasons, but also in the realm of natural philosophy each time observed data and scientific truth require it. This is established by A. Zimmerman in an article titled "Albert le Grand et l'étude scientifique de la nature," published in *Archives de philosophie*, Oct.–Dec. 1980, 695–711. The author attributes the Aristotelian dogmatism that Galileo struggled against to the Latin Averroism at the Paris Faculty of Arts. Combatted by St. Thomas, Averroës (twelfth century) is known as an ultra-Aristotelian interpreter of Aristotle.

composed of an aetheric or crystalline matter, is in perfect accord with a metaphorical use of the term. And it might even be said that this accords all too well. For—and this is the conclusion toward which our entire analysis is leading—the consistency that philosophy attributes to the world of appearances is much stronger than the consistency it attributes to religious symbolism when it makes use of these same appearances. What is rationally true, is universally and timelessly so. By making the intelligibility and topological structure of the cosmos interdependent, Aristotle conferred on this structure a *definitive and necessary reality*. The order of the visible is ontologically justified. Whereas sacred symbolism, prior to justifying the order of appearances by conferring on it the dignity of 'divine word', begins with a dismissal of this reality's claim to the title of initial and unique reality, positing it from the start as a reflection, that is, as ontologically deficient. Certainly visible beings receive, through their appointment to the title of sacred symbols by religious discourse, a kind of divine enhancement. But this is based on the 'vestigial' character of their mode of existence: "Heaven and earth will pass away, but my words will not pass away." The symbolic word saves precisely what there is of the essential in creation by trans-forming things into expressions of the divine. But this occurs through a passing beyond their existential form exactly because, if they remain in this form, they are lost. In appearance, sacred scripture's symbolism is *based* on the visible universe. In reality, only insofar as they serve as a mode of expression for the revelatory Word will the things of this world—doomed by nature to annihilation—not pass away.

So, it should be abundantly clear now that the Aristotelian ontology of physical reality is directly opposed to such a perspective. Although superficially conducive to biblical symbolism, deep down this ontology contradicts it. The day when the Aristotelian universe collapses, it will bring sacred symbolism along with it in its fall. And this is why we have spoken of a misunderstanding, because the implicit ontology of sacred symbolism is incompatible with the explicit ontology of Aristotelian cosmology.[1]

Did St. Thomas have a certain 'foreknowledge' of this collapse? One would think so in reading this text, which today conveys an almost prophetic tone (it deals with knowing whether the firmament was created on the second day): "In discussing questions of this kind two rules are to be observed, as Augustine teaches [*Gen. ad lit.* i, 18]. The first is, to hold the truth of Scripture without wavering. The second is that since Holy

1. On objections raised by our theses, cf. Appendix 4: *St. Thomas Aquinas: a theophanic cosmos?*, 139–142.

Scripture can be explained in a multiplicity of senses, one should adhere to a particular explanation, only in such measure as to be ready to abandon it, if it be proved with certainty to be false; lest Holy Scripture be exposed to the ridicule of unbelievers, and obstacles be placed to their believing." And St. Thomas continues by examining the various possible interpretations. The gist of what he says is this: if the firmament designates the starry heaven, "we must reconcile our differing modes of explanation according to the variety of opinions concerning the firmament," that is according to the variety of cosmological doctrines to which we intend to refer as warranted.[1] From such a perspective, we see that St. Thomas does not hesitate to declare that, on the whole—with the truth of Scripture safeguarded—we ought to subject its interpretation to the rational jurisdiction of the natural philosophies.

We might think it regrettable that Galileo's judges had forgotten these wise words at the most critical moment of this 'affair'. This would be wrong. They would have changed nothing, for the peculiarity of a misunderstanding is to remain imperceptible, insofar as its consequences do not give rise to any flagrant contradictions. Now there is a certain contradiction between Aristotle's explicit cosmology and biblical symbolism's implicit cosmology (just as in theology "[t]here is almost no similarity between the Christian's God of love and Aristotle's kind of God"),[2] a contradiction St. Thomas does not seem to perceive when he tends to favor what would later be called 'the equivalence of hypotheses', as pointed out above in connection with the Ptolemaic system; while the Stagirite's physics is basically realist, and also as little conjectural as possible.[3] It is not enough, then, to be pleased with the theologian's foresight; it should also be asked if it is compatible with the Aristotelian

1. *Summa theologiae* I, q. 68, a. 1.

2. Pierre Aubenque, "Aristote et le Lycée," published in *Histoire de la philosophie*, Encyclopédie de la Pléiade, tome 1 (Paris: Gallimard, 1969), 655.

3. The theory of natural place is a physical theory: it defines the nature of physical realities. The movement of the stars (and their distribution in space) necessarily flows from this. Aristotle's world is a world subject to as strict a necessity in its structure as in its movement. "Such a physics," writes Duhem, "would necessarily impose a clearly defined form on astronomical theory" (*Le système du monde*, tome II, 57). But, as Duhem shows, true astronomers are rather quickly obliged, taking observed appearances into account, to break with this physicist necessitarianism. We are in this case witness to an actual conflict between Aristotelian physics and Ptolemean astronomy, which required the decentering of certain planetary orbits, a conflict which spans Hellenic science and continues with the Middle Ages, the Arabs, and the Renaissance until Copernicus (ibid., 60–67). St. Thomas himself, who is clearly Aristotelian when he comments on the *Metaphysics* around 1265, is more sympathetic to Ptolemy's arguments (How do we explain the variations in the apparent orbits of the planets?) when he comments about

principles of his philosophy of nature. This (relatively) conventional view betrays in any case, for St. Thomas, an almost Platonic indifference towards a realist cosmology, a view which has its sufficient reasons in his religious faith.[1]

And so, to formulate the conclusion of our analyses as clearly as possible, Aristotle's natural philosophy taught the medieval West two things: overtly and superficially, it gave a certain structure to the cosmos, a certain astronomical order, corrected somewhat by Ptolemy, and open to being treated symbolically. But, more profoundly, it taught a wholly physical realism that, *epistemologically*, underlies the astronomical order. With Aristotle, cosmology definitively ceases being a myth in order to become an objective science and realist description of the world. Because it was Christian, the medieval West generally lost sight of this physical realism through the power of a religious faith that knew the figure of this world will pass away, enabling it to imbue this realism with a kind of 'conventionality'. But, once a search began for the truth about the right order of nature, the realist tendency came to the fore and became stronger and stronger.[2]

THE INEVITABLE 'FAILURE' OF NICHOLAS OF CUSA

Two facts bear out this conclusion. The first consists in the debates stirred up among Peripatetics themselves over the problems of Aristotelian physics, above all from the second half of the thirteenth century. The second is represented by the new cosmology proposed by Nicholas of Cusa.

As satisfying as Aristotle's physics might have been, it did not, however, allow for an intelligible answer to certain major problems, espe-

On the Heavens (1273), and leans, as a consequence, towards a somewhat 'conventional' view (ibid., tome III, 349–355).

1. It seems we can interpret in the same sense the formal condemnation issued (1277) by Etienne Tempier, bishop of Paris, in the name of the doctors of the Sorbonne and at the request of Pope John XXI, against those (the Peripatetics) who would deny there might be a plurality of worlds, and that the stars might move according to a rectilinear motion (Denifle and Chatelain, *Chartularium Universitatis Parisiensis*, tome I, pièce n° 473, 546). This condemnation actually rests on the *a priori* bias for the primacy of Divine Omnipotence over scientific certainties. It supposes that these certainties should be limited by the awareness that, after all, God is more knowing and can do what He wishes. It therefore basically destroys the underlying principle of Aristotelian physics.

2. As surprising as it may seem, St. Thomas *never* supports his astronomical or cosmological considerations with a text from Scripture. Occasionally a text will be used as a confirmation, but never as a basis. The significant text from *Joshua* (10:12), where the leader of the Hebrews stops the Sun, was never cited by him to prove the movement of that luminary, whereas it would come to be at the center of the quarrel of Galileism.

cially those involving violent motion, movements other than those caused by the necessity of regaining a natural position, since an effect demands the presence of a cause.[1] How, then, are projectiles moved? Once a stone leaves the hand that threw it, how can it continue to move according to a violent and non-natural movement? *A quo moveantur projecta?* By what are projectiles moved? John Philopon, a sixth-century Peripatetic, thought that force "is not imparted to the air by the initial mover (Aristotle's thesis), but to the body in motion."[2] This famous doctrine of *impetus* would be taken up again by St. Albert the Great and St. Thomas Aquinas, and is already a notable improvement on Aristotle. But it was above all in the fourteenth century that critiques and hypotheses, often quite modern in character, were formulated. So much so that this century seems today to be the most important between Aristotle and Galileo.[3] At Oxford, the 'Mertonians' (teachers at Merton College), the best known of whom is Thomas Bradwardine, and in France the 'Parisians', among whom we find Jean Buridan and Nicholas Oresme, set up methods for analyzing motion and its various types (e.g., qualitative and quantitative variations of intensity), methods that included arithmetic, calculations with letter symbols (*a, b,* etc.), and movements represented by means of geometric tracings. Although we are inclined to conclude, with Clavelin, that this was not really a scientific revolution,[4] we should acknowledge that, even so, science or, at the very least, a pronounced effort to grasp the reality of facts in a more and more precise way is involved. From this point of view Mertonians and Parisians, although distancing themselves from Aristotle, did not stop following his concept of scientific knowledge.

The case of Nicholas of Cusa is even more exemplary, but for another reason. Numerous historians have seen in him an inspired forerunner of the new physics' cosmological infinitism, for he proposed in advance a model of the universe whose center, as well as outer limits, have disap-

1. Aristotle declares: "the mover is mover of the mobile, the mobile is mobile under the action of the mover" (*Physics* III, 1, 200b). One can therefore, with this doctrine, consider movement in itself, outside of the cause that makes something move. This is why Aristotle imagines that "projectiles move themselves out of the hand, in fact, either by the return of a counter-blow (antiperistalsis) according to certain theories, or by the thrust of the air pushed which impresses on the projectile a movement more rapid than its transport toward a natural place" (*Physics* IV, 8, 215a).

2. A. Koyré, *Galileo Studies,* 12. This has to do with a Latin text of Bonamico, Galileo's teacher, in which the author sets forth the historical background of the question.

3. M. Clavelin, *The Natural Philosophy of Galileo,* 61.

4. Ibid., 85.

peared.[1] And yet it is altogether impossible to draw from the Cusan even a vague Copernicanism or Galileism. His character as a forerunner corresponds to a kind of illusion in retrospect: on many occasions we find Nicholas of Cusa discovering truths explained by the later history of Western science from Copernicus to Einstein. But they are not contemplated in the same spirit, the 'scientific spirit'; at least in his treatise *On Learned Ignorance,* the cardinal is essentially a metaphysician. Nor should we attribute to chance the fact he may have had 'lucky guesses' on some points, since, to the contrary, he developed his doctrine with exceptional vigor and daring. Were these lucky guesses due, then, to his being the first one to reject the medieval idea of the cosmos, as Koyré thinks;[2] and, by going contrary to it, could he not help touching upon the truth now and again? This break is not, however, truly 'scientific' in nature. This same author explains that in no way did Nicholas of Cusa seek to criticize the astronomical theories of his time,[3] that "it is impossible" to take his cosmological concepts as a basis for a "reformation of astronomy,"[4] and that, after all, he, like the people of the Middle Ages, *believed* "in the existence and also the motion of the heavenly spheres ... as well as in the existence of a central region of the universe around which it moves as a whole."[5]

1. The essential texts have been collected by Koyré in his study *From the Closed World to the Infinite Universe* (Baltimore, MD: Johns Hopkins Press, 1957), 6–23. The Mertonians likewise treated of the infinity of the world, and debated the metaphysical and theological consequences its acceptance entails (in particular, Koyré, "Le vide et l'espace infini au XIVe siècle," in *Etudes d'histoire de la pensée philosophique* [Paris: Gallimard, 1971], 51–90). But the doctrine of Nicholas of Cusa seems to be more cosmologically significant. After all, the infinite space of Bradwardine is called 'imaginary' (Koyré, 84). This is an uncreated space existing before the creation of the world, and is not, therefore, physical (or concrete). It simply expresses, in our opinion, the limitlessness of God's creative power. On this topic, however, the question seems to be completely vitiated by the Aristotelian manner in which it is posed. When they speak of the world, its expanse, the problems arising from its creation, Mertonians and Parisians are actually seeing only the visible and corporeal world, the only one that exists for Aristotle. But this world never seems to include the creation of invisible things (animic and spiritual). This is why the assertion of spatial infinity, which can have a *symbolic* sense (designating, according to the Kabbalah, the *tsim-tsum,* the "intradivine emptiness" into which God projects the archetypes of all things), runs up against the contradiction of being merely conceivable and yet extensive and measurable.

2. Koyré, *From the Closed World to the Infinite Universe,* 6.

3. Ibid., 8.

4. Ibid., 18.

5. Ibid., 19. Do we need to join Koyré in adding that he "denies the very possibility of the mathematical treatment of nature" (ibid., 19)? Gandillac has shown that there was nothing to this by publishing an integral translation of the dialogue *The Layman,* the

We must, therefore, reverse the relationship and consider that, in reality, it is modern science that has discovered, more or less tangentially, certain declarations of Nicholas of Cusa, but starting from a viewpoint completely foreign to them. If these tangential contacts are possible, they are so by virtue of the *freedom* of the metaphysical perspective and its universality. Obeying only the necessity of pure intellective evidence, metaphysics unfolds the order of intuitively perceived truths regardless of constraints imposed by cultural fashions and widely accepted ideas; which does not rule out that a metaphysician might be, in extrametaphysical realms, strongly influenced by them. Cusan cosmology is less a break than a reaction and a reminder: a reaction against the realist physics of Aristotle; a reminder that there is no certain and perfectly fixed knowledge of what is, in itself, uncertain and necessarily interminable (*interminatum*). The beings of the world are never 'terminated' (or fixed) in themselves, because they are always able to receive another termination; they are always capable of admitting more or less. In a way it is only from the viewpoint of God (of the "negatively Infinite," as the Cusan puts it, that which is limited by absolutely nothing) that the finite is really finite, terminated. The concrete universe is only limited by God, and is thus infinitely limited or infinitely finite. This is why, "since the universe encompasses all things which are not God, it cannot be negatively infinite, although it is unbounded and thus privatively infinite. And, in this respect, it is neither finite nor infinite."[1]

What do these considerations mean? They simply mean that the existence of a Supreme Reality is taken seriously and its every consequence deduced. Lazy theological thinking readily admits that God is the only 'substance', 'unity', 'life', or 'measure' of all things, the only 'center' that is truly Center, and so forth. But once this axiom is posed, this kind of thinking will also accept the existence of the created substance, unity, life, measure, and centrality of all kinds of relative realities. Never denied, this theological axiom remains as a scenic backdrop. As the need arises, some thinkers will even seek to make a specific relationship Primary Being can maintain with the order of secondary beings more precise, analogical relationships in particular. They will see, for exam-

last part of which is dedicated to a praise of balance, and where Nicholas of Cusa develops notable considerations on various measuring instruments capable of treating even biological or psychological data mathematically. Cf. *Œuvres de Nicolas de Cues* (Paris: Aubier, 1942), 328–354; Eng. trans. J. Hopkins, in *Wisdom and Knowledge* (Minneapolis: Banning, 1996).

1. Book II, chap. 1; Nicholas of Cusa, *On Learned Ignorance*, trans. J. Hopkins (Minneapolis: Banning, 1981), 90.

ple, an image of divine centrality in a given spatial center. But we need to go further. Of course a center can be such only by participating in the supreme centrality—the basis of cosmic symbolism; but, even more, the unicity of the divine Center proves there is no physical center in cosmic space, there cannot be one. Nicholas draws a *physical* conclusion from a *metaphysical* certitude, and this presupposes—something specifically rejected by Aristotelianism—that there is no autonomy of the physical, even within its own order; there is, at the very least, no autonomy that does not give precedence to the metaphysical.

Obviously Nicholas of Cusa does not deny that the world in its entirety is a symbol of God. This is even a major theme of his doctrine, the doctrine of God as an enfolding (*complicatio*) out of which the created universe is only an unfolding (*explicatio*): "It is therefore through the very unity of the Eternal Word, which enfolds all things, that each creature participates through a development in form and in various ways."[1] But Nicholas felt obliged to react in the face of a *cosmologized Platonism,* of a realist physics based more and more on Peripatetism, an outlook that intended to 'localize the intelligible'. Certainly he felt obliged to react in the name of truth, but he also wanted to rectify a philosophy of nature that is a dead end for Christian thinking. And this is why he so insistently denounced the illusion of what we have called the 'topology of the intelligible', for the intelligible (the essence) is not truly in a thing or a place, contrary to what Aristotle asserts; it is there only insofar as a thing or place participates in it.[2] "It is no less false that the center of the world is within the Earth than that it is outside the Earth; nor does the Earth or any other sphere even have a center. For, since the center is a point equidistant from the circumference and since there cannot exist a sphere or a circle so completely true that a truer one

1. Letter to Rodriguez Sanchez of Trier, May 20, 1442, in *Œuvres choisies,* 172; and elsewhere: "You appear to me, my God, as the exemplar of all men . . . in all species You appear to me as the Idea and Exemplar. . . . You are the truest and most adequate Exemplar of each and every thing that can be formed. . . ." *The Vision of God* (Jasper Hopkins, *Nicholas of Cusa's Dialectical Mysticism: text, translation, and interpretive study of de visione dei* [Minneapolis MN: Banning, 1985], 157).

2. Nicholas of Cusa is steeped in Scholastic, and therefore Aristotelian, vocabulary. Nevertheless he is able to criticize the Stagirite (*De Beryllo,* chaps. 25–27 in *Nicholas of Cusa: Metaphysical Speculations,* trans. J. Hopkins [Minneapolis: Banning, 1998], 46–47). Often, like the Arabs, his intent is to reconcile Plato with Aristotle, but always by showing that the latter said the same thing as the former, and not the reverse (*De Mente,* chaps. 13, 15; *Complete Philosophical and Theological Treatises of Nicholas of Cusa,* trans. J. Hopkins [Minneapolis MN: Banning, 2001], 580 ff and 586 ff).

could not be posited, it is obvious that there cannot be posited a center (which is so true and precise) that a still truer and more precise center could not be posited. Precise equidistance to different things cannot be found except in the case of God, because God alone is Infinite Equality. Therefore, he who is the center of the world, viz., the Blessed God, is also the center of the Earth, of all spheres, and of all things in the world. Likewise, he is the infinite circumference of all things."[1] Nicholas of Cusa answered realist physics with a realist metaphysics: if God is real, and even *supreme* Reality, there should be no strict ontological determination for what is not God.

One last consideration should sufficiently convince us that such is indeed the significance of the Cusan endeavor. This involves those texts from *On Learned Ignorance* in which the cardinal opposes the cosmological inferiorizing of the Earth, such as we encounter in Aristotle and his disciples. One would be very much mistaken to see in this a denial of the theory of natural place that underlies, as we have seen, the whole of Peripatetic mechanics. Nicholas admits this doctrine, but with a slight modification that changes its import considerably: "The entire motion of the part tends, *in order to gain perfection,* toward a *likeness* with the Whole. For example, heavy things (are moved) toward the earth and light things upwards; earth (is moved) toward the earth, water toward water, air toward air, fire toward fire. And the motion of the whole tends toward circular motion as best it can, and all shape (tends toward) spherical shape...."[2] The italicized passages clearly show that the Cusan's 'natural place' is no longer Aristotle's. An intelligibility identified with the shape and structure of the sense world as such is no longer involved here. What there is of the intelligible in the physical world is tied to its symbolic function, or, if preferred, to its *iconic* function. The image cannot help but imitate its Model, and out of this need we must seek the raison d'être of appearances, the appearances of forms as well as motion: "Every motion tends toward likeness." But precisely because this only has to do with an image or a likeness, the intelligibility of appearances can only be approximate: "as best it can." Hence, each natural being participates in supreme Perfection in its own way, and there should be, then, no 'more vile', no maximum of baseness or ignominy in cosmic space. Nicholas of Cusa is not opposed to a 'scalar' ontology, to borrow a fortuitous expression that designates a vision of degrees of

1. *On Learned Ignorance,* book II, chap. 11, 115.

2. Ibid., book II, 117. Emphasis added. [Trans.— The French version of this text is followed where it differs from the English.]

reality divided according to a scale of ascending perfections.[1] But, as such, this scale does not include the lowest or the highest possible degrees. Thus it avoids the purely cosmological translation of Platonism effected by Aristotle when he divides the world into two parts, in accordance with a truly topological axiology. The Earth, in its own order, is as noble as the Sun: "Therefore, the shape of the Earth is noble and spherical, and the motion of the Earth is circular; but there could be a more perfect [shape or motion]. And because in the world there is no maximum or minimum with regard to perfections, motions, and shapes (as is evident from what was just said), it is not true that the Earth is the lowliest and the lowest.... Therefore, the Earth is a noble star which has a light and a heat and an influence that are distinct and different...."[2] Without doubt an Aristotelian would have difficulty understanding that bodies do not all have the same nobility and yet no body is *most* vile or *most* noble, since for Aristotle, *and here we must pay close attention,* a first term and a last term are indeed necessary. But this is what Nicholas denies and rightly so. *In the relative order* there is no end. René Guénon has said exactly the same thing in a formula that is one of the major keys of metaphysics: "the indefinite is analytically inexhaustible."[3] To find an end to the finite, we must 'go outside' its order, but we must not seek to analytically exhaust it within its own order. The end of time is outside of time; the end of space is outside of space. *By itself,* no reality is limited. Every reality is limited synthetically by an immediately superior reality—such is the metaphysical realism of the Cusan. We repeat—he is much more opposed to the metaphysics than to the cosmology of Aristotle, even if the stakes of the battle at first seem to involve the overall shape of our world.

To profess at the same time, and logically, the basic ontological endlessness of the created and its symbolic or iconic dimension in no way signifies that, for Nicholas of Cusa, mere scientific knowledge of the

1. We encounter this expression in the work of Pierre Magnard, who used it to define Charles de Bovelle's 'serial approach' ("L'infini pascalien," in *Revue de l'enseignement philosophique,* num. 1 [October–November 1980]: 15). Undoubtedly he is restating a formula of Pierre Quillet: "L'Ontologie scalaire de Charles de Bovelles," a contribution to *Charles de Bovelles en son cinquième centenaire* 1479–1979, Actes du Colloque International, held at Noyon, September 14–16, 1979; Guy Trédaniel, 1982, 171–179.

2. *On Learned Ignorance,* book II, chap. 12, 117–118.

3. *The Metaphysical Principles of the Infinitesimal Calculus* (Hillsdale, NY: Sophia Perennis, 2003), 111. Wanting to analytically exhaust the finite is Zeno's 'error' in particular. We put *error* in quotation marks, for it is possible that Zeno may have only wanted to show, *a contrario,* the absurdity of this undertaking.

world is illegitimate in itself. But it necessarily becomes so when it wants to reach a maximum of exactness, because the object of its efforts is truly unable to withstand the test of such precision. Here we cannot help but be reminded of Werner Heisenberg's 'uncertainty principle': by very virtue of the Cusan principle relating to the ontological endlessness of the world, every intensified effort at precision will only *reveal* the indeterminacy of phenomena. But neither should this mean that the world is only an illusion, a nothingness of being, and lead to some form of idealism. To say that the universe is a symbol does not deny its reality, but qualifies it. Quite the contrary, and a metaphysics of the symbol will demonstrate this clearly, it is even the only way to preserve the ontology of the cosmos—the history of science is there to prove this. It is physical realism that leads to idealism, and to the philosophical impasse in which contemporary epistemology has become mired. But, from the neo-Aristotelians to the formalists by way of the materialists, is there still anyone who understands this today? Besides, it was the same during the cardinal's time.

One sure law of history is that "when the wine is tapped, drink you must." The 'Aristotelian vintage' drew off our knowledge of the world onto a course of a most strict physical realism. To combat Aristotle—which Galileo had expressly proposed to do—amounts to placing oneself on his own terrain, that of cosmological realism, since it was upon this very terrain that his contradictions and limitations were most obvious. The attempt of the cardinal, which tended to displace the question and situate it on another level, was, from this point of view, bound to fail.[1] Surely, it offered in advance the possibility of combining, in one and the same metaphysical synthesis, both the vision of a symbolic cosmos and the later constructs of science—of which the cardinal was obviously unaware. But only the rarest type of intellect would have been aware of such a theoretical possibility—one aware of the ontological demands of the new physics.[2] Galileo was not such a one. We might even go so far as to say that the Keplers and the Newtons themselves formulated such a mythico-metaphysical doctrine of the cosmos. And yet, in the history of science, everything transpires as if the Newtonian

1. At the very least as far as the 'general trend of ideas' was concerned. For there are also those who, in reading him, 'recognize' in him that truth they bear, more or less consciously, within themselves.

2. In his article "L'infini pascalien," Magnard points out how different are the reactions of Nicholas of Cusa and Pascal when confronted with the world's infinity. For the cardinal, "The (infinite) universe is a figure of God's superabundance, which a person

concept of space as *sensorium Dei* was purely and simply a theological curiosity unrelated to his concepts of physics and astronomy. And they see in his famous *hypotheses non fingo* ("I feign no hypotheses") the inaugural charter for scientific positivism;[1] but his real concern was to set aside the 'false hypotheses' of Cartesian mechanics which, by rejecting empty space and by asserting the physics of vortices, "deny God's action in the world and push him out of it," whereas "the true and ultimate cause of gravity is the action of the 'spirit' of God."[2] For Newton, space is the mode according to which God is *universally present* to all things, and outside of this 'function' it is meaningless and quite simply does not exist.[3] In return, if (universal) space is indeed the Divine Organ by which all things are known, then their existence is established since, by it, they both exist for God and, God being one, his being universally present to things accounts for their mutual relationships and therefore for an otherwise unimaginable 'activity at a distance'. Therefore, both the being and the motion of physical realities demand this theological basis. How remarkable, then, that Newtonianism, through a profound alteration of its nature, will progressively become the model

wherever he finds himself, experiences as a kind of jubilation" (3). "When Pascal cries out that 'the eternal silence of those infinite spaces terrifies' him, is he just expressing the sense of emptiness left by the withdrawal of God?" (p. 4). The reason for this difference is that Galileo and the denial of the theophanic world falls between them. We do not think that Pascal is expressing his personal sentiment here, but that of anyone aware of the cosmological consequences of the Galilean revolution: *cf. infra*, 85–86, nn 1, 3.

1. "I have not been able to discover the cause of those properties of gravity from phenomena, and I feign no hypotheses" *(Mathematical Principles of Natural Philosophy,* General Scholium). Koyré (*From the Closed World to the Infinite Universe,* 228–229) observes that *non fingo* should be translated as: "I do not feign," that is: "I do not surmise *false* hypotheses." Cf. likewise the fine study by J. Zafiropoulo and C. Monod, *Sensorium Dei dans l'hermétisme et la science* (Paris: Les Belles Lettres, 1975).

2. Koyré, *From the Closed World to the Infinite Universe*, 234. Newton writes (*Mathematical Principles*): "Now we might add something concerning a certain most subtle spirit which pervades and lies hid in all gross bodies; by the force and action of which spirit the particles of bodies attract one another at near distances, and cohere, if contiguous, etc."

3. Henry More agreed: "Space is an attribute of God and His *Sensorium*" (cited by Zafiropoulo and Monod, *Sensorium Dei dans l'hermétisme et la science,* [Paris: Les Belles Lettres, 1976], 137). Perhaps it was More who provided Newton with the idea of the *Sensorium Dei.* The term is found for the first time in a commentary by Boethius on Aristotle (*Ar. Top.,* 8,5). In the words of Zafiropoulo (ibid., 11): he is designating the *organ* of a particular sense (Aristotle, *On the Parts of Animals,* III, 3, 665 a 12); the *sensorium commun* is "the central organ where sensations, coming from different directions, coalesce in such a way as to give the mind a representation of the object" (Lalande, *Vocabulaire,* s.v.).

for 'celestial mechanics' in the eighteenth century! A profound alteration, but (and we need to recognize this) an alteration that finds justifiable grounds in Newton himself: the concept of space as *sensorium Dei* suffers from a basic ambiguity. Such a formula cannot be understood in a grossly realist sense; otherwise space as commonly perceived is ontologized. It is truly acceptable only if one specifies that *what* appears to be 'space' is, in reality, the *sensorium Dei.* In other words, and to get a clear grasp of the matter, we need to bring about a conversion of the intellectual gaze analogous to that which Plato strives to instill in his reader when speaking of the 'receptacle' in the *Timaeus*. One can indeed proceed from metaphysics to physics; but not from physics to metaphysics, for that is impossible. We need to notice these discontinuities, then, and decry a certain commonly perceived illusion of space. In short, we must distinguish between "an infinite expanse not susceptible to movement and immaterial, on the one hand, and, on the other, the material expanse, mobile and hence finite."[1] This distinction, which comes from Henry More, will be taken up again by a disciple of Newton, the young Joseph Raphson, and will be at the heart of the debates that will put Leibniz and Clarke on opposite sides of this question. To Leibniz, who criticized the Newtonian ontologization of space, Clarke retorted: "Space is not a Substance, and eternal and infinite Being; but a Property or a consequence of the existence of an infinite and eternal Being."[2] Such a space is absolutely indivisible. But, if Leibniz was misled, this is because there was sufficient reason to be misled. This whole question of the nature of space, which troubled the most eminent seventeenth-century minds, does not appear to be posed in its true perspective, that of mytho-cosmology (which does not mean that we are not speaking of true realities here, but simply that one can only speak of them symbolically). Lacking this perspective, the discussion seesaws endlessly from one side to the other, from metaphysical symbolism to the conceptualism of physics, without the adversaries seeming to realize it and, often, in an inextricable manner.[3] And surely we should also see here a sec-

1. Koyré, *From the Closed World to the Infinite Universe*, 191.

2. Cited by Koyré, ibid., 247

3. We shall return to the question of space in the next chapter. But for the moment, and to give a synopsis, we will say that space—an existential *condition* of corporeal manifestation—is the reflection of *Prakriti* (or *materia prima*), and, even more profoundly, of Universal Possibility or of the very essence of supreme and absolute reality, when viewed as the matrix of all that is and all that can be. This Universal Possibility, or intrinsic infinity of the Absolute, is the *basis* of all that theologians call the 'presence of immensity', or God's presence in the totality of the created by virtue of divine Immensity, which is to say by the fact that divine Reality is without measure and therefore measures all

ondary but quite real effect of Aristotle's physical realism paradoxically 'confirmed' by Galileo.

things. Since nothing is able to limit or circumscribe It, It is necessarily present in all that is: "Wherever the virtue of God is, there is the substance and essence of God. . . . Let us refrain from seeing God's immensity as a vast expanse, like the rays from the Sun, for example. God is everywhere, not by a part of his substance, but with his whole substance, just as the soul is entirely within the body" (Abbé Berthier, *Abrégé de théologie,* num. 317). Likewise for Clarke: "God, being everywhere, is actually present to all *essentially* and *substantially*" (Clarke's third reply, §12). Thus we see total agreement between the Catholic theologian and the Newtonian philosopher, provided that we clearly understand both and pose the necessary distinctions; the presence of God is not local, or spatial, and only thus can it account for the infinity of space. We do not get the impression that Clarke was always aware of these distinctions—there is a certain confusion of the levels of reality in all this. Witness for instance this text by Joseph Raphson who, so to make of space an attribute of God, identifies it with the *En-sof* of certain Kabbalists: "It is to this Infinite that assuredly must be referred a great number of passages of the Holy Scripture as well as the hidden wisdom of the old Hebrews about the highest and incomprehensible amplitude of the *Ensof*" (*De ente infinito,* cap. V; Koyré, *From the Closed World to the Infinite Universe,* 195). *En-sof* literally means 'limitless'. Not of ordinary derivation and of relatively late usage (Scholem, *Origins of the Kabbalah* [Princeton, 1987], 265 ff.), it designates "the absolute infinity of the supreme essence" (Schaya, *The Universal Meaning of the Kabbalah* [Baltimore: Penguin Books, 1973], 36), the Godhead 'envisaged' within Itself, independent of all relationship to the created. As we see, we are far from the space of the physical world that is only its most distant reflection. See also what we have written (51, n. 1) concerning the infinite space of Bradwardine, whose idea, moreover, anticipates Newton's.

2

The Destruction of the Mythocosm

'The Galileo Affair' or the Rupture Completed

LEGEND AND TRUTH

Reading so many authors, some of whom, such as Leibniz and Newton, are truly among the most potent minds known to the western world, one cannot help but think that, in a certain manner, the Galilean crisis was avoidable. Basically, those interested in showing how (what we have called) a mythocosmology was still possible and even necessary were numerous.[1] Had they succeeded the evolution of the West would have been truly different. In any case, humanity would not have known the most terrible blow inflicted on its cultural soul in the course of its (known) history. The more-or-less episodic actors who play a role in this drama without doubt had some vague awareness of the unique importance of this struggle and its stakes. The spectators themselves perceived its gravity; otherwise there would be no explanation for the violence of the passions unleashed, or the repercussions from the condemnation still felt today, three centuries later. After all, there were more important discoveries than those of Galileo in European cultural history. It was Copernicus who truly inaugurated the astronomical revolution (Galileo was only a disciple).[2] And it was Newton who carried out the synthesis of physics and astronomy by his universal gravitation

1. We will gain some idea of the importance of this effort once we see that, among the body of manuscripts left by Newton, half concern theology, a quarter alchemy (121 treatises), and a quarter physics.

2. Jean Bernhardt, in an inaugural and important article, "L'originalité de Copernic et la naissance de la science moderne" (*Revue de l'enseignement philosophique*, 23e année, N^o^6, août–septembre 1973, 1–35), is intent on showing how Copernicus is the first to

theory, a theory based on 'the greatest scientific discovery of all times', that of the elliptical orbit of the Earth by Johannes Kepler. Yet it was in the person of the Florentine scholar that the cultural crisis found its eponymous hero. And it was the Catholic Church, the actual (more or less willing) heir of ancient humanity, delegated by all of the Earth's traditional civilizations, that pathetically clashed with the new science: in condemning Galileo, it seems to have condemned itself.

All the same, the disproportion between the mildness of the punishment inflicted on Galileo and the extent of the reactions aroused by his condemnation just confirms, if necessary, that what was at work surpassed by far the anecdotal detail of events. The imagination very quickly seizes on this affair to confer on it those features more in keeping with the demands of its exemplary nature. A legend has rights to the very extent that what is expressed by it is of greater value than historical accuracy.[1] However, in Europe, there is no lack of victims of intolerance whose philosophical or scientific activities were much more costly than a house arrest! Giordano Bruno, burned alive as a heretic by the Catholic Church; Michel Servet, burned alive as a heretic by the Calvinists; without speaking of Lavoisier, guillotined in the name of Republican morality. But they are forgotten for the benefit of Galileo alone, who by himself symbolizes science martyred by faith.

Under these conditions, we understand why works devoted to this question almost always give the appearance of a plea for or against Gali-

---

construct a cosmological model: "What began with Copernicus was a new metatheory light years removed from Platonic contemplation, the metatheory of a speculative, mechanical demiurgy or the completely new audacity of a conceptual shaping of physical objectivity" (34).

1. Today the historical truth is much better established (cf. especially Giorgio di Santillana, *The Crime of Galileo*; E. Namer, *L'affaire Galilée*; Pierre de Vrégille, 'Galilée', an article in *Dictionnaire apologétique de la foi catholique*, tome II, col. 147–192; etc.). But certain points that will undoubtedly long remain mysterious are also underscored. This last development was due to the discovery in the Vatican archives, by Pietro Redondi (*Galileo Heretic*), of an unlisted document denouncing Galileo to the Holy Office for heresy in the matter of physics: the atomic theory, professed by Galileo, making Eucharistic transubstantiation impossible. By denouncing Galileo, a personal friend of the Pope and figurehead of the aristocratic intelligentsia, the Jesuits, presumed authors of the denunciation and favorable to a Spanish alliance, wanted to compromise a papacy favorable to France. Urban VIII, to foil the maneuver and save the Vatican's official scholar, diverted the accusation to a much less burning topic: the Copernican theory, and set up a fake trial in connivance with Galileo. Since its publication in Italy, Redondo's fascinating book has aroused lively polemics, and even an attempt at refutation on the part of the Jesuits.

leo. This does not only proceed from the nature of the event itself, but also from a desire to rectify the historical perspectives to which it has given rise in the course of three centuries, and thus to respond to what seems calumny or slander. In short, the trial continues. The merit of Arthur Koestler's book *The Sleepwalkers* is that he frankly recognizes this.[1] But this is not the case with all studies, even those reputed to be the most serious. Thus the work of Emile Namer from which we have borrowed our article's title and the author of which is regarded as a great specialist of this period constantly presents the facts as if Galileo had actually verified, established, and demonstrated the system of Copernicus, so that it is obviously unforgivable that the Church had condemned him, while Galileo seemed just a person anxious for objectivity colliding with the most incomprehensible of negations: the denial of plain facts.[2] But this is untrue and sets us on the path to our first point to be remembered in this complex affair (the second will concern the Church).

THE FAITH OF A PHYSICIST WITHOUT PROOF

Among all the scholars of his time, Galileo was the first, not to have subscribed to the Copernican system, but to have insisted on imposing it as an official truth guaranteed by the Church. Galileo was a hyper-Copernican—which constitutes moreover, on the topic of planetary orbits, a regression with respect to Kepler. But *he was never able to prove* this truth in which he believed.

Why did he so ardently adopt the Copernican system as a special cause seventy years after its author's death in 1543, whereas, during this whole time, the Copernican theory aroused no official reaction, and was even taught and supported by church scholars?[3] As we know, Galileo

1. Pages 425–426.

2. *Cf. L'affaire Galilée*, presented by E. Namer (57, 59–61, 75–76), which is summarized on page 114: "How could the system of Copernicus be condemned, now that it was verified through rigorous and proven experience?"

3. Recall that *De revolutionibus* was dedicated to the pope and prefaced by a cardinal. In 1533, the chancellor Johann Albrecht of Widmanstadt had already given an account of the *Commentariolus* before Clement VII and his retinue. The pope was enchanted by it and rewarded him for it. (The *Commentariolus*, a short manuscript treatise, contains the first explanation of the doctrine of Copernicus. Cf. Koyré, *The Astronomical Revolution*, 28; likewise *Dictionnaire apologétique de la foi catholique*, tome II, col. 155.) That the publication of Copernicus' book had no repercussions is generally asserted. But this is not altogether exact. Prior to Giordano Bruno, recognized as the first Copernican (the *Ash Wednesday Supper* dates from 1583), Thomas Digges must be named, whose work (discovered in 1934) entitled *A Perfit Description of the Celestiall Orbes according to the most*

was not originally a Copernican, at least overtly. He even intended to demonstrate that the Earth is indeed at the center of the world.[1] Such was his teaching from 1593 to 1610. There is a letter, however, which informs us that he held "the opinion of Pythagoras and Copernicus with respect to the motion of the Earth . . . as more probable than that of Aristotle and Ptolemy."[2] But everything changed with the discovery of the astronomical telescope. From that moment he crossed from physical realism, which he supported because of his studies on the movement of the pendulum and the fall of bodies, to an astronomical realism, which he turned to because of his observations with the telescope on the earth-like natures of the heavenly bodies and the nonexistence of the sphere of the fixed stars. Henceforth the Copernican system no longer seemed a probable opinion, but an absolute and proven certainty. He was not just content with pointing the optical tube invented by a Dutchman toward the sky, and thus to discover "things that no one has ever seen," as the title of his *Siderus Nuncius* (*Sidereal Messenger*) announced in 1610. When he came to Rome, in the following year, to have his triumph officially recognized, he clearly intended to link his discoveries to the system of Copernicus: "By his presence at Rome, and his contacts with the Jesuits at the Roman College, Galileo expected a double acknowledgement, the official recognition, by the experts in mathematics and astronomy, of the accuracy of his observation, and the confirmation, in light of sensory experience, of the Copernican system."[3]

This desire for recognition even went so far as aspiring to dictate to the religious authorities their duty and enjoining them to declare themselves in his favor theologically. Evoking the truth of both Scripture and nature, he concludes: ". . . since manifestly two truths can never contradict each other, the *duty* of wise exegetes is to determine the true meanings of the passages of Scripture, so to harmonize them with natural

aunciente doctrine of the Pythagoreans, latelye revived by Copernicus and by Geometricall Demonstrations approved dates from 1576. On the other hand, we know that Strasbourg's famous astronomical clock, built in 1547, is decorated with a portrait of Nicholas Copernicus and exhibits an orrery consistent with his system, which is thought to have been created around the time of Galileo's condemnation as a sign of protest on behalf of the university (Flamarion, *Copernic*, 81).

1. *Treatise on the Sphere or Cosmography*, *Ed. Nat.*, vol. II, 205–06; Namer, op. cit., 46.

2. *Ed. Nat.*, vol. X, letter no. 4. Galileo to J. Mazzoni, May 30, 1597; Namer, op. cit., 46.

3. Namer, ibid., 70.

conclusions; conclusions that prior sensory experience and the necessary demonstrations have rendered sure and certain in our eyes."[1]

As we see, what mattered to Galileo the physicist was not so much the astronomical truth in itself, as the support that it could bring to his physical realism. Galileo—and the historians have stressed this—was not a great astronomer. He can be compared neither to Copernicus, Kepler, or Newton. But he was a physicist, a 'philosopher of nature'. And he was convinced that what is revealed beneath the scientific gaze is the world's very reality, such as it is in itself. His confidence in the ontological value of his analysis was total, and hence too its demands. In astronomy, however convincing the 'proofs', it is always possible to explain appearances equally well by different hypotheses. Not so for physics, which cannot help but pose questions about the objective value of its hypotheses, precisely to the extent that it is not, like astronomy, only a science of the visible. In this sense, Galileo is the direct heir of the physicist Aristotle. Would he have encountered so violent an opposition from the Peripatetics if he had situated himself on their plane? Besides, with respect to modern science, the real master is Aristotle, not Plato. Listen to what an historian of science, J. Bernhardt, has to say on this subject: "A precise astronomy, careful to account for the details of phenomena, was not then possible in Platonism," whereas the realist concern of Aristotle, even though not always successfully realized, as Ptolemaic astronomy shows, bears witness to its scientific seriousness, even in its failures and by the very conventionalism to which it is resigned: "A hardly contestable scientific progress was thus realized," Bernhardt continues, "and we see how, despite their limitations, Aristotelianism and the phenomenism derived from it prevail, as to scientific value, over Platonism, contrary to a widespread legend."[2]

Wherever Galileism breaks with the Aristotelian concept of knowledge, this is through the introduction of mathematics into the treatment of the real. "Philosophy," declares Galileo in a famous text, "is

1. *Ed. Nat.*, vol. V, 281–288; letter to Father Castelli, Dec. 21, 1613. He would recapitulate the terms of this letter almost literally in his major epistle to Madame Christine de Lorraine. *Ed. Nat.*, vol. V, 309–348; Namer, ibid., 112.

2. "L'originalité de Copernic et la naissance de la science moderne," *Revue de l'enseignement philosophique*, Aug.–Sept. 1973, n°6, 30. The author is alluding here to the thesis of Koyré in particular. We have already said that Galileo's Platonism seems to be of a rather strategic order. This is also the opinion of Clavelin. It is true that Koyré distinguishes two Platonisms: "the Neoplatonism, of the Florentine Academy, [is] a mixture of mysticism, numerology and magic; and the Platonism of the mathematicians, that of a Tartaglia and a Galileo, a Platonism which is a commitment to the role of mathematics in science but without all these additional doctrines" (*Galileo Studies*, 222). We think that the second Platonism has nothing of the Platonic.

written in this grand book—I mean the universe—which stands continually open to our gaze, but it cannot be understood unless one first learns to comprehend the language and interpret the characters in which it is written. It is written in the language of mathematics, and its characters are triangles, circles, and other geometrical figures, without which it is humanly impossible to understand a single word of it."[1] But this mathematizing of the real—we must not be led astray here—is also a reification of mathematics. We say it again: Galileo truly intends to identify words and things here. Mathematical language is not a metaphor, it expresses the very intelligibility of the real. The real, in its very substance, is no longer governed by Aristotelian causality, but by a rational necessity that adequately expresses geometrical relationships. "Eschewing *a priori* speculation no less than pure description, Galileo set himself the task of elaborating a conceptual system in which rational necessity took the place of physical causality; as the clearest expression of that necessity, geometry became the language of scientific research, it was transformed from technical aid into the master key to the door of experience . . . as we read his works, we are struck above all by the remarkable way in which he impressed the features of classical science upon a two-thousand-year-old picture of scientific rationality."[2]

Neither could Galileo accept, under these conditions, the 'Neo-Platonic' criticisms of Aristotelian cosmology, which all derived from what we have called mytho-cosmology. In fact, it is surprising to see how, in his major work, the *Dialogue*,[3] where he carries out the most systematic criticism of Aristotle's system, Galileo makes no mention of the critiques that Giordano Bruno and Nicholas of Cusa had already directed at it. Some historians conjecture that, if Galileo ignores them, this is because, for him, they remain prisoners of the same concept of knowledge: they legislate *a priori* on the world and bodies and next ask experi-

1. This text is drawn from *Il Saggiatore* (*The Assayer*), published by Galileo in 1623, with the encouragement of Cardinal Barberini (Urban VIII), in reply to his detractors; *Ed. Nat.*, vol. VI, 232. René Taton makes clear that this text glorifying mathematical language expresses a wish rather than a finding based on established results; *Galilée. Aspects de sa vie et de son œuvre* (Paris: PUF, 1968), 42.

2. Clavelin, *The Natural Philosophy of Galileo*, 383.

3. *Dialogue Concerning the Two Chief World Systems, the Ptolemaic and the Copernican* (in four days), Florence, 1632; *Ed. Nat.*, tome VII, 1–546. This is Galileo's greatest work. It is written in Italian (Galileo, an excellent Latinist, could have written it in Latin). It is a masterpiece of style, pedagogy . . . and polemic. Galileo displays there his genius as a physicist and even more as an astronomer. The book was known in Europe above all through the Strasbourgian Mattia Bernegger's Latin translation (but Descartes had read it in Italian).

ence to justify theories concocted without it.[1] This is possible. But we think that, even more profoundly, what disqualified the Cusan criticisms of Aristotelian cosmology from Galileo's consideration was that they lead to the rejection of every realist cosmology; they invalidate, in a certain manner, the whole scientific project of Galileism. Do not forget that the *Dialogue*, which brought Galileo's second condemnation, is not a book "about astronomy, nor even about physics."[2] A combative work addressed to reasonable men, its essential aim was to prove the reality of the Copernican system. So true is this that, in its beginnings, this work that would cause Galileo so much hardship was conceived "as a resumption of his work *On the Flux and Reflux of the Sea*: the movement of the tides should constitute (in an erroneous manner it is true) the essential proof for the rotational movement and revolution of the Earth."[3]

Lastly we come to this most curious fact, hardly to be ascribed to the realist passion of Galileo the physicist: all his life he sought a physical proof for the double Copernican movement of the Earth, but all he found was the movement of the tides. Now, by very virtue of the Galilean principle of relativity, such a proof is completely worthless. And yet Galileo held fast to it more with an obstinacy of faith than of knowledge. It is worthless because, as he abundantly demonstrated, both many times in the *Dialogue* and as implied by the principle of inertia, it is impossible to register the motion of a body if one does not have at one's disposal a coordinate system independent of the moving object.[4] Galileo's idea is the following. The Earth turning about itself and around the Sun, during one of the phases of its own rotation (the nocturnal phase), its two movements are in the same direction and their

1. This is the explanation offered by Clavelin, op. cit., 158, and this is basically what Galileo denounced at the beginning of the third day of the *Dialogue*, *Ed. Nat.*, 300.

2. Koyré, *Galileo Studies*, 158: "The astronomical part of the *Dialogue* is particularly weak. Galileo completely ignores not only Kepler's discoveries but also even the concrete content of the works of Copernicus" (222).

3. Namer, *L'affaire Galilée*, 175.

4. *Dialogo*, II, 169 ff; Koyré, *Galileo Studies*, 165ff. Galileo shows that, if a stone is dropped from the height of a mast on a moving ship, the stone falls to the foot of the mast, even though this foot has left the 'place' where it was when the stone was dropped. In other words, the stone drops vertically from a point which, with respect to a supposedly immobile space, was displaced by the speed of the ship during the time the stone was falling. And yet, for anyone on board the ship, the stone always falls vertically. The reason for this is that the stone is not only propelled by a rectilinear movement from above to below, but also by a 'horizontal' movement, the motion of the ship itself (obviously it must be supposed that the ship's speed is constant; otherwise, an abrupt change of speed would skew the trajectory of the body). For an observer standing on the shore, the stone does not trace a vertical, but a parabola; cf. I. Bernard Cohen, *The Birth of a New Physics* (New York & London: W.W. Norton Company, 1991), 83–85. It is not

speeds are added; while, in the diurnal phase, the movement of rotation is contrary to the direction of orbital movement: "as a result, the water is 'left behind' at night, and rushes ahead of the land in daytime. This causes the water to get heaped up in a high tide every twenty-four hours, always around noon."[1] This is a completely untenable explanation and, moreover, contrary to what is commonly observed, since there is not one but two tides in twenty-four hours. The heap of water could be propelled by a speed differing from the continental mass only if the Earth abruptly changed its speed of rotation or orbital translation. But then there would no longer be any Galileo to perceive it. However, he was so convinced of the strength of his argument that, after having presented it as his unique and decisive proof in the years 1615–1616, at the time of his first condemnation, he took it up again and developed it fifteen years later in the *Dialogue* by having it preceded by, it is true, two other arguments, the first concerned with the course of the planets, more simply explained by Copernicus, the second with the movement of solar spots.[2] But it was always the phenomenon of the tides that remained in his eyes, against the advice of both his friends and the best astronomers of the time, the irrefutable argument.[3]

It is not possible, then, to underestimate the importance of 'faith' in the struggle that Galileo waged in favor of a hyper-Copernican system. It is the importance of this factor that explains, in part, the attitude of the

possible to prove then the motion of a body, the Earth for example, by means of elements integral to this body, and therefore all hope of proving in an absolute manner, on Earth, the Earth's motion should be abandoned. Even Foucault's pendulum experiment, in 1851, cannot be considered decisive. Besides, all this is open to a different interpretation in an Einsteinian perspective.

1. A. Koestler, *The Sleepwalkers*, 465, summarizing Galileo, *Dialogue*, Ed. Nat., tome VIII, 472ff.

2. Neither of these two proofs is decisive. The second led Galileo to deny that the Sun could move while remaining parallel to its axis of rotation, whereas he admitted this same parallelism for the Earth! The first argument is more interesting and is, in itself, incontestable. But, convincing as this might be, it does not prevail over the more complicated, but also more precise, Ptolemaic explanation. To this it must be added that Galileo, by presenting this argument, made the Sun the geometric center of the planetary orbits, which does not correspond at all to the Copernican system in which, as we stated, the center of the planetary orbits is the center of the earthly orbit (or *orbis magnus*), which is not the Sun, but is found to be one twenty-fifth of its radius to the side, to account for the eccentricity of the Sun's apparent motion. This anomaly was resolved by Kepler's discovery of the elliptical orbit, which Galileo rejected. In short, Galileo is more heliocentric than Copernicus.

3. *Works, Ed. Nat.*, bk. XIV, 289. This is why Clavelin, in a study devoted to "Galilée et le refus de l'équivalence des hypothèses," published in *Galilée. Aspects de sa vie...*, concludes: "Surely then, by asserting the *de facto* truth of Copernicanism, Galileo has clearly exceeded what his discoveries or his own progress in the science of motion warranted" (129).

Catholic Church. It was almost the rivalry of one faith with another, a faith in the invisible clashing with a faith in the visible. This is what one Galilean enthusiast recognized: "Galileo does not just give astronomy its instrument; he gives it much more by giving birth to the 'faith' needed for being able to make use of it successfully.... The struggle that Galileo undertook [that is, with respect to the objective value of simple sensory vision] and won, represents an extraordinarily dramatic chapter in the history of science. Only a matter of 'faith' is involved here. Galileo had no philosophical arguments to undermine the verdict: science cannot be created by means of a single insight; and above all he had no technical arguments to support with confidence what he saw through the telescope.... He was only asserting his faith...."[1] And this is why, as M.D. Grmek states, "We believe in Galileo's sincerity, as we do in his inability, of a psychological and not intellectual order, to understand the subtlety of Bellarmine's and Urban VIII's reasoning, and even less to truly grasp their policies."[2] What then were the reasoning of Bellarmine and the reasons of Urban VIII? These questions lead us to the second point to be examined in this 'affair', the attitude of the Catholic Church.

A CHURCH ENSNARED AND WITHOUT METAPHYSICAL VIGOR

We are unable to take up and examine here all items of documentation.[3] We will only give the conclusions reached, while the reasons for them

1. Vasco Ronchi, "Galilée et l'astronomie," in Taton, *Galilée. Aspects de sa vie et de son oeuvre*, op. cit., 160.

2. "Le personnalité de Galilée et l'influence de son oeuvre sur sa vie," in ibid., 64.

3. On the juridical aspect of the condemnation, cf. Lucien Choupin, S.J., *Valeur des décisions doctrinales et disciplinaires du saint-siège* (Paris: Beauchesne, 1929), 161–186. The history of Galileo's relationship with the Catholic Church is yet to be concluded. In 1979, on the occasion of the centenary of Albert Einstein's birth, Pope John Paul II asked that an interdisciplinary inquiry about Galileo's trial be undertaken. A pontifical commission was appointed on July 3, 1981 to study the Ptolemy-Copernicus controversy in the sixteenth and seventeenth centuries, the work and conclusions of which were presented eleven years later, on October 31, 1991 in Rome, at the time of a plenary session of the Pontifical Academy of Sciences. Did this involve a rehabilitation of the physicist, as hoped for by many? Even today we hear about the erecting of his statue in the Vatican gardens in 2009. In fact, as stressed by Cardinal Poupard, the president of the pontifical commission, the work accomplished shows above all the extreme complexity of an affair in which the participants were overtaken by the importance of the stakes in play. Today the Church has recognized that this condemnation was a juridical mistake, but also that the 'good faith' of the protagonists is not open to suspicion (cf. Paul Poupard, *L'affaire Galilée*, [Édition de Paris, 2005]). As for the heart of the question, although the Earth's rotation upon itself is beyond doubt, its universally agreed orbital translation around the Sun seems, however, hard to reconcile with the results of Michelson's experiment (1881), unless we interpret them within the framework of Einsteinian relativity.

will be found in the body of cited works. Those conclusions that we hold as proven are the following:

- The Church was wrong, objectively, for officially condemning heliocentrism. It would have also committed another error in solemnly confirming it since, from the viewpoint of relativity, there is no *absolute* truth for the movement of the Earth about the Sun.[1]

- The Church, however, felt such a condemnation repellent. It did everything it could to avoid it. After all, the *dogmatic* scope of this condemnation, given its canonical forms, is quite narrow, and has always been considered as such by the theologians.

- The Church was caught in a pincers between the unrelenting hatred of the Peripatetics—truly responsible for the condemnation—and the inflexible and sometimes provocative attitude of Galileo, a situation that constrained the hierarchy (in the person of his friend the Pope) to withdraw its so long-accorded protection.

Now we will briefly comment on each of these points.

1. — When we say that the Church had committed an objective error by condemning heliocentrism, we are stating a fact, since it seems relatively true, from the viewpoint of the solar system, that the Earth and other planets revolve around the Sun. In all respects it might be thought that the Church had better things to do than become involved in this conflict. However, we need to realize that to take literally the terms of the first (1616) condemnation, the only one that concerns heliocentrism directly[2]—since the second (1633) concerns only Galileo and basically punishes him for not obeying the first—contains no error: the Sun is not, in fact, at the center of the world (nor of the planetary orbits, nor of the galaxy) and is not motionless. On the other hand—and without con-

1. "By virtue of this general principle, no physical experience enables an observer linked to a system in accelerated movement, that is linked to a moving body whose speed or direction is not uniform, to know if this system is or is not at rest. In fact, phenomena which, from within a system, might be attributed to a change of speed or direction, might also be logically attributed to the existence of a gravitational field external to the system" (Lenis Blanche, "Les paradoxes du système du monde," *Revue de l'enseignement philosophique*, 18e année, n^o^5, 16); cf. also Albert Einstein, *Relativity: The Special and General Theory*, trans. R.W. Lawson (Rahway, NJ: Henry Holt and Company, 1920), 71.

2. As we know, the official text of this condemnation (Feb. 24, 1616) does not mention Galileo by name. It dealt with only two propositions, the first: "The Sun is the center of the world and hence immovable for local motion," the second: "The Earth is not the center of the world, nor immovable, but moves according to the whole of itself [moves about itself]." The first proposition was declared philosophically absurd and formally heretical as contrary to Scripture; the second received the same philosophic notice and was judged *theologically* "at least as erroneous according to the Faith" (*Ed. Nat.* vol. XIX, 320–321).

sidering the 'cultural' aspect of the question to be mentioned in the next article—we must indeed recognize that the Church had some excuses to refuse to make the Copernican system canonical as Galileo wished.[1] This system, in the account of *De Revolutionibus*, was highly complex. Superior to Ptolemy, although it explains all celestial movements by a single principle, the Earth's motion, it is more complicated than it in the detail with which it is carried out.[2] Moreover, for Copernicus himself, the superiority of his system resides above all in its structural unity, and not in its better agreement with observed data.[3] And lastly he provides no physical proof for the Earth's motion: "The proofs that Copernicus advances for his doctrine are most curious. Properly speaking, they prove nothing at all."[4] In particular, Church astronomers knew quite well that, if the Earth was propelled by an orbital movement, variations in the parallax would be registered, since, with the Earth moving about, the angle by which a fixed star is viewed (this is the parallax) should also vary. Now such a variation had never been measured.[5] Under these conditions it was hard for a religious institution to decide in favor of any set astronomical system. It could only see it as an hypothesis.

2. — Precisely all of these reasons explain the Church's visible reluctance to take sides in this 'affair'. This attitude is often presented as that of a 'power of the past', a survival from the dark times of the Middle Ages resolutely hostile to all innovation, and bound to the letter of a Scripture that compelled it to deny clear proofs: in short the triumph of superstition, intolerance, and stupidity. But this view does not hold up under examination of the facts. We will cite only one passage from the

1. Ultradogmatic by nature, Galileo wanted the Church to recognize the truth of scientific propositions when proven and subject the apparently contradictory interpretation of scriptural passages to them (which is, as we have stated, the attitude of St. Thomas). But he also wanted the unproven propositions in contradiction to the Bible condemned: "Propositions taught, but not necessarily demonstrated, should be judged to be undoubtedly false as soon as something in them is found to be contrary to Scripture" ("Letter to Christine of Lorraine," E. Namer, op. cit., 113). Now such is not the position of the Church, which considers them to be only hypotheses, and this is precisely the case with the Copernican system, which it has never condemned.

2. A. Koyré, *The Astronomical Revolution*, 45.

3. Ibid., 108.

4. A. Koyré, Introduction to his edition of *De revolutionibus*, 19.

5. Such a variation would only be measured in 1837 by Bessel. The difficulty with this measurement is due to the extreme distance between a star and the Earth, and so the angle by which the ray between Earth and star is viewed is extremely small. The Ptolemaic world was already quite large (240 millions of kilometers), but it was small compared to the Copernican and Galilean universe. They were not accustomed to imagining such dimensions (Koyré, *The Astronomical Revolution*, 55).

letter that St. Robert Bellarmine addressed to Galileo and Father Paolo Antonio Foscarini. Foscarini had sided with the Copernican system in a work, written in Italian, and published in Naples in 1615.[1] This letter is important because it defines the Church's position exactly.

"It seems to me," writes the cardinal, "that your Reverence and Signor Galileo act prudently when you content yourselves with speaking hypothetically and not absolutely, as I have always understood that is how Copernicus spoke. To say that on the supposition of the Earth's motion and the Sun's quiescence all the celestial appearances are explained better than by the theory of eccentrics and epicycles, is to speak with excellent good sense and to run no risk whatever. Such a manner of speaking is enough for a mathematician." But to wish to assert the physical reality of Copernicanism is to be in opposition to "all the scholastic philosophers and theologians" and to Holy Scripture, such as the Church interprets it as it has the right to do; this is then to go against the faith, not as to the object of faith, but as to the one who speaks for it. And Bellarmine continues: "*If there were real proof that the Sun is in the center of the universe*, that the earth is in the third heaven, and that the Sun does not go round the Earth but the Earth round the Sun, then we should have to proceed with great circumspection in explaining passages of Scripture which appear to teach the contrary, and rather *admit that we did not understand them, than declare an opinion to be false which is proved to be true*. But as for myself, I shall not believe that there are such proofs until they are shown to me."[2]

And so the Church's attitude is clear, at least in keeping to declarations of principle, for it is obvious that the imperfections of human nature must also be taken into account. This attitude does not at all consist in unconditionally subjecting reason to Revelation, but in postulating their necessary agreement: truths of faith cannot contradict truths of reason, God being the author of both. Hence, if reason establishes a truth that seems to contradict Scripture, the interpretation of Scripture must be subject to this truth. With the whole crux of the matter being to know if the truth is established, it is fitting then, lacking such a certainty, to remain prudent.

This is why, at the time of the Holy Office's 1616 decree, cardinals Caetani and Mafeo Barberini stood openly against the Pope (Paul V

1. *Lettera del R.P.M. Foscarini Carmelitano sopra l'opinione de'Pittagorici e del Copernico della mobilità della Terra a stabilità del Sole e del nuovo pittagorico sistema del mondo.* A Latin translation of this work was added to the first Latin edition of the *Dialogue*.

2. Cited by A.C. Crombie, *Augustine to Galileo: the history of science AD 400–1650*, (Cambridge, MA: Harvard University Press, 1953), vol. II, 215–216. Our emphasis.

wanted to condemn Copernicus as contrary to the faith), and he ended up yielding to their good judgment.[1] And so Copernicus was placed on the Index (although without the sanctions of a higher authority, that is without the Church enlisting its highest power of authority) only *donec corrigatur*, "until it be corrected" in the sense of his system's hypothetical character.[2] Geocentrism (or heliocentrism) was never dogmatically defined as an article of faith. And although it is true that "Urban VIII [the former Cardinal Barberini become pope] and Bellarmine were taken at unawares by an intellectual revolution that they could not even imagine . . . they always acted in a lordly way. . . . The commissary general of the Inquisition himself did what he could to stifle an affair that he deemed absurd and save Galileo *in extremis*."[3]

As for the manner in which the 1616 decree was received by theologians, it should be mentioned that it was viewed as a 'disciplinary' decree, which it was. Father Tanner, a Jesuit, simply affirmed in 1626 that heliocentrism could not be taught as a certainty. In 1631, the theologian Libertus Fromont declared that no decision had been definitively made upon the Copernican system. Father Riccioli, S.J. wrote in 1651: "Since there is not in this matter a definition by the Supreme Pontiff, nor by a council led and approved by him, by no means is it *de fidei* that the Sun turns and the Earth is immobile, at least by very virtue of a decree, but at most and only by reason of the authority of Holy Scripture, for those who are morally certain that God has so revealed it." In short, concludes P. de Vrégille, who has collected these testimonies, "Not a single official theologian considered the decrees in question as definitive and irreformable decrees."[4]

3. — This is why it is reasonable to admit that "the action organized against Galileo was rather the work of lay and university interests,"[5] as well as Galileo's ambiguous and intractable attitude, as Koestler has

1. Giorgio de Santillana, *The Crime of Galileo*, 315.

2. This correction happened four years later—the work of Cardinal Gaetani—and had to do with nine sentences of the *De revolutionibus*. Starting in 1620 the book (corrected) was then no longer on the *Index*. It could be read and taught. And this, moreover, was what transpired. Citing a study by B. Szczesniak, *The Penetration of the Copernican Theory into Feudal Japan*, Koestler writes: "the type of astronomy they [the Jesuits] taught, from the end of the seventeenth century onward, was the *Copernican* system of the world; and that the rapid spreading through China and Japan, of the doctrine of the Earth's motion was thus primarily due to the Society of Jesus, working under the supervision of the Sacred Congregation of the Propaganda in Rome" (*The Sleepwalkers*, 495).

3. De Santillana, *The Crime of Galileo* (French ed. [Paris: Club du Meilleur Livre, 1955]), 420.

4. *Dict. ap. de la foi cath.*, s. v. "Galilée," col. 188.

5. De Santillana, *The Crime of Galileo* (French ed.), 9.

endeavored to establish. It is useless to go back over the history of the Aristotelian opposition, which is incontestable. The ambiguity of Galileo's attitude basically stems from his own declarations, wherein he seems to "be successively and contradictorily sincere."[1] For many times, and on his own authority, Galileo protested over his submission to Church decisions, its rejection of Copernicanism at a time when no precise threat hung over him;[2] but for all that he rejected the compromises, rejected the principle of the equivalence of hypotheses proposed by Bellarmine, and proclaimed himself an ultra-Copernican at the most critical and most difficult moments of his existence. He went even further. He knowingly and without good reason challenged papal authority by ridiculing Urban VIII in his famous *Dialogue*. Urban VIII, who had been his friend when Cardinal Barberini, who had aided and protected him, and who had shown a certain penchant on several occasions for the Copernican system, yet could be recognized, and rightly so, in certain statements of Simplicio, a daft character in the *Dialogue*.[3]

Such appear to be the chief features of the Galileo affair. It would be unnecessary to mention them if the historical truth were better known. This is not the case. Now, for want of a just appreciation of exactly what happened during this extraordinary period in which one of the major events of modern civilization occurred, we would not understand its consequences for European culture, consequences still being felt.

We can recognize now that, as important as this affair may be, it does not amount to much. If the immense literature that it has given and continues to give rise to is disregarded, it is reduced to a bad quarrel made with an especially 'coleric' scholar, one moreover incapable of proving his astronomical theses, by a vacillating religious institution giving way to some pressure groups, and constrained to disciplinarily condemn what it no longer knew how to integrate speculatively. The immediate consequences of this condemnation were extremely limited: Galileo suffered very little, except morally; as for science and its pro-

1. Talleyrand with respect to Alexander I.

2. "I would rather tear out my eye than resist my superiors by upholding against them, to the prejudice of my soul, what today seems certain to me...." (*Ed. Nat.*, bk. XII, 28; letter of Feb. 16, 1614). "If you could see with what submission and respect I agree to treat with dreams, chimeras, equivocation, paralogism and falsehood all the proofs and arguments which seem to my superiors to shore up the system they disapprove of...." (*Ed. Nat.*, bk. XIV, 258; letter of 3 May 1631). Cited by Father de Vrégille, *Dict. ap. de la foi cath.*, col. 179.

3. Not that Urban VIII was irreproachable. This humanist pope entertained at times some strange notions that made one doubt that he was truly of the faith (Koestler, 471). Mostly he was a politician with an irenic turn of mind who sincerely admired Galileo. He even wrote an ode to the glory of the learned Florentine!

gress, all has transpired for heliocentrism as if this affair never happened! And yet it would be correct to see here an exceptional moment in the history of our culture. But what was truly at stake surpassed by far the account of a trial. The historians have lingered exclusively over the circumstances of the event, its least details scrutinized, and it seems that what has happened to them is what happened to the contemporaries and actors in this drama: they cannot see the forest for the trees. Considerable efforts have been devoted to establishing the least fact, defining responsibilities, and sifting testimonies, as if it were preferable not to ask about the profound reasons, precisely those that incite us to a deep interest in this inquiry. Likewise, we see the actors in the drama argue, cite the ancients and the Fathers, oppose scientific reasonings with other reasonings, and compete with juridical subtleties. But they are so absorbed by the urgency of their task that none of them take time to reflect on the cultural significance of the battle joined against them. In short, philosophy was absent at the appointed rendezvous between faith and science. Nevertheless it has had its word to say. Not that the protagonists were wholly unaware of the importance of the stakes. Quite the contrary, to read the items in the documentation is to discover among them a mighty feeling, however confused, of living through something major. But it is obvious that they were unable to formulate it. This is precisely what we will attempt to do now.

The Effects of the Cosmological Revolution

AN IRREMEDIABLY 'FAKED' UNIVERSE

If we seek to characterize the effects produced by the Galilean revolution on the idea of the world and the upheavals it has brought, an initial feature should be stressed: contrary to what is most often stated, and as our historical brief has shown, the 'cosmological humiliation' that people should have suffered from the fact of heliocentrism seems to have played no role.[1] It was never a concern, at least to our knowledge; there was even a totally opposite concern, that is to say having to do with a

1. Freud, as we know, asserted that "the universal narcissism of men . . . has up to the present suffered three severe blows from the researches of science"; the first being due specifically to the astronomical revolution of Copernicanism driving the Earth from a central position, which seemed to man to "fit in very well with his inclination to regard himself as lord of the world"; the other two being Darwinism (biological humiliation) and psychoanalysis (psychological humiliation); *The Standard Edition of the Complete Psychological Works*, vol. XVII (London: The Hogarth Press, 1955), 139–140. A recent study has shown that Freud borrowed this outline linking Copernicus and Darwin from Ernst Haeckel; cf. the article by Paul-Laurent Assoun, "Freud, Copernic et Darwin," in

rehabilitation of the Earth: "... that she is movable and surpasses the moon in brightness, and that she is not the dung heap of the filth and dregs of the universe, and we will confirm this with innumerable arguments from nature." This somewhat lyrical declaration of Galileo, drawn from the *Sidereal Messenger* (1610),[1] and which heralds the grand cosmological presentation of the *Dialogue Concerning the Two Chief World Systems* (1632), quite clearly expresses the idea that he has of the cosmological significance of his theses.

It is harder to appreciate the consequences of the abrupt contradiction brought by the new astronomy to the order of appearances, and therefore to the powerful data of sensory experience. That the Sun, in actuality, neither rises or sets, while we see it accomplish these two movements every day, and that so many meanings and sentiments tied to the most profound rhythms of our psycho-corporeal life are attached to this—such a knowledge could not but introduce, into the existential awareness of Europeans, a latent divorce and a irreparable breach between the order of being and appearances. Without any doubt one can, in a very general overview, see this divorce as the origin of that philosophy of duality which, from Descartes to Kant, runs through all European thought, except for Spinoza and Hegel. With Descartes, the very potent idea that the world is 'mechanized' conveys such a divorce. The devil is not only in the thinking substance (by way of hypothesis); he is also in the extended substance: *diabolus in machina*. Just as animals are only mechanisms hidden beneath the living thing, so the marvelous effects offered us by the 'fable of the world' are wholly explained by a complex machinery of pulleys, cords, bellows, pipes, wheels, jointed levers, weights, and other devices which always come back to extension and motion. The world is truly a 'fabulous opera' that a machinist God has constructed, perhaps for our astonishment, and this 'new philosophy' enables us to pierce through to its secret.

The stage-setting for this opera presents us with a pleasant and colorful spectacle where the most beautiful and most harmonious forms are blossoming: the mountains, forests, carpets of flowers, varied animals, astonishing meteors, hail and rainbows, clouds drifting on an azure

---

Ornicar, n° 22–23 (1981), 33–56. In a vigorous and very scholarly article, Rémi Brague has done justice to this judgment of Freud (via Haeckel): "Le géocentrisme comme humiliation de l'homme," in *Herméneutique et ontologie*, Hommage à Pierre Aubenque, (Paris: PUF, 1990), 203–223; this study was republished (revised and enlarged) in *Au moyen du Moyen-âge—Philosophies médiévales en chrétienté, judaïsme et islam*, (Paris: Flammarion, 2008), 361–396.

1. *Siderus Nuncius, or Sidereal Messenger*, trans. A. Van Helden (Chicago & London: University of Chicago Press, 1989), 57.

background, leaping torrents and lazy rivers, that our eye (another natural marvel) ceaselessly gazes on with rapture, all our senses are appealed to. But, if we look behind the scenes, all this beauty vanishes, those qualities that seem so real are blotted out, and there remains nothing but cleverly arranged wheels. And yet these wheels are not as subtle as the 'method' that can see through them.

Such is the watch that ticks
With ever equal, blind and purposeless tread.
Open and read within: many a wheel
Has entirely displaced the spirit from the world,
A first moves a second, a third follows
and, at length, it chimes.[1]

What characterizes such an etiology is not only a mechanical determinism, but also, and even above all, the idea of a radical heteromorphism of cause and effects; a heteromorphism bound up with what must be called a *physics of suspicion.* This suspicion towards appearances, this belief in a 'faked' world seems to be a constant feature of the Cartesian approach. Extraordinarily distrustful by nature, Descartes lived in the midst of a physical, social, and *even mental* world that deployed all its resources to cheat and deceive his belief. Everywhere he sees snares to be eluded at the cost of the greatest vigilance. To acquire any certainty demands that we surround ourselves with a thousand precautions and develop the immense apparatus of the *Meditations on First Philosophy.* And among the innumerable multitude of these intellectual 'effects' that are ideas, there is but a single one to escape this etiological heteromorphism, a single one that resembles the Cause that produced it, "the mark of the workman impressed on his work,"[2] the idea of God.

As we see, and this is the conclusion to which we would like to come, such a divorce of appearance and being ends up destroying all symbolism, insofar as the symbol is rightly a mediator between visible and the Invisible orders. Or at least there remains, for Descartes, only a single mediating reality: the objective reality of the idea of God; a single vestigial being that can be, in its very immanence, a sign of the Transcendent. It does not, however, seem possible to see in this etiological heteromorphism a simple consequence of the Copernican discovery of the geostatic illusion. The reason for this is that, as we have shown, cul-

1. La Fontaine, "Discourse for Madame de la Sablière," *Fables*, Book 9.

2. "... and it is not also necessary that the mark should be something different from the work itself; but considering only that God is my creator, it is highly probable that he in some way fashioned me after his own image and likeness" (*Third Meditation*, §38, Veitch trans. [Washington: M. Walter Dunne, 1901]).

tures and philosophies have easily combined a more or less 'Copernican' cosmology (the Pythagoreans and Platonists) with a basically symbolic mindset. It is not then the *optical* problem of the Sun's apparent movements that destroys the possibility of religious symbolism. Perhaps in a way this makes it more difficult for the average person—besides, the truth of heliocentrism never seems to have been spread among the people—but not impossible. There must be something else.

We have already shown[1] that sacred symbolism implies both the 'vestigial' character of cosmic realities and their nature as divine words. In other words, the ontological dimension of symbolism is correlative to a dismissal of the sensible world by a more inclusive reality. Or again, sacred symbolism necessarily joins together the two following axioms:

- every symbol only signifies through presentification, that is through a participative correspondence with the realities that it signifies;

- the visible world, the forms of which constitute the various symbolic signifiers, only possesses a deficient being, an image or manifestation of a world invisible and alone truly real, or at the very least nearer to the unconditioned Source of being and truth.

As a result the traditional cosmos also presents a distinction between being and appearance, but a distinction that is not a sundering and entails no heteromorphism. Quite the contrary, appearance is the image and revelation of being. It conceals this only if we idolize it, attributing to it a reality for which it is unsuited. But symbolism is there precisely both to awaken us to an awareness of the Other World, and to save this world by revealing its theomorphism. The world is a theater, as the symbol of the cave teaches, but the setting and the players are likenesses: existentially distinct, essentially one. As we see, the traditional cosmos is multiple: it contains in itself a hierarchy of planes of existence or degrees of reality and, since we cannot do otherwise than take our own state for a point of departure, there is first of all an ontological distinction between what is visible and what is not-visible, that is to say everything else. This is why the Platonic distinction between the sensible and intelligible entails no irremediable sundering: it does not cut the world in two, it vertically distinguishes two worlds, two degrees of reality, one of which is the manifestation of the other. The cosmic theater, which the symbol of the cave specifically has as its end to *reveal*, is not the work of a *deceiver*, an evil spirit, but of a demiurgic power subject to the order of essences. The cosmos viewed in its total expanse as the 'sum' of all the

1. Chap. 1, art. 3, "The Scholastic Misunderstanding."

worlds[1] is of an inexhaustible *depth*; heavens rise tier upon tier above heavens; one behind the other perspectives open out; man is surrounded by an indefiniteness of 'underlying worlds' that, in each thing, open onto an immense and generous reality, mysteriously united with the shoreless ocean of the divine substance.

Quite to the contrary, the Cartesian contradiction comes from the new philosophy wanting to 'house' its cosmic theater and all its machinery in a world without density, 'inside' a visible universe that is pure exteriority and the only reality. That the world of bodies refers to a incorporeal cosmos, for which it is the image and sign—nothing is more logical: reality is at least dual, if not multiple, but, when we do not see something, we can no longer see it. Our senses do not deceive us, it is we who deceive ourselves when we take perceived being for absolute being. The sensible is all that it ought to be, but it does not exhaust the field of the real. In return, for a visible and lying corporeal state to *still* refer to a truthful but invisible corporeal state ... this introduces an illusion with no other option. It is the corporeal world as a whole that is irremediably faked. We will never 'see' its truth. Following Plato, we can at least hope to know one day, through the 'eye of the soul', the truth of the sensible world in a directly experienced contemplation of its intelligible model. But, in the Cartesian universe, there is no possibility of ever seeing the visible—however deceptive—other than as we see it. The truth of perception exists only 'in' human reason. It can only be reconstructed through abstraction, this is the only 'experience' that we can have of it, and yet it is objectively true and real. In the Platonic mythocosm, the ontological deficiency of the sensible world is ransomed by the reality of the *topos noetos*. All is not lost. There is a cosmic and objective salvation. The last word of the tragi-comedy of the universe has not been said, the 'game is not over'. But, in the Galileo-Cartesian universe, all is said, all accomplished, all is present, indefinitely present. With loss of sight, there is only a spreading out of an expanse without ontological depth and without mystery: space, always space beginning again. And, within this irremediably faked space, the trompe-l'oeil of the setting conceals only another setting, which we will never see however. The truth of appearances is only 'in the head' of the spectator. The soul is truly separated from the body, the *logos* no longer dwells in things, no other-world can save the world.

In Sartre's *No Exit*[2] Garcin asks the valet what is 'outside', and the

1. We are speaking of 'sum' as the integral of all possible worlds. For, as innumerable as their multiplicity might be, this is nonetheless contained *integrally* within the manifested (or the created), which thus constitutes its unsurpassable limit.

2. Translated literally, the French title *Huis clos* = 'closed doors'. —trans.

waiter does not understand: "Outside?" he asks. "There is no outside." "Everywhere there are only other rooms, passages and stairs." "And beyond them?" "That's all." The closed door is therefore not only the one for the room in which a man and two women are shut up for eternity. The closed door is the 'door' of cosmic reality as a whole, forever closed upon itself, forever cut in two, a falsified world without hope of a cosmic salvation, a hell. We understand why then such a world is radically destructive to all symbolism, since precisely, as we have said many times, the sacred symbol is the 'salvation of the world'. Not only does it save it from its spatio-temporal contingency, from its crumbling away into indefinite multiplicity, by conferring on it a raison d'être and a unity as 'word of God' and presentification of the Invisible, but again it saves it, in its 'phenomenality', from its 'specular illusion' by revealing the iconic character of appearances and signifying by the same stroke their intelligible model. A world without 'outside', without 'beyond', this is what the Galilean revolution has imposed, but this is also an unendurable idea. Basically, and without knowing it, it is against this cosmic damnation that the Catholic Church protested, and such is, in our opinion, the true stakes of the Galileo affair.

AN INDEFINITELY PHYSICAL UNIVERSE

These analyses have shown us, in fact, the most profound significance of the change that a geometrizing of the real brought to cosmology. All the historians of science agree in seeing in this geometrization Galileo's essential work. But it must not be forgotten that it was conditioned first by a spatialization of reality. We forget this because, for modern man, this spatialization is self-evident. Descartes has taught us that the totality of the corporeal world is only extension and movement. This seems so natural and so obvious that we no longer think about it. In other words, the mathematical treatment of the real in Galilean science is supported by an ontological reduction of the world to the sole dimension of space. Time only intervenes to measure movement in space, but is not, properly speaking, a dimension *of* reality itself, a component of its *substance*.[1] Extension is the essence of corporeal substance. This is precisely what Cartesian ontology affirms, and what Newton attempted to

1. Thus, as Koyré remarks, one "slips into a thorough-going geometrisation." Even phenomena of the temporal order, like the acceleration of speed in uniformly accelerated motion, are evaluated as a function of space and not of time (*Galileo Studies*, 73). It is true that Galileo, who had at first conceived of the dynamic in geometric terms, subsequently endeavored to return to the 'cinematic stage' and formulate the acceleration of the fall of heavy bodies as a function of time (ibid., 109), contrary to Descartes, for whom extension alone enjoyed substantiality, while time was only a mode (ibid., 91).

endow with a metaphysical and religious sense by posing absolute space as a *sensorium Dei*. This topological reduction also exists for Galileo, even if not isolated and formulated in itself. In any case, he knows quite well that it is somehow presupposed by Copernican astronomy. In other words, as Clavelin shows, Copernicanism implies premises about space and the world that Galileo would strive to establish.[1]

These premises define the anti-symbolic universe of modern science. And first of all, this space is not in six directions, as for Aristotle: there is neither above, nor below, nor cardinal points. To reiterate an expression of René Guénon (in *The Reign of Quantity and the Signs of the Times*),[2] there is no *qualified space*. Hence the world has no center and bodies have no natural place. As a result the order of the world depends on no qualitative determination inherent to space itself. It is simply a container, neutral with respect to movements unfolding within it. In traditional cosmologies this is the 'ordering structure' of the world, these are the qualitative determinations inherent to space itself which are logically primary and which account for the need for a center and natural movements. For Galileo, movement constitutes the ordering structure: "Having identified motion with order, [Galileo] could dispense with its ontological function. At the same time he severed the link between local motion and changes in respect of quality and quantity, and thus robbed the concept of κίνησις [*kinesis*] of all meaning."[3] Since that time motion, no longer having any end, can be regarded as a *state*, and it can therefore be maintained indefinitely. This is why, in such a perfectly homogeneous space, it is impossible to accommodate any quality. "For my own part . . . never having read the pedigrees and patents of nobility of shapes, I do not know which are more and which are less noble, nor do I know their rank in perfection. I believe that in a way all shapes are ancient and noble; or to put it better, that none of them are noble and perfect, or ignoble and imperfect, except insofar as for building walls a square shape is more perfect than a circular, and for wagon wheels, the circle is more perfect than the triangle."[4] This declaration clearly expresses this qualitative neutralization of the cosmos, correlative to the neutralization of symbolism.

Thus space-matter is everywhere the same. "This postulate," affirms Jacques Merleau-Ponty, "is almost unbelievably audacious in the extent

1. *The Natural Philosophy of Galileo*, 207–208.
2. Trans. Lord Northbourne (Hillsdale, NY: Sophia Perennis, 2002), 38–41.
3. Clavelin, op. cit., 216.
4. *The Sidereal Messenger*, *Ed. Nat.*, book VI, 319; quoted by Clavelin, op. cit., 212.

to which it extrapolates experience."[1] Euclid's geometry assumes a *physical truth*. Insofar as viewed as a container, this geometrico-physical space is perfectly indifferent to what occurs within it. There is no relationship between its structure, identical throughout, and moving bodies. And this is also why it is infinite.

To tell the truth, Galileo himself was not much interested in this question, and seemed hesitant to take sides. "For Galileo the question of a finite or infinite macrocosm is insoluble. But here is a curious point. He was not overly interested in this question. The universe is perfectly fine for Galileo except for this: at its center is the Sun and not the Earth and, moreover, the confines and center of this universe bear no system of natural placements."[2] The reason for this relative indifference (for Galileo tended rather to favor the infinitist thesis),[3] in the words of Koyré, is that the problems of cosmology and celestial mechanics hardly interested him: "It may be, therefore, that, like Copernicus himself, he never took up the question *a quo moventur planetae* [by what are the planets moved], and thus never made the decision—though it is implied in the geometrization of space of which he was one of the foremost promoters."[4]

The correlation between the geometrization of physical space, that is the affirmation of its complete homogeneity, and its indefiniteness is in some manner illustrated by Kepler's counter-example. Kepler was at once an adversary of a purely homogenous space and of an indefinite space. He was also, among the founders of modern astronomy, the only one who strove to build, or restore, a relatively symbolic cosmos.[5] As for the first point, Kepler's reasoning was as follows: if the world is infinite, it is homogenous, for there is then no *sufficient reason* for the Creator to have distinguished one celestial region from another, it being given that an infinite universe has no center with respect to which differences

1. *The Rebirth of Cosmology* (New York: Knopf, 1976), 74.

2. Boris Kouznetzov, "L'infinité du macrocosme et du microcosme chez Galilée," a paper given at the 1968 International Conference at Brussels and published in *L'univers à la Renaissance. Microcosme et macrocosme* (Paris: PUF, 1970), 55. In this paper Kouznetzov gives 'microcosm' the unusual meaning of the 'infinitely small'.

3. The argument which, as Galileo himself tells us, swayed him "rather toward the infinite and unterminated than toward the terminated" was that, if the world was finite, we would not hesitate to declare it such (Letter to Liceti, Feb. 10, 1640, *Œuvres*, XVIII, 293ff.).

4. *From the Closed World*, 99. The question of a physical cause for planetary movements is actually posed in a geometrized universe, since there is no longer any structuring order for cosmic space to account for them.

5. And yet we will soon see how this evaluation must be qualified, for Kepler's symbolism is no longer altogether traditional.

might be measured: "How can we find in infinity a *centrum* which, in infinity, is everywhere? For every point taken in the infinity is equally, that is, infinitely, separated from the extremities which are infinitely distant."[1] Now there is at least one region to be distinguished from all the others, and that is the solar system. This fact proves that the stars are not equally distributed throughout the universe, which has then a 'regional structure' and is not infinite.[2] As for the second point, we will only recall the Pythagorico-Platonic doctrines that Kepler set forth in the first part of *Mysterium Cosmographicum*,[3] not only the one concerned with the role played by the five perfect solids (the five 'Platonic' bodies) in the structure of planetary orbits,[4] and the sympathy that joins each solid with each planet, but also the astrological, numerological, and zodiacal doctrines, as well as the rapport uniting musical harmony with that of the spheres, the heptacord, and the celestial septenary.[5]

1. *De stella nova in pede Serpentarii* (Prague, 1606), cap. XXI; *Gesammelte Werke*, ed. Max Caspar, vol. I (Munich: C.H. Beck, 1938), 254; trans. Koyré, *From the Closed World*, 53.

2. Kepler never varied on this question, even after the discovery of the telescope (Koyré, *From the Closed World*, 62–87). Besides, as Koyré states (ibid., 65), "modern science seems rather to have discarded than to have solved the problem" and even, because of the impossibility of imagining a world that is actually infinite, "has returned to a finitist conception." Kepler's argument can be likewise compared to the famous 'paradox of Olbers': this German astronomer remarked, at the beginning of the nineteenth century, that, if stars were equally distributed in the universe, assuming that they arranged as indefinitely deep tiers, the depth of the sky should appear uniformly luminous instead of black. Classical infinitist physics cannot resolve this paradox.

3. *Gesammelte Werke*, vol. I (1597). Kepler was twenty-five when he published this first work. Cf. *Mysterium Cosmographicum—The Secret of the Universe*, trans. A.M. Duncan (New York: Abaris Books, 1981).

4. There are five perfectly regular solids, that is to say inscribable within a sphere (*Timaeus*, 54d–55c). This is why, according to Kepler, there are only six planets. He imagines in fact that the sphere (or orbit) of Saturn is circumscribed by a cube in which is inscribed the sphere of Jupiter, and so forth: tetrahedron, the sphere of Mars, dodecahedron, the sphere of the Earth, icosahedron, the sphere of Venus, octahedron, the sphere of Mercury. This arrangement provides the sufficient reason for the respective magnitudes of the planetary orbits, while the number of facets of the Platonic bodies (4, 6, 8, 12, 20) correspond to the order of planetary succession, from most distant to nearest the Sun. This correspondence is Kepler's invention (a discovery that filled him with an inexpressible joy) and is not to be met with in traditional symbolism. For Plato, the perfect solids correspond to the elements (cube-earth, icosahedron-water, octahedron-air, tetrahedron-fire). Kepler also made use of this correspondence: *On the Harmony of the World*, book II, *G. W.*, book 6, 81.

5. The correspondence between the celestial septenary and the musical scale is universal: Macrobius and Martianus Capella, for example; Duhem, *Le système du monde*, book II, 10–13.

Nevertheless, whatever Galileo's attitude about this, it is a fact that the cosmological mutation effected by Galilean Copernicanism can be described, it seems, as a passage "from closed world to infinite universe," to borrow the title of Koyré's book. Contemporary philosophy, and modern man in general, do not remark this infinitization of the cosmos without a certain pride. A pride felt so much more vividly when in stark contrast to the fear or confusion of a Pascal or a John Donne before this shattering of the world:

> And new philosophy calls all in doubt,
> The element of fire is quite put out;
> The sun is lost, and the earth, and no man's wit
> Can well direct him where to look for it.
> And freely men confess that this world's spent,
> When in the planets and the firmament
> They seek so many new; they see that this
> Is crumbled out again to his atomies.
> 'Tis all in pieces, all coherence gone.[1]

All historians have in fact noted that "the immediate effect of the Copernican revolution was to spread skepticism and bewilderment."[2] It is facilely explained that, as prisoners of the narrow medieval cosmos, the Europeans of the time stood in fright of the spaces that the new knowledge displayed before them. Their timid souls took fright before such audacious perspectives, like children who could not imagine that the world was so large. Quite happily, our soul has grown with the universe. Henceforth we are adults. We no longer dread "the eternal silence of the infinite spaces." Fearless, we confront this cosmic indefiniteness, in which we even see a direct image of human development; for the indefinite enlargement of space is witness to the indefinite enlargement of our science and hence our power, the supreme form of which power is called today the 'conquest of space'. Thus Valéry reprimands Pascal severely for having stayed a stranger, by "lack of faith in research insofar as it hopes in the unforeseen,"[3] to the enthusiasm that the spectacle of

1. Donne, *An Anatomy of the World* (1611), ll. 205–213.

2. Koyré, op. cit., 29. The same author notes, it is true, that the historians who have recently studied "the disastrous effects of the seventeenth-century spiritual revolution" do so with "some nostalgic regret" (281). We even think that the Western soul has never recovered from the cosmological shock suffered at that time. But this bruise was quickly forgotten for the sake of the scientific and technical enthusiasm that the medieval cosmos seemed to thwart.

3. *Variétés* (Paris: Gallimard, 1924), 147.

cosmic immensities should not fail to arouse in every truly scientific or even religious spirit.

However—and without entering into the problem of the exact interpretation of Pascal's fragment[1]—it should be said that not only is infinity objectively terrifying, but so is the silence of the universe, a silence of the universe pointed out by Pascal on several occasions: "When I see the blindness and the wretchedness of man, when I regard the whole silent universe. . . ."[2] Now it is precisely this silence that makes clear the metaphysical truth of the cosmological revolution, a truth for which Pascal is, we think, the greatest, most lucid and, in a certain manner, the only witness. For, with creation no longer the other Book of divine revelation, with heaven and earth no longer singing the glory of God, this is not only a calamity for religion, it is also a terrible condemnation for the universe itself, the being of which ceases by this very fact to have any meaning. A Laplace or a Valéry, through metaphysical ignorance, might well be unaware of what the 'realist' geometrization of the universe implies. And yet its effects were constantly unfolding right down to the Sartrean proclamation of the world's existential absurdity, which is the strict consequence of this. It is only registering the irreparable divorce of the flesh from the Word. The universe is no longer anything but a cosmic cadaver.[3]

And let no one object that some of the most Christian thinkers, such as Bradwardine in the fourteenth century, were perfectly capable of reconciling an affirmation of infinity with the Christian faith, and manifested no pessimism. This is because the situation was altogether different. We have already shown the true significance that must be attributed to the infinitism of a Nicholas of Cusa. Likewise we have stressed, in connection with the infinity of space for Bradwardine, that this does not involve a physical infinity. We are far removed from the

1. Fragment 206 of the Brunschvicg edition, 201 of the Lafuma [Trans.—Trotter translation]. Actually it is almost certain that this saying, far from expressing an anguished confidence of Pascal, ought to be presented, in the mouth of a libertine, as someone becoming aware of his true cosmic situation. Moreover, if Valéry had read his author more attentively, he would have seen that he also writes that "those who have the living faith in their heart see at once that all existence is none other than the work of the God whom they adore" (Br. frag. 242, L. 781.); M. de Gandillac, "Pascal et le silence du monde," in *Blaise Pascal*, Cahiers de Royaumont, 1 (Paris: Ed. de Minuit, 1956), 342–365.

2. Br. 693, L. 385.

3. "The problem—for Pascal—is not of knowing what remains, after sin, of cosmic harmony, but what the world of the new science might mean for man if it is still a *true symbol*, an interpretable sign. The answer is beyond doubt: this world is 'silent', and it is still more so for the believer than for the libertine." This reflection of de Gandillac ("Pascal et le silence du monde," 349) clearly identifies the significance of Pascal's attitude.

reification of Euclidian geometry. This is why, contrary to what Pierre Magnard[1] asserts, for example, in an already cited study, we should be by no means "justified in expecting a religious interpretation" of the shattering of the closed world of the ancients, despite "the persistence of a Dionysian and Cusan tradition." The exclusively physical and realist character of the Galilean universe deprived this tradition of its efficacy, at least for common cultural awareness, that of the honest libertine to whom Pascal rightly intended to speak.

A UNIVERSE INDEFINITELY LIMITED BY THE SPATIAL CONDITION

If Pascal is 'right' then, if he expresses the truth about a moment in European culture, we must, behind the apparent enthusiasm arising from the infinitization of space (and not ignored by Pascal since he can make it the source of a certain cosmic lyricism[2]), grasp the true significance of this cosmological revolution witnessed by seventeenth-century man. Our thesis is as follows: this infinitization of the physical expanse

---

However, we must make this distinction: the believer has a more acute awareness of the world's silence than the libertine. But the believer is also the only one for whom this silence can be eventually transformed into words: "God, wishing to show that He could form a people holy with an invisible holiness, and fill them with an eternal glory, made visible things. As nature is an image of grace, He has done in the bounties of nature what he would do in those of grace. . . ." (Br. 642, L. 274). Basically, Pascal's thesis is that God is at once present and absent: "All appearance indicates neither a total exclusion nor a manifest presence of divinity, but the presence of a God who hides Himself. Everything bears this character" (Br. 556, L. 449); "Nature has some perfections to show that she is the image of God, and some defects to show that she is only His image" (Br. 580, L. 934).

1. "L'infini pascalien," *Revue de l'enseignement philosophique*, Oct.–Nov. 1980, 11. To be precise, our reserve bears upon this secondary point alone, for we have learned much from this scholarly study.

2. When one sees only cosmic pessimism in Pascal, as do Valéry and some others, they have forgotten to ask how the same author can propose that we contemplate "the whole of nature in her full and grand majesty," a nature so simple that our imagination is made weary with it: "No idea approaches it. We may enlarge our conceptions beyond all imaginable space; we only produce atoms in comparison with the reality of things. It is an infinite sphere, the center of which is everywhere, the circumference nowhere. In short it is the greatest sensible mark of the almighty power of God, that imagination loses itself in that thought." And, deploying next the 'prodigy' of the infinitely small, he concludes that whoever "regards himself in this light will be afraid of himself, and observing himself sustained in the body given him by nature between those two abysses of the Infinite and Nothing, will tremble at the sight of these marvels; and I think that, as *his curiosity changes into admiration*, he will be more disposed to contemplate them in silence than to examine them with presumption" (Br. 72, L. 199; our emphasis). We do not think then that Pascal's attitude is so totally foreign to the Dionysian and Cusan tradition. After all his 'epistemological' conclusion rejoins the Cardinal's: our knowledge has no sure grasp on the indefinite.

is illusory. Far from enlarging the world's reality, it only serves to *mask* its irremediable enclosure in the corporeal modality alone. Everything actually transpires as if this infinitizing, by the fascination it had on minds, had caused this 'shutting out' or 'shutting in' to be completely forgotten. The title of Koyré's book is misleading: *only with respect to the Galilean world is the medieval universe closed.* Everybody repeats this formula, everybody restates this opposition as if it exhausted the question. But there is nothing to it. Ancient and medieval people did not have the feeling of living in a cosmic prison. First of all, and we have already mentioned this,[1] the Ptolemaic universe is not small. But this is not what is essential. The important thing is that this universe is prodigiously rich and profound. Within it there is a profusion of innumerable degrees of being and it therefore possesses inexhaustibly varied forms of reality. In short, its 'infinity' is of the qualitative order, whereas post-Galilean infinity is purely quantitative. The famous words of Hamlet to Horatio—"There are more things in heaven and on earth than are dreamt of in your philosophy"—apply perfectly to the medieval cosmos. But it would not suit the Cartesian. Reduced solely to the ontological dimension of extension, wholly spread out before us, *partes extra partes*, a mechanism able to be dismantled and dismantled, without residue or mystery, without interiority, *its bounds are everywhere, its center nowhere.* And Descartes, basically, would agree with this, he who so clearly distinguished infinity (applied only to God) from the indefinite.[2] For this indefiniteness of space is for him rather an imperfection than a perfection.

To say that in a particular space the center is nowhere is readily conceded, since in fact nothing in the universe has the quality of a center.[3] This space is no longer qualified and no longer *qualifiable*. In the medi-

1. See 71, n5.

2. *Principles of Philosophy*, First Part, §§26 and 27.

3. We are not sure we can apply to space, as Pascal does, the celebrated formula first found, it seems, in the *Book of the Twenty-Four Philosophers*, a pseudo-hermetic writing composed in the twelfth century of which it is the second proposition. This formula does not mean just anything. To say that the "center is everywhere" is to say that everything has the quality of a center: the center is not *no matter where* but omnipresent; this is not an indefinite state but an infinite one. To say that the circumference is nowhere is to say that this 'metaphysical' circle (or sphere) is contained or encompassed by nothing, that it is integrally unlimited. Clearly the circular form takes on a symbolic meaning here, and therefore such a circle is no longer of the spatial order. But we can suppose that, for Pascal, an application to space is possible precisely to the extent that this is the universe as a 'locus' of Divine Omnipotence. For, "if we are well informed, we understand that, as nature has graven her image and that of her Author on all things, they almost all partake of her double infinity" (Br. 72, L. 199).

eval universe the determination of the center is geometric in appearance only. Surely, the center of a space can be fixed in an extrinsic manner, by starting with its extreme limits and taking the middle of a diameter, therefore by dividing it. But, in reality, the center is an intrinsic property of space itself, a peculiar quality rather than a mathematically calculable and locatable point in space. The universe has a center because it is turned toward the One. When Nicholas of Cusa denounces the illusion of a geometric, exactly assignable center of the world, he is calling into question a certain reifying tendency of Latin Averroism that had lost the sense of analogy, but he does not destroy the notion of a qualitative center of the universe, of a certain centrality of cosmic space. This is possible because basically space is more than space, because it is the most immediately obvious image of divine Omnipotence. In return, the Euclidean homogeneity of Galilean space excludes all qualitative centrality and, by the same stroke, every geometric determination of any center whatsoever, since there are no longer any extreme limits with respect to which it would be mathematically determinable. In other words, according to the perspective of ancient metaphysics, the physical centrality of the cosmos is only an image and is based on an intrinsic qualitative centrality of space, which is a direct consequence of the divine omnipresence in creation. It is God, the unique center, who confers on every point in space its capacity to be taken equally for a center. This is why, when the analogic and symbolic nature of cosmic centrality is lost sight of under the effect of an over-insistent realist physicism, there always remains the possibility of making an appeal to its metaphysical basis, and of showing at the same time: no point is absolutely a center, there is only a single (metacosmic) Center, but every point can be a center relatively by participating in divine Centrality. This is even more so when the centrality of space is denied in the name of physical truth. In such a case, if all points of space can be taken indifferently for a center, this is because the center is no longer anywhere, and because every center is illusory or, at best, a geometric construction. And it is certainly possible to see a symbolic proof of the disappearance of the center, and therefore also of the world's qualitative sphericity, in the fact that "while ancient and medieval physics contrasted natural circular motion with violent rectilinear motion, this contrast is inverted by classical physics, for which it is rectilinear motion which is natural, and circular motion which henceforth is considered to be violent."[1]

Perhaps it will be less well understood when we declare that the limit of such a universe is everywhere. And yet this is no less evident, since

1. Koyré, *Galileo Studies*, 130.

here it is only a strict consequence of the spatialization of corporeal reality. The indefiniteness of space should not mislead us; it is always just the extension of finiteness as such. Indefinitely repeated, the finite remains no less the finite; and, even if we follow this line of thinking right to the end, such a repetition only makes this finiteness increase indefinitely, that is, only brings it to its maximum. The truth about the infinitizing of cosmic space is, then, solely the truth about the universal limitation of such a cosmos. As long as the world is something other than space (and, with Aristotle for example, as long as it is populated by souls and intelligible forms) its spatial finiteness, its *spatial closure* if you like, is of little import: this world is inwardly open to the infinite. But when the world's very substance is reduced to a space, then it is definitively enclosed within itself; even more, it is identified with the limitation itself. For ultimately, what is geometric space thus reified, what is this extension spoken about by Descartes?[1]

We will define space by exteriority, or again as that which determines the mode of existence *partes extra partes*. Actually, in our perspective space is not viewed as a thing in itself, nor as an *a priori* form of sensibility, but is regarded as a condition of existence which, among other conditions (time, quantified matter, form, energy or movement), affects all beings and all things that belong to the realm of corporeal existence.[2] This means that for a particular being to exist corporeally is to be subject to the spatial condition (as well as other conditions). Now, to know exactly in what this condition consists, what is its essence, it must be asked in what manner it affects the corporeal being, what is the effect produced in this being. We will say that the clear effect of the spatial condition on a corporeal being—or spatialization—is to render it indefinitely divisible. Divisible because, for any being, to be extended makes it capable of being divided: only the unextended is insecable. Indefinitely, on the other hand, because a non-indefinite divisibility would lead to atoms of space, in other words to spatial unities, and because such a notion, which also corresponds exactly to the definition of the geometric point, is a contradiction *in terminis*, since it is a consequence of the preceding proposition. To give a meaning to the notion of point, we are led then to consider the point as the *inaccessible limit* of divisibil-

1. *Principles of Philosophy*, Second Part, §§ 9–15.

2. In what follows, we are relying in part on the article by René Guénon, "The Conditions of Corporeal Existence," published in *La Gnose* (Jan. 1912) and reprinted in *Miscellanea* (Hillsdale, NY: Sophia Perennis, 2004); this article is unfortunately unfinished. As for space, this must be completed by the considerations set forth by Guénon in *The Symbolism of the Cross* (Ghent, NY: Sophia Perennis, 2001), *passim* and above all Chapter 16, "The Relationship between the Point and Space," 86–90.

ity (we do not say of division), a limit presupposed by divisibility (which it posits implicitly), for otherwise divisibility would not be divisibility, that is to say the continual possibility of separating parts from each other, with every part in this operation always considered to be as if a *relative unity* distinct from all the others. Now the notion of a relative unity requires, for a basis, that of absolute unity which, with respect to space, is called a point. This notion of point clearly then 'orients', as its unrealizable term, the operation of divisibility. Space is not therefore, properly speaking, composed of points, but of a single *potentially omnipresent* point, the inaccessible aim of all divisibility. Space thus appears as the very 'movement' by which this aim is reached, the tension by which divisibility strives to approach the point. But the aim or end is also, and necessarily, the principle. The movement of divisibility not only has in view the point as its aim, it also starts from the point as from its origin; this tension is equally an ex-tension. The point-principle is therefore also the original potentially omnipresent limit of every process of divisibility. Obviously point-of-origin and end-point are only one and the same point, insofar as principle or end. But if space is no longer considered in either its root or its (non-spatial) completion, but in the very process of its divisibility, matters no longer proceed in the same way; the point-of-origin is temporally distinct from the end-point, for otherwise there would not even be any actual space. On the other hand, we see by this that time is indeed a fourth dimension of space or, more generally, that space implies time as an intrinsic condition of its own actuality.[1] Time is at the core of the movement of extension by which space is actualized. On the other hand, to the very extent

1. The converse is not true: time does not imply space; or rather—for there is an undeniable correlation between physical time and space—there are modes of duration in which no space is implied. This is the case with a whole category of psychic phenomena: a sentiment, love for example, possesses a real becoming, its own duration, its own rhythm (as Proust has so admirably shown), which do not require, by themselves, any spatiality. In return, if physical time is considered, it must be said that space is as if a fourth, *extrinsic* dimension of time, a possibility that time conceals within itself (in a non-spatial state) and is produced as a residue that time casts off behind itself, in the manner of a spider spinning its web. From this point of view, time is transformed into space. Time completed is space entirely realized, succession is transformed into simultaneity. The point is no longer omnipresent then in a solely potential manner. However, neither is this actual omnipresence any longer that of a point. But it is a pure and simple, contradictorily realized bursting asunder. No longer is there any point. This is then also the end of space, which implies, as we have seen, the temporal (infinitesimal) isolation of origin- and the end-points. The abolishing of this temporal isolation (in pure simultaneity) therefore also abolishes space. The end of time (as a condition of existence for corporeal beings) is also the end of space.

that we consider space in its actuality, or rather in its actualization (for space can never *be* perfectly actualized), the potential omnipresence of the unique point gives way to the indefinite exteriority of every point with respect to all the others, that is, to the indefinite potential multiplicity of points that mutually limit each other. For to be on the outside of a thing is necessarily to limit it. In spatialization every point is the limit of all other points, and every point is limited by all the others. Therefore there is, in space, only limitation: space is the sum of all the mutual limitations of all points. This is truly limitation as such; from the ontological point of view, this constitutes the ultimate limitation for a reality, beyond which there is no longer anything. To be in space, for a body, is therefore to participate in this exteriority: the spatiality of a body signifies that every point of this body is *distant* from every other point of the body, and that the body itself—which is comparable to a point by virtue of the relative unity conferred on it by its form—is distant from all the world's other bodies. By this, then, is the body limited, not only as a form upon a background, but ontologically, in its very substance. To make the body identical to space is to make it identical to its own omnipresent limitation.

The opening of cosmic space must not mislead us by imposing the image of an enlarging, a widening of the world taking the place of a universe timidly turned in upon itself. This so-called widening is in fact an annihilation. Also, it is in some manner only the cultural projection of the spatialization of physical reality. This is why our whole analysis sheds light on a dynamic conception of space. Space in its very nature, in its 'internal' structure, we venture to say, is a spacing out, an *almost-realized explosion*. 'Almost' because an absolute explosion would be equivalent to nothingness, and because space would be annihilated by its realization. Space is therefore, by nature, extension or, again, expansion.[1] It is this expansive component of space that, following from the reduction to extension of corporeal reality, is projected culturally, in the seventeenth century, under the form of an indefinite enlargement. However, to this dynamic conception might be objected the observable

1. These purely philosophical considerations are in accord—at least in a certain sense—with the conclusions of Einsteinian cosmology. But even there an *enlargement* of the universe can only be spoken of (as proven by the *red-shift*, or displacement of the yellow spectrum line of sodium toward the red, interpreted, in terms of the Doppler-Fizeau effect, as giving evidence of a radial displacement of the stars) in a relative manner, for there is rightly no absolute space with respect to which the magnitude of the universe could be measured. That said, our account leads to the notion of a cosmic space, a function of the temporal process, for which it is as it were the terminal mode. And so we interpret the *red-shift* and Hubble constant as the trace of this dynamism of

fact of the stability of corporeal forms: in general, bodies do not explode. But this is exactly what they should do if composed of extension alone, as Descartes would have it. For, if we view the spatial condition no longer insofar as it affects a body, which truly confers on it its whole reality, but so to say in itself, it no longer appears as just exteriority, distancing, scattering. If bodies do not explode, this is because they are somehow held back 'from within', because their expansive movement is held in check or 'reined in' by the transpatial unity of their essence. As we will see when we broach the ontology of the symbol, the fundamental 'tenon' that secures the substantial reality of a corporeal being, its ontological consistency, should not be in space, which, as such, is pure dispersion and indefinite divisibility.[1]

Have we now answered every objection? Surely not, for what remains evident to our mind is the indefiniteness of geometric space that seems altogether indispensable. Some will ask: why not an empty space to 'house' your explosion of space? Here is truly the most subtle snare of the space question: *space always seems to presuppose space*. We spontaneously try to produce space. We begin with the non-spatial, with the unextended, with the point for example, which may be deployed and repositioned, and then we perceive that this repositioning presupposes precisely the existence of what we wanted to produce. We speak of the world's expanse, of cosmic space, and unconsciously situate this space

exteriorization that is space. This does not mean that the expanding universe occupies more space, that it is dilated *in* space, but rather that—if we might hazard something in this realm—the light traversing space, as shown by the red-shift, is subject to this centrifugal dynamism and expresses it after its fashion by a change in its wavelength.

1. The most acute analyses of recent physics even lead to a paradoxical reversing of ordinary conceptions. No longer is it the separation or distance between two points that is primary and obvious, and therefore the possible correlation of these two points that must be explained, but, to the contrary, it is 'inseparability' that becomes 'principal', and distance or remoteness that causes a problem: "In an intrinsic way, there *is* no distance between such or such an element of independent reality. In a certain manner, we are the ones who put it between such or such elements of *empirical reality*, or, in other words, of the image of reality constructed for our exchanges and our use" (Bernard d'Espagnat, *A la recherche du réel*, (Paris: Gauthier-Villars, 1979], 46). This work is open to some criticism, especially because of the somewhat summary character (to our eyes) of certain philosophical considerations (what then might a physicist think of our own considerations on scientific matters!). But this would not be enough to discredit his fundamental thesis: objective reality is not material in nature. The question of a non-spatial principle required for the physical unity of a body is the object of a new and expanded treatment in *Amour et Vérité* (rev. ed. of *La charité profanée*), chap. VII, sect. 2 (Paris: L'Harmattan, 2011): "Le détour cosmologique". We have explained the principle of non-separability in *Problèmes de gnose* (Paris: L'Harmattan, 2008), 131–139.

in a greater space. We are told of an expanding universe, and we cannot help but imagine a preexisting space which makes this expansion possible. Now, it is true that this pure, isotropic, and completely empty space is the condition for the possibility of space as an extensive dynamic, as exteriority on the way to realization. But it is a passive condition, even preeminent passivity. And whoever says 'passive condition' is speaking of a condition inseparable from an active condition. Just as with Aristotle matter should not exist without form, so empty, static space should not be isolated from space-expansion. This is however what we do when we think about space geometrically. We 'realize' it conceptually, separating it from that which is its condition; but only with respect to this condition does it have meaning. We complete in thought the constitutive movement of the distancing of space, of its indefinite divisibility; we deploy it at a single stroke in its total and absolute exteriority, forgetting that it would be then a pure and simple contradiction, since, with exteriority, there can be only something of the relative, only something of one part relative to another, and therefore that, in reality, the true condition of the possibility of space is the potential existence of the point-limit, which is the preeminent 'part', the absolute term with respect to which alone everything is relation. Besides, it is inevitable that this is so, for there is a particular affinity between thinking activity and geometrical extension. To think in fact is to think about possibility, at least implicitly, since pure impossibility is inconceivable.[1] To think about space is therefore to think about the possibility of space. But, on the

1. For example, we cannot really think the concept of a circle-square, we can only name it. If someone objects that we can indeed think about in a certain manner since we can speak of it, we reply that we do not think about it in itself, but we are only thinking about its impossibility. And if it is objected that we are then contradicting ourselves, since we have called an impossibility inconceivable, we will point out that there is indeed a difference between impossibility understood absolutely—which is inconceivable—and a relative impossibility, that is the impossibility *of* something. This relative impossibility is altogether conceivable since in reality it is only the thought of two mutually exclusive possibilities. This discussion proves precisely that we only think positively or really about the possible. And so true is this that a concept envelops its own possibility in a generally implicit manner: we think of the concept directly in itself; possibility as such is explicitly conceived and posited only by reflecting on the potential impossibility of what we are thinking about. But this impossibility is not always easy to perceive. And that is the chief cause of our errors. It consists precisely in what thought is always thinking, directly and spontaneously, what it thinks as possible. But this fact does not always correspond to what is correct. We should not, however, blame thought for this; thought can no more think 'outside' possibility than the eye can look straight at something that breaks up refracted light. Hence, for thought, the duty to examine critically: the only faulty thought is the lazy or arrogant thought.

other hand, space itself, so conceived, is nothing but the idea of pure corporeal possibility: the possibility of space is the space of possibility, since geometric space is the idea of a pure, universal, neutral, and isotropic container. This is the locus of the possibility of bodies. Such a 'space' is then as if an objective projection of thought's own conceptive ability. Geometry confers on this 'conceivable' space a kind of objective consistency because of the rigor of the relationships that it brings about there. But, in reality, this space is much more 'conceivable' than spatial: its spatiality, its indefinite extension, is a consequence or a translation of the very idea of pure possibility; which is to say, not the possibility of something, but the possibility of everything, and therefore the absence of all determination. We can certainly give it a metaphysical and therefore altogether 'realist' sense, by seeing in it the symbol of universal Possibility (Guénon) or the Infinite, which refers us back, in a certain sense, to the uncreated space of the Kabbalah and, more assuredly, to the doctrines of Bradwardine, Henry More, and Newton. Unable to conceive of anything beyond this universal Possibility, which is also Immaculate Conception,[1] our understanding necessarily perceives in it the reflection of everything it knows. But it also understands that Possibility is universal only in the Universal and the Absolute, beyond every determination and even the first of all, the ontological determination, there where complete interiority is identical to perfect exteriority. Short of this thearchic Super-Essence, all possibility is relative. It is the possibility of something and cannot be separated from this except by abstraction. In other words, physically, space is not a container for bodies. We too readily imagine bodies as unextended points penetrating into space so to be deployed, as if space offered them precisely the possibility of this deployment. But bodies are not in space as contents in a container. Bodies *are* spatial, or rather are affected by the spatial condition, and it is the body's own spatiality that in some manner confers on it the capacity to be received in space. Because the body is spatial, space manifests as a possibility and not the reverse. And it is not spatial possibility that explains the substantiality of bodies, since it is 'possibility' within itself only because it is nothing. Now certainly: *ex nihilo nihil fit*.[2]

1. We have explained the metaphysics of the 'Immaculate Conception' (in rather theological language) in our *Amour et vérité*, 293–303.

2. On the level of the relative, we need to say: something *is*, and therefore something is *possible*; for we know the possibility of something, that is to say basically its intelligible structure, its 'logic', its non-contradiction, only from its reality. But, at the level of the Absolute, we need to say: something is possible, and therefore something is (without specifying the mode of being), for, *in divinis*, everything that is possible is real (but

Such considerations enable us to grasp the true significance of the infinitization of the cosmos at the beginning of the seventeenth century, and to justify what we are saying on this subject about an indefinite limitation. It is precisely this concept of an indefinite or universal limitation that enables us to explain, paradoxically, both the apparent enlargement of cosmic space, and the *imprisoned* feeling that seized minds and hearts at that time. We repeat—for, in our eyes, here lies the essence of the cultural change undergone by the European soul—the limitlessness or *openness* of cosmic space is fundamentally correlative to the reduction of the world to geometric extension, that is to its ontological *closure*. Conversely, as we have sufficiently established, the spatially enclosed ancient and medieval world is correlative to an ontological limitlessness, because the order of things is mysteriously tied to the divine order that it symbolizes. This intrinsic limitlessness of the world is in harmony with the *depths* of human life, for human life always develops in a universe at once finite and profound, never in a unlimited space composed of surfaces. Life is finitist, it demands a frame, a 'milieu' (*Umwelt*), an ecological niche, a matrix, a paradise. But at the same time it adds to this closed three-dimensional space (symbolized by the cross of rivers in the Garden of Eden) a fourth dimension, the only one that is infinite, or nearly infinite: the inner depths, geometrically unrepresentable, the concealed presence of a hinter-world in all things, a world that prolongs their secret being toward infinite Being. This is why Galilean science might indeed offer us the intoxication of a limitless expanse into which our gaze plummets, but the collective subconscious is not deceived by this. It feels this opening to be a loss, this enlargement to be a trap. Language lies in this respect: have we not lately—although the

everything possible is not 'realized' in the relative, which would exhaust universal Possibility: there are more things in God than can be contained in heaven and earth). Thus the viewpoint of the possible, and therefore the viewpoint of knowledge, surpasses that of being, since God is identical to universal Possibility, which, in order to be absolute, implies the surpassing of a non-contradiction conceived of in a simply ontological manner. In other words, the possible is non-contradictory; but this non-contradiction can be interpreted in two ways: ontologically and super-ontologically. The first interpretation: it is impossible for a thing to be and not be; being is, non-being is not. The second, super-ontological interpretation: absolute Possibility is absolute Non-contradiction, that is, the Non-contradiction of a Reality that can be contradicted by nothing, or again, That to which nothing can be opposed, and therefore That which includes in Itself Its own contradiction, or again, That which eludes every opposition because It eludes every *position*, and the initial one first, the ontological position or affirmation. For a more complete explanation, cf. the chapter 'La Possibilité de l'Être' in *Penser l'analogie* (Geneva: Ad Solem, 2000), 89–117, as well as *Problèmes de gnose*, 167–211.

enthusiasm is somewhat spent today—spoken of the 'conquest of space'? As if we could 'conquer' space! While, to the contrary, every movement towards the indefinite subjects us more and more 'narrowly' to our finiteness, and only serves to bring out the irremediable disproportion, not of our very being, but of our enterprise, of the 'ends' pursued. Spatial conquest is not a major step for humanity. It is (despite its technical or military usefulness) a major defeat, as are all strictly vain acts. For man, in his spiritual reality, is greater than space, and, to conquer it, it is enough to return to himself. But he loses the secret of the interior path in wanting to confront the pure exteriority of the universe, and instead of conqueror becomes the decisively conquered.

A UNIVERSE WITHOUT SEMANTIC CAUSALITY

Our examination of the cosmological revolution's effects would not be complete if we did not pause over the case of a recently mentioned contemporary of Galileo: Johann Kepler.[1] Now it seems to us that cosmic symbolism is as if the hidden stakes of the Galileo affair, so well hidden moreover that it actually seems to play no role in the violent assaults that the new physics waged against Aristotelian philosophy. And this is not at all surprising, since, as we have shown, both are in principle foreign to it. Doing this, are we not privileging a certain line of Western cultural history to the detriment of another, a contemporary of the previous one, and which merits to be taken into consideration just as much: the line that proceeds from the magico-symbolist *episteme* of the Renaissance to the scientifico-astrological astronomy of Kepler? But this evolution does not seem to present the same characteristics as the previous one; scientific progress in particular does not seem to run afoul of the symbolist mentality, quite the contrary. At the very time we witness to the manifestation of a strong magico-symbolist current, about which we have said that its *episteme* is as a whole founded on the likeness of "the form making a sign and the form being signalized,"[2] in reaction to Aristotelian conceptualism, we also notice a tremendous curiosity about the 'Book of Nature', which is set, they say, above the Book of Revelation,[3] and which will give rise to the progress of knowledge in a

1. Kepler, born in 1571, was seven years younger than Galileo, and died in 1630, twelve years before him.

2. Foucault, *The Order of Things* (New York: Random House, 1970), 29.

3. This is what Ernst Bloch declares in his study on *The Philosophy of the Renaissance* (1972), 130 of the French translation (Paris: Payot, 1974). In truth the author forgets, in this rather deceiving publication, that the comparison of the world to a book is extremely widespread, and that in particular it was in use with the great doctors of the

throng of domains. Under these conditions is it still possible to uphold our conclusions? We think so. This is precisely what we want to establish, first by studying the attitude of Kepler with respect to the etiology that the astrology of his predecessors brought into play, next by examining the Renaissance concept of knowledge that Foucault presents in chapter two of *The Order of Things*.

This attitude of Kepler has been thoroughly studied by Gérard Simon in *Kepler. Astronome astrologue.*[1] It is extremely critical with respect to traditional astrology and certain reasonings that the imperial mathematician considers to be superstitions. Now, and this is already a first confirmation of our theses, this denial of the etiology of astrology is altogether independent of the Copernican revolution: "It has been often asserted that the Copernican theory destroyed astrology by doing away with the absolute privilege of the situation that the Earth had with respect to the other stars: when the Earth became a planet like any other, the heavens could no longer be viewed as being ordered in terms of the center of the world. This is in fact an overly hasty conclusion: we fear it is still something of a retrospective rationalization of history."[2] What is actually in play is a certain idea of causality, the very one implied by sacred symbolism understood in its most integral sense, and

Middle Ages. Derived from Origen (*De principiis*, IV, 1, 7) as far as Christianity is concerned, this image is to be found again in John the Scot, who informs us that "the eternal light reveals itself in a twofold manner through Scripture and through creature" (*Homily on the Prologue to the Gospel of St. John*, XI, trans. C. Bamford [Hudson, NY: Lindisfarne Press, 1990], 37), Scripture for which "the human mind was not made, and of which it would have no need if it had not sinned" (*Super Hierarchiam caelestum*, l. III; *P. L.*, t. CXXII, col. 146 C). It is taken up again by the Victorines, for whom the world "is as if a book written by the finger of God" (*Didascalion*, l. VII, c. 4; *P.L.*, t. CLXXVI, 814), and is continued without interruption until our own time (Henri de Lubac, *Medieval Exegesis*, trans. M. Sebanc [Grand Rapids, MI: William Eerdmans/Edinburgh: T&T Clark, 1998] vol. 1, 76–77). To this must be added examples offered by world literature, that of Islam in particular. We will cite only this text of Muhyi 'd-Din ibn al-'Arabi drawn from the *Al-Futuhat al-Mekkiyah* (*Meccan Revelations*): "The Universe is a vast book; the characters of this book are all written, in principle, with the same ink and transcribed on to the eternal Tablet by the Divine Pen; all are transcribed simultaneously and inseparably; for that reason the essential phenomena hidden in the 'secret of secrets' were given the name of 'transcendent letters'. And these same transcendent letters, that is to say all creatures, after having been virtually condensed in the Divine Omniscience, were carried down on the Divine Breath to the lower lines, and composed and formed the manifested Universe" (according to the translation of R. Guénon, *The Symbolism of the Cross* [4th rev. ed., Ghent, NY: Sophia Perennis, 2001], 78).

1. Paris: Gallimard, 1979.
2. Ibid., 94.

which we have named semantic causality.[1] This remark should be enough to prove, if there were still any need, that the major crisis that then shook western culture, with effects continuing down to our own time, is indeed a crisis of the symbolic sign. For ultimately Kepler is in no way situated outside of the astrological *episteme*. This knowledge—devoid of any interest for Galileo—was in Kepler's eyes something eminently serious and altogether real. In a certain manner, Galileo, that truly modern spirit, ignored symbolism rather than struggled with it. But such is not the case for Kepler: he believed deeply and without reticence in astrology, and the territory that he meant to occupy was the very one where traditional symbolism seemed to hold undivided sway. He was dealing with it directly then, but he no longer believed in it. What Kepler denied was the efficacy of a significance, that is to say an etiology positing a causal relationship between signs, insofar as signs. And, if he no longer believed in this, it was because he no longer understood it. The mode of a constellation's action on those subject to it is not at all dependent on the name that it has received, which by no means expresses its nature, and which should in no way account for any effect whatsoever.

Such a semantic causality is however at the heart of traditional symbolism. We need to understand this once more. There was surely superstition, in the etymological sense of the term, as Guénon says, in attributing an efficacy to the sign reduced to itself and detached from its ontological root, and in wanting to produce real effects through a simple semiotic manipulation: a little like someone who would imagine himself changing the temperature by acting upon a thermometer. But such is not the case with a true symbol that realizes the basic identity of signifier and referent, or again that signifies by presentification. And so it is not the 'form' of the sign which, by itself, is efficacious; it is only so by virtue of the referent's presence, which it realizes at the same time that it expresses and signifies it. The symbolic form that tradition makes into a sign and consecrates, that form apprehended by hermeneutic understanding as the simple words of one discourse among others, are rightly not—are never—'simply words' for traditional symbolism. Otherwise it would be obviously absurd to attribute to a simple designation, or to any other symbolic mark (color, material, etc.), any power whatsoever. But true form is always the necessary expression of a nature. Between symbolic signifier and referent there is the meaning that unifies them because it is their common identity, and because it surpasses every determinate ontological degree. Between words and things there is the

1. Supra, chap. 1, art. 1, "The problematic of semantic causality."

intelligible referent, the archetype in which resides the true foundation of semantic causality, that is to say, basically, the divine Word, the Author of both worlds and traditional symbolism. This is why it is not unimportant that the matter of Baptism is water, that the form of the baptistery is octagonal, that a particular constellation is designated by a particular name, etc. In this perspective it can be said that everything is a sign, and that all these signs become 'intelligent signs'. The symbolic universe, the traditional mythocosm, is in reality *the intelligible world showing through every sensible form*, and in reality these causal relationships are effected from essence to essence: these are *relationships of comprehension*. It is this intelligible, subjacent sparkling of the cosmos that Johann Kepler no longer grasped, and this is why he wanted to substitute for it both a purely mathematical *semantics* (because it is thus rationally verifiable) and a purely physical *causality* (although not necessarily material, since it can be equally psychic). "It indeed seems," he writes, "that the signs of the zodiac have received the name of elements[1] only by reason of arbitrary inventions; in reality, they are not connected through any particular kinship with the elements that have been used to name them."[2] "This is why," G. Simon comments, "the technical arguments issuing from the Copernican revolution find so little place in his critique: the change is not effected on this level. The discussion deals with something much more profound—the origin and value of words. It is about knowing whether words are modeled on things: whether there exists a similarity between the figures of the constellations, the animal that they evoke, and the names that designate them."[3] The answer, for Kepler, being evidently negative, there followed the need to search for a properly 'physical' causality for the influence that they in fact exercised upon earth's humanity. According to the expression of G. Simon, his intent was to substitute for a 'cause astrology' a 'sign astrology.'[4] But that clearly proves that he no longer understood what a truly semantic causality was, that is a causality which—the unity of the Principle being reflected in creation—results in an analogical correspondence of all the degrees of reality among themselves, by virtue of which every degree of reality can be considered an expression of higher degrees, and therefore a reading-system enabling them to be deciphered. From this point of view, astrology is nothing but a logic of the stars. The excellence of this

1. The twelve signs of the zodiac are distributed, in order, between the four elements succeeding each other in this way: fire, earth, air, water, etc.

2. *On the New Star*, vol. 1, 179.

3. *Kepler*, 102.

4. Ibid., 125.

logic, of this system of writing, only arises from the extreme remoteness of its semiotic elements, which, by reducing them to the essential, confers on their forms and relationships a maximum of simplicity and clarity: an eminently readable and universal cosmic text. Hence, if such is indeed the true nature of traditional astrology, there are no grounds to speak on this subject, as G. Simon does, about a sign-cause circularity: in the astrology combated by Kepler, he asserts, the star is sign because it is cause, and it is cause because it is sign.[1] But this is untrue. In reality, the second relationship is the foundation and standard for the first, which basically means that every causality (and not only astrological causality) leads back to semantic causality.

This last remark directly introduces the second point, which is concerned with the *episteme* of the Renaissance. It must in fact be said that true astrology is so lofty that it escapes ordinary understanding, and that this semantic causality tends to be degraded into a gross physical determinism. But this degradation is not new. Already Plotinus was obliged to recall that "the movements of the stars announce future events, but do not produce them, as is too often believed.... The stars are like letters written at each moment in the sky...; consequently, while accomplishing other functions, they also have a signifying power. Everything transpires in the universe as in an animal where one can, thanks to the unity of its principle, know one part from another part."[2] Now the Stoic corruption of astrology, which Plotinus is taking aim at here, is even more pronounced at the end of the Middle Ages. The Platonic semantism of the cosmos[3] being progressively replaced by Aristotle's physicism, one was led to conceive of astral causality in a physical manner only, even at the cost of a rather fanciful etiology. This tendency was only accentuated in the sixteenth century. It is altogether significant that a master as competent as Paracelsus felt the need to protest against such a conception and to recall the traditional doctrine. No astral influx is impressed in the body or acts upon it: "Nothing impresses; neither an *astrum* that necessitates, nor one that governs or acts by inclination...."[4] But the 'power of the stars' is nothing but the bond of constitutive analogy (microcosm-macrocosm) that connects them to the human being: "The conjunction between heaven and man is as follows.... There is a double firmament,

1. Ibid., 124.

2. *The Enneads*, II, 3, "On the Influence of the Stars," § 7; trans. Bréhier. t. II, 33.

3. "Everything teems with signs. To be wise is to know one thing by another," says Plotinus, ibid., 34.

4. *Fragmenta medica. Fragmenta ad Paramirum de V Entibus referenda. De Ente astrali*; cited by Walter Pagel. *Paracelsus: An introduction to philosophical medicine in the era of the Renaissance* (Basel, New York: S. Karger, 1958), 67.

one in heaven and one in each body, and these are linked by mutual concordance and not by unilateral dependence of the body upon the firmament."[1] This is why the heavens, that form as if a person's portrait, express a 'prophecy' of the person rather than act on him as a cause.[2]

From this perspective, we are then led to consider the *episteme* of likeness and analogy (by which Foucault wants to define the Renaissance concept of knowledge) rather as a Platonic and anti-Aristotelian resurgence of a very ancient current before its almost total disappearance or complete marginalization by Galilean science. Far from constituting a stable and precise *episteme*, this analogical riot represents the swan song of a cultural era in full transformation. Besides, if we were to accept such a definition, this would be not at all the Renaissance *episteme* that is characterized in this way, but, whatever the various degrees, the *episteme* of *all pre-Galilean humanity*. It could be perfectly applied to those vast systems of correspondences displayed by the *Upanishads* as well as to those symbolic connections established by *The Key of Meliton of Sardis*, the *Gemma animae* of Honorius of Autun, or the *Rationale* of Durand of Mende.

But what characterizes the Renaissance for us was, first, a massive disoccultation of doctrines rather esoteric in nature: it witnessed the publication of multiple works—Henry Cornelius Agrippa of Nettesheim's *De occulta philosophia* being a typical example[3]—which dealt not only with a knowledge mysterious in itself, in its own nature, but even hidden from the unqualified masses, according to both senses of the term 'occult'. This disoccultation should be clearly understood however. There is actually a paradox in making public what, by nature, should remain unknown, and in declaring *openly* that a philosophy is *occult*. But this hidden knowledge was, in fact, somehow constrained, because of the progressive change in *episteme*, to show itself as such to the very extent that it was inexorably 'marginalized' by the cultural revolution. In reality, the Renaissance was the period when western culture 'rid' itself of its 'esoteric impurities', and therefore when these, like weeping sores on a face, appeared most visibly. But this appearance should not mislead us: the epistemic tissue forming just below the surface is a tissue of mathematical rationality. Whereas this same esoteric knowledge, in the course of the preceding period, is much harder to glimpse to the very extent that it formed part—much more closely than we are told—of the official culture, and even to such a point that it is not always easy

1. Pagel, op. cit., 67–68.
2. *Opus Paramirum*, lib. II, cap. 7; ibid., 68.
3. *Three Books of Occult Philosophy*, trans. J. Freake (St. Paul, MN: Llewellyn, 1993).

to distinguish between official and esoteric.[1] Thus, to cite only the most illustrious case, the reputation of St. Albert the Great as a magician is based on the exceptional importance that the Master himself attributes to the hermetic tradition, in the person of its eponymous 'founder': Hermes Trismegistus.[2]

But, and this will be our second remark on the Renaissance *episteme*, the close relationship that the esoteric traditions harbored with theology and religion maintained them in a properly spiritual sphere: the hermetic intellect is, for Albert the Great, the principle of deification, because, as Hermes ('*pater philosophorum*' Albert calls him) teaches, in agreement with Holy Scripture, it is "the divine that is within us."[3] While the marginalization of these same traditions in the Renaissance, by distancing them from the Christian spiritual tradition, caused them to lose, in part, their supernatural character, and risked allowing them to be imbued with the ambient *vitalist naturalism*, so that the hermetic, alchemical, kabbalistic, astrological, etc. doctrines were degraded into a common *magia naturalis*. For Albert the Great, only the deified man, the one who "lives according to the divine intellect ... as Hermes testifies," through a lengthy ascesis that leads to the culminating point of the spirit (*perseverante homine in mentis culmine*), becomes master of the world (*gubernator*), transmutes bodies (*agit ad corporum mundi transmutationem*) and accomplishes, so to say, prodigies (*ita ut miracula facere dicatur*).[4] Of all that the Renaissance would often retain only recipes for increasing one's power, and would occasionally end up in the lowest kind of sorcery.

On Cosmological Demythification

TWO CONCEPTS OF MATHEMATICAL LANGUAGE: KEPLER AND GALILEO

Now we can fully understand what we have called the ontological neutralization of sacred symbolism. The destruction of the medieval mytho-

1. Such, in particular, are the conclusions that the historians Chiara Crisciani and Claude Gagnon have arrived at and express in *Alchemie et philosophie au Moyen-Age. Perspectives et problèmes.* (Québec: éd. L'Aurore-Univers, 1980), under the form of a critical bibliography on the subject. The opposition between the thought-system of alchemy and that of scholastic metaphysics "is perhaps not so systematic as has been thought, and above all not always uniform" (19).

2. Loris Sturlese, "Saints et magiciens: Albert le Grand en face d'Hermès Trimégiste," in *Archives de philosophie*, vol. 43, n°4, Oct.–Dec. 1980, 615–634. The role and importance of Hermeticism would be rediscovered later with Meister Eckhart and Tauler.

3. *Super Matth.*, 10; Sturlese, 626.

4. *De animalibus* XXII, 1, 5; Sturlese, 627.

cosm does in fact deprive symbolism of its ontological reference. This was an event of considerable import and all its consequences are hard to gauge. Henceforth the world is shut up within itself; henceforth the world is reduced to the state of a cosmic corpse; no longer does it sing the glory of God. It is a silent universe, a mute, indefinite space where everything is meaningless. Not absurdity, but a semantically neutral cosmic reality: quite simply it no longer signifies anything.

Now is the time to stress the correlation that unites the ontological closure of the universe with the epistemological closure of the concept:[1] in the order of being the former corresponds to what the latter is in the order of knowledge, both being essentially realized by an exhaustive recourse to mathematics. We will start by showing that the philosophical openness of the concept is itself correlative to an open ontology.

Insofar as the concept is open to being and is nurtured by it, there is within it something of the unthinkable, of the mentally incomplete, or, to say it more nobly, there is mystery: the concept's hidden face, its ontological face, is extended to the being that informs it and confers on it its inexhaustible intelligibility; it is an umbilical cord attached to the being's ontological matrix, leaving its mark on the concept's surface like a scar and the sign of its incompleteness, of its *origin*.[2] In other words the philosophical concept is never entirely definable, or hence entirely able to be formulated or stated; its idea, its cognitive content, is mentally inexhaustible. It includes within itself a solution or break in conti-

1. This notion is explained in *Histoire et théorie du symbole,* chapter 4, article 1.

2. What we are saying here about the concept is valid for all beings that have an origin. To have an origin, cause, parents, etc., is an ontological situation whose metaphysical significance merits more ample development. In particular, it implies that such a being is not truly self-possessed, since its being is its unconsciousness: insofar as conscious beings we are not only born of our parents, but also of that being that we will become, a being that is, from the cellular stage to the age of two or three, completely ignorant as subject and does not remember *itself.* Its identity is therefore purely 'fiduciary' or traditional: we think that we are this being because someone has told us so. Hence the truth of our being (and not just our being) is given to us, so to say, 'ready-made'. This is why we precede ourselves; this is why tradition is, ultimately, always right; this is why we are Being's children. Every philosophy that would forget these principles is like a navigator who, forgetting the ship that carries him, imagines himself walking on the water: it speaks by denying that out of which all speech is possible. We would have to lay aside three quarters of modern philosophy if we seriously considered what this fact of having parents signifies. (Among others we might think about Descartes' project to kill the child within him, and about the Christic injunction: if you do not become like to a little child—that is, if you do not consciously identify yourself with your origin [= your essence or truth, with what is not your individual self and yet is only one with you]—you will not enter into the kingdom of heaven. Perhaps it was basically the desire to be for himself his own origin that led Descartes to define God as the cause of oneself.)

nuity, contemplative and intuitive by nature, that is to say an attentiveness to being, a quest for its light, that *silence* of thought by which alone can be heard the word of being. Thus being speaks a word to the attentive intelligence, a word that it has perhaps never heard; it reveals a knowledge about which it is perhaps ignorant. Even more, a being is unforeseeable: unforeseen and essentially new because, being timeless, a being is also without past. Being is surprise, its objectivity implies every possible objection. In short, it is literally true for philosophical knowledge that, by definition, there is more to an object than can be conceived. Clearly, the constitutive openness of a philosophical concept is correlative to an open ontology or again, as we have endeavored to show, to a *deep* ontology.[1] Conversely, the epistemological closure of the concept posits its object as essentially exhaustible, having an end, finished: basically, there is no more to the scientific object than what we conceive about it. Hence it is a closed or flat or surface ontology: bodies (and all of macrocosmic reality) are solely composed of extension and movement.[2]

Now, the only science that can actually realize a true closure of the concepts that it elaborates and with which it works is mathematical science: a mathematical being is identical (or nearly so)[3] to its definition. It even exists insofar as it is constructed by the concept that it thinks. This is why, mathematics being *the* model for what is scientific, every science capable of a mathematical translation demonstrates by that very fact how scientific it is. This is quite precisely what Galileo asserts in an already cited text: "Philosophy is written in this immense book held constantly open before our eyes—the universe—and which cannot be understood if one has not previously learned how to understand its language and know the letters used to write it. This book has been written in a mathematical language; its letters are triangles, circles and other geometric figures, without the intermediary of which it is humanly impossible to understand a single word of it."[4]

1. Likewise we could speak of a *scalar* ontology, one that allows for a scale or hierarchy of degrees of reality.

2. In the very same way an epistemological closure of the concept equally realizes a kind of ontological neutralization. The concept is no longer a mental symbol, a psychic reflection of immutable essences and therefore a participation in their being, but an autonomous rational operation. To tell the truth, the tendency toward such a theory of the concept is already to be met with in Aristotle. We will return to this in a moment.

3. We say 'nearly', for there are unconstructible mathematical beings, the primary numbers for example, or the notion of the continuity of space.

4. Ed. Naz., vol. VI, 232 (*Il Saggiatore*, 1623).

It might be objected, however, that the use of mathematics in cosmology does not sufficiently account for the symbol's ontological neutralization, since a scholar like Kepler uses it constantly while remaining within a symbolist mode of thought. We might go even further: for him mathematics becomes the only symbolic reality, to the exclusion of all cultural symbolism, because its significance resides in the intelligible necessity of its own structure, a significance not bestowed on it by an uncertain and fanciful tradition.[1] But mathematics actually plays a rather different role in Kepler and Galileo, and without doubt this has not been sufficiently noticed. For the imperial mathematician, and within a globally Pythagorean perspective, mathematical entities retain their own semantism, a semantism only revealed to an esthetico-mystical contemplation: they constitute the 'architectonics of the world'[2] and reveal, because of their intrinsic intelligibility, the cosmic order's raison d'etre, everything that makes of the world "a marvelously organized work of art."[3] Thanks then to their own esthetic intelligibility the sym-

1. Kepler in particular seems to be the first to have explicitly symbolized the trinitarian mystery with the aid of a sphere: we find "the image of God the Three and One in a spherical surface, that is [the image] of the Father in the center, the Son in the surface, and the Spirit in the regularity of the relationship between the [central] point and the circumference" (*Mysterium Cosmographicum*, chap. II, *G.W.*, t. 1, 23; English trans., *The Secret of the Universe* [New York: Abaris, 1981], 93). This analogy is inspired, it seems, by Nicholas of Cusa: *Complementary Theological Considerations* (1453), chap. 6; *Nicholas of Cusa: Metaphysical Speculatios*, trans. J. Hopkins (Minneapolis, MN: Arthur J. Banning Press, 1998), 14–15. That Nicholas of Cusa speaks of a circle and Kepler of a sphere makes little difference. Of greater importance is their divergence in the representation of Son and Spirit, which, contrary to Kepler, the Cardinal figures by the radius and circumference respectively. This figuration respects the traditional order of the Trinitarian processions and seems more in conformity with the 'straight' nature of the *Logos* and the 'circular' or rather 'spiral' nature of the *Pneuma*. Cf. *Un homme une femme au Paradis* (Geneva: Ad Solem, 2008), 119, n. 2. Later he speaks of the sphere as the "most excellent figure of all. . . . For, by forming it, the omniscient Creator was pleased to reproduce the image of his veritable Trinity. This is why the central point is, so to say, the sphere's origin; and the surface, the image of the innermost point and the way leading to it; it can be understood as an infinite emanation (*egressus*) of the center out of itself, which would be obedient to a kind of equality: the center is communicated to the surface in such a way that, once the relationship of density to extension is reversed between center and surface, there is equality. This is why there is everywhere between center and surface the most perfect equality, the closest union, the most beautiful agreement, intimacy, relationship, proportion and commensurability. And although the center, surface and interval are manifestly three, yet they are only one, to the point that no one can even conceive of the lack of one of them without all being destroyed" (*Compliments to Vitellion,* chap. 1, *G.W.*, t. II, 19; G. Simon, op. cit., 134).

2. G. Simon, op. cit., 282–3.

3. Ibid., 436.

bolism of the universe is spared and heaven and earth continue to tell the glory of God: "Praise Him heavens, praise Him sun, moon and planets with, so to perceive your Creator and address Him, the sense and the tongue that are yours, whatever they may be; praise Him Celestial Harmonies, praise Him you who think to discover these Harmonies...."[1] For the Italian physicist the idea of an intrinsically esthetic intelligibility for mathematical entities does not make the least sense. As pointed out, in his eyes mathematics constitutes a language in the most profane and modern sense of the term: a translating system, an intermediary. There is no intrinsic intelligibility to geometric bodies capable, by their very *nature,* of giving a sufficient reason for the structure of the world; there is only an extrinsic intelligibility, an intelligibility for us and for physical reality. What there is of the intelligible within them are those conceptual operations that they represent: mathematics is already, for Galileo, a science of operational diagrams. For Kepler, to extract the geometric architectonics of the cosmos is enough to explain its necessity: *the world is justified in its very being.* For Galileo such a justification is excluded. With him the order of the world remains contingent, it is observed and described, to be explained only in its functioning. Intelligibility has a bearing on process, not on structure; we might say on the 'how', not on the 'why'. In other words, the Galilean approach necessarily implies the following axiom: in physics the question of the world's meaning has no meaning. The world is what it is; we only ask how it functions. And the answer can only be formulated in mathematical language.

A comparative study of mathematical use in Kepler and Galileo has thus spotlighted the truly radical character of the symbol's ontological neutralization, that is to say the total negation of its naturalness, through a no less total disappearance of nature symbolism. Actually, cosmological symbolism is not only rendered *impossible* by physicist mechanics, it is even rendered philosophically *useless,* that is to say unnecessary, by the Galilean mathematization of knowledge. The question of the world's meaning is branded as nonsense.

NO COSMOS WITHOUT MYTHOCOSM

But we have to ask: what is the price to be paid for such an eradication of the reason-for-being or sufficient-reason principle? A truly cosmological intellection is the price. Up until that time none of the known

1. This is the conclusion to *Harmonia Mundi, in five books* (1619), V, chap. X, *G.W.,* t. 6, 368; G. Simon, op. cit., 440.

western philosophies considered themselves able to renounce this principle that defines the very intelligibility of being: a reason for *being*[1] is being become reason, a being justified, a being somehow saved. For Platonism, this reason for being is in the palpable world only according to an iconic mode, by a participative image, and this is, we repeat, the viewpoint of sacred symbolism. For Aristotelianism, this reason for being is immanent to the world itself; it is that of the intelligible form, and this is why we have said that, for him, the visible world as such somehow becomes the intelligible world. For Kepler, this reason for being is not identified with the visible as such; it resides in the mathematical structures that, from 'beneath' and invisibly, give order to cosmic reality. To some extent this is a return to Plato, since reason-for-being orders the world to its image, and yet the finality of this justification is much more cosmological than metaphysical: where the physical world finally reaches the locus of its true reality, Kepler's mathematical models are not eternally living essences; their entire importance is exhausted in their functioning as cosmic intelligibility. But, with the advent of Galilean physics, such a principle is quite simply lost from sight. However, and correlatively, what is lost from sight at the same time is the very idea of a cosmos. Doubtless there is still no awareness of this other than in a negative and indirect way, as with John Donne or Pascal; we have to wait for Kant for the thesis to be articulated philosophically. All the same the conclusion was implicitly established: the disappearance of the mythocosm is the death of the cosmos. In other words: for cosmology there is only myth or symbol. And contemporary science will certainly not contradict us. Rather, there is even the impression that contemporary science has still not clearly understood

1. Properly speaking, we need to distinguish 'reason for being' from 'reason for existing'. 'Reason for being' is concerned with essence, 'reason for existing' with existential accidentality. The former is clearly a principle that no true philosophy can do without. The latter remains contingent. *The* rose has no why because the answer to this why, its 'reason for being' a rose, is in the obviousness of its essence, its beauty: the absence of why is correlative to a superabundance of intelligibility. Conversely, a *particular* rose has no why because no one can provide any response to the question: "Why is such or such a flower here and now a rose?" This is an absence of why by lack of intelligibility. But the answer is: 'At God's behest'. This complete contingency (or inverse reflection of the Absolute's superintelligibility) by no means excludes however all kinds of justifications. To say that a particular rose, at a particular place or time, concurs with the harmony of all is a certainty and obvious because an *a priori* and universal truth; but this does not grant a determinate why to a particular determinate rose.

this, and that it will resolve those difficulties debated since 1927[1] only when it will have grasped the conditions for a true thinking about the cosmos.

These conditions are the following: 1. — We cannot but think about the world since man is precisely that being for whom the world *exists*, that is for whom the sum total of existing things constitutes an objective order, independent and existing within itself. 2. — To think about the world is also to think about a unified whole: unity, totality, existence, such are the marks of the idea of 'world'. 3. — As we have shown in *Histoire et théorie du symbole*, the emergence of thinking about the world takes place in the discovery of language: to give birth to an awareness of the sign is to give birth at the same time to an awareness of the objective and subjective universe; 4. — as a result, thinking about the world is initially a thinking about a world 'to be expressed', in other words about a being whose meaning has to be revealed. 5. — Hence, a concept of the world that excludes the thought of its unity, its totality, its meaning, and retains only the thought of its existence, and therefore reduces the cosmos to a purely physical 'being-there', is not even a thinking about the world, or is so only by surreptitiously and unconsciously reintroducing those conditions that it has deliberately set aside. Now obviously we cannot think about the unity of a world in which isotropism is incompatible with any ordered structure whatsoever. We cannot think about the totality of a world in which physical reality has been framed by a spatial indefiniteness, while the notion of totality implies the finite. Lastly, we cannot think about the meaning of a world that, by epistemic definition, is devoid of it.

Therefore, imagining the physical world as a mass of indefinitely divided bodies in an indefinitely extended space needs to be decisively renounced. The world is not in space, it is space that is in the world. This point is decisive, and this is why we have examined it at such length and from so many different vantage points. Cosmological thinking that posits an indefinitely extended physical reality 'prior' to itself is instantly a prisoner of its own representation: no longer can it 'depart' from this universal expanse surrounding it on all sides and 'wherever it goes'. And

1. This is the date when the Fifth Congress of Solvay convened, with the world's greatest physicists in attendance (Einstein, Heisenberg, de Broglie, Lorenz, Planck, Marie Curie, Dirac, Schrödinger, etc.), who decided to accept, for lack of another theory, Bohr's doctrine and the quantum postulate: "This postulate has as consequence a renunciation of the causal description of atomic phenomena in time and space." This doctrine is said to be of the 'Copenhagen school' that ran afoul of Einstein, de Broglie and Schrödinger, and which, unless we are mistaken, has yet to be superseded (J. L. Andrade e Silva and G. Lochak, *Quanta, grains et champs* [Paris: Hachette, 1969], 147 ff).

yet this is the representation that invades minds at the beginning of the seventeenth century as a veritable collective suggestion. The cultural subconscious is endowed with a new 'mental image', an image that functions in an automatic and unreflecting manner as spontaneously evident and accompanies all thinking about the world.[1] By this image, to be found as the basis of all speculative activity, thought becomes a scene, a representation. Now the illusion proper to the thought of space is a self-supposition, or, what amounts to the same, our being led to think that it is self-supposing; that is, that space supposes a space for existing, that 'in' and 'where' are *in* and *where*. Whoever pauses over the thought of space will be convinced that it is, in fact, always implicitly affirming that beneath space there is always space. Hardly have we imagined such a container than we imagine the container's container and so forth indefinitely. But this is a mistake. Space is not in space. Space is nowhere.[2] And, even though we have some difficulty imagining this, we should have none conceptualizing it. Hence, if we hold firmly to this conclusion, we see that the traditional figurations of the cosmos are, in reality, the only ones possible.

Classical Galilean cosmography provides a visual geometric representation of the universe that destroys the very notion of a universe. As for the conceptions of modern Einsteinian cosmology, they are non-figurable.[3] And yet such a representation is quite simply inevitable. What is left to be recognized, then, is that the symbolic representations of the mythocosm alone remain: to picture the world as a series of concentric spheres, bounded by a last Empyrean Heaven, that is, the sphere of the divine metacosm or something analogous,[4] that from which nothing

1. Seeing time under the form of a 'historical awareness', preconditioning and accompanying all our conceptualizings so as to temporalize them, will only come at the beginning of the nineteenth century. From the seventeenth to the eighteenth centuries, the European cultural soul is conditioned by the 'thought of space'. While today it seems that the thought of time is losing its importance and that we are being dominated by the thought of relationship (hence the importance of computers that treat information only from the viewpoint of relationship and not of meaning). This is why we have so much difficulty thinking about things in themselves.

2. This is, moreover, what the Theory of Relativity affirms by conceptualizing a curved space and therefore a finite world.

3. "Like most concepts of modern science, Einstein's finite, spherical universe cannot be visualized—any more than a photon or an electron can be visualized" (Lincoln Barnett, *The Universe and Dr. Einstein* [New York: William Sloan Associates, 1948], 94).

4. In particular, we are referring to Muhyiddin Ibn 'Arabi's cosmology as explained in a previously cited book by Titus Burckhardt. Its general plan is that of a succession of fifteen concentric spheres with seven below and seven above that of the Sun: earth, water, air, ether (fire is absent), Moon, Mercury, Venus, Sun, Mars, Jupiter, Saturn, the

can escape. What is important in this cosmography is not only the hierarchic and concentric order of the universe, but also that its ontological finiteness is spatially represented by the bounds of the last heaven. A cosmification of the divine will perhaps be seen in this, but it is actually just the opposite—it is a symbolically signified surpassing of the visible world. This means that the frontier of the visible and corporeal, the confines of the world, are necessarily the 'locus' of a passage to the Infinite. Only the Absolute can limit the relative, only the Infinite can enclose the finite. Space will never be able to limit space. Therefore, in all objectivity, it is the Empyrean Heaven that marks the limit, not only of the visible world, but of all creation. And so such a representation is much more rigorous and truer than the one presented by Galilean cosmography (not to mention the Einsteinian, which has none).

SYMBOLIC REALISM

Without doubt, some will say, but what truth is involved here? A symbolic truth, an allegorical figure and not a scientific truth or a realist and photographic picturing. We can cross space indefinitely and never encounter the Empyrean Heaven. Even though the ancients themselves sincerely and naïvely believed that things were indeed objectively so, we cannot. But this objection is much stronger in appearance than in reality, for it wholly rests on belief in the evidence and the unicity of scientific truth: science describes things such as they are. And this evidence is, in fact, the evidence of materialism: before us there is a world of concrete, palpable and measurable things, situated *wholly* in space and time, and there is nothing else. This is the true 'cosmological principle' of Galilean physics.[1] And it is this cosmological principle, which might be characterized as that of the ontological homogeneity of the real,

heaven of the fixed stars, the starless heaven, the Sphere of the Divine Pedestal and the Sphere of the Divine Throne. As Burckhardt indicates, only the planetary spheres and that of the fixed stars correspond to sensory experience. The elemental spheres are invisible. As for the supreme spheres (according to the *Qur'an* the Pedestal contains Heaven and Earth and the Divine Throne contains everything), they are the symbolic spheres: "They mark the passage from astronomy to metaphysical and integral cosmology," a discontinuity marked by the starless heaven or "the 'end' of space" (*Mystical Astrology according to Ibn 'Arabi*, 14).

1. We know that, around 1930, the American physicist Milne coined the expression "the Einsteinian cosmological principle"; this involves "the dual hypothesis that the distribution of matter in space is homogeneous and that, at every point, the appearances of the universe are isotropic for every observer at rest with respect to the cosmic *substratum* taken all together" (Jacques Merleau-Ponty, *Cosmologie du xxe siècle* [Paris: Gallimard, 1965], 273. Milne seems to have declined this paternity; ibid., 147, n.1).

which modern physics has forsworn and should be no longer upheld today.[1] For nearly three centuries the European mindset has become addicted to the invincible belief that the materialist concept of physical being apprehends it as it is in its reality, as a perfectly isolatable individual entity, an entirely locatable existential point, that is, enveloped by space on all sides and strictly reducible to spatial presence, in such a way that to be real, to really exist, is equivalent to existing materially. Everything else (ideas, sentiments, etc.) is almost synonymous with something nonexistent. But this is untrue. That our first notion of the real arises within us as occasioned by our perception of the corporeal world is incontestable, and therefore that we are led to identify reality with corporeality is understandable. However it is no less certain that the raising of this spontaneous attitude into a philosophical thesis is untenable. Not only modern physics, but also philosophical reflection shows that there is nothing to this: for space, being purely exterior (if we identify the real with the spatial), all interiority becomes impossible and, at the same stroke, all existence. As Leibniz says, in a letter to Arnauld (April 30, 1687), "a being that is not *a* being is no longer a *being*": *esse et unum convertuntur.* To be, in one way or another, is to be oneself. There is no ontology without interiority, no presence to the world without presence to self. Total Reality is deployed between an absolute interiority beyond all creation, which is the interiority of the Principle that India terms the Self (*Atma*), and absolute exteriority, an inaccessible boundary this side of creation that is equivalent to pure nothingness, to non-existence. Between these two extremes are to be found all possible degrees of interiority, from the minimal interiority of the corporeal world to the maximal interiority of the semantic world,[2] leaving aside

1. Recall that, in large part, we are referring to Raymond Ruyer's cosmology (*Néo-finalisme* [Paris: PUF, 1952], *Cybernétique et origine de l'information,* Flammarion, 1954, *La gnose de Princeton,* Pluriels, 1977, etc.). This cosmology rejects the determinist materialism of classical physics, as well as the probabilist idealism of contemporary physics. This separating into blocks, or alternativism, is due solely to an enduring materialist prejudice: either the physical world is material, or else one no longer knows anything about the real. But there is also, and even above all, the incorporeal and perfectly objective real of the psychic (or subtle), spiritual (either intelligible or semantic) and divine orders, with the first two orders being just as cosmic as the corporeal. This thesis should be considered normal and rational, and not taken, as is feared, for out-and-out metaphysical insanity.

2. Theologically, this semantic function of unification in presence-to-self, whereby multiple entities gain access to proximity with each other, can be related to the Word considered as prototypical Relation. We have examined this point of view in *Amour et vérité*, chap. VII, sect. 2: "Le détour cosmologique."

the pure interiority of metacosmic Being.[1] This signifies, as already shown, that space does not truly realize exteriority; otherwise it would not exist at all: as we have stated, it is a *half*-realized explosion or *out*-burst. There is, then, already some interiority 'in' space—the point-like possibility—and hence something of the non-spatial: it is the permanent presence of the non-spatial at all 'possible' points in space that founds its reality. Now, every interiority is of a semantic, intelligible or spiritual nature. One of the best examples for proving this is provided by a consideration of a text's meaning: the signs of a word are semiotic entities immediately present to each other—fused but not confused—only in the meaning 'realized' by a word. Words themselves are united and present to each other only in the meaning realized by a sentence. And sentences would remain irreducibly strange to each other without *the* meaning of the text that they form and within which they are 'contained' in a unity beyond multiplicity. Only the 'meaning' of the point, the non-spatial limit of space, we might almost say the 'intelligible point', can account then for the reality of the spatial condition, it being of course that which is valid for this existential condition, and *a fortiori* valid for the corporeal beings affected by it.

And what is this intelligible point? It is that which can always be seen as the trace of a vertical transecting a plane. The plane is in fact the most direct image of spatial extension starting from a point. The vertical in this point is also then the most direct image of an emergence 'outside' space. But, if anyone has clearly grasped what is meant, it will be understood that it is not solely an image of it: the vertical also somehow 'realizes' this emergence. A comparison will be helpful. For a being that lives in a two-dimensional universe, a flat being for example, a rectangle would be enough to enclose a treasure *on all sides*, and no one would be able to take it. A being lending itself to the third dimension would steal this treasure in an inexplicable way; it would have passed into an extra-cosmic 'dimension'. Thus the vertical really constitutes an emergence outside the plane of space, and therefore 'spatially' symbolizes the exceeding of the spatial condition both in general and for the corporeal world.[2]

1. The Supreme Principle is also Exteriority and not only Interiority. This is what the Qur'anic mention of the two divine Names *Ah-Zahir* and *Al-Batin* expresses. "He is the First (*Al-Awwal*) and the Last (*Al-Akhir*), the Outward (*Az-Zahir*) and the Inward (*Al-Batin*), and He knows all things infinitely" (*Qur'an* 57:3). On the subject of this verse, Frithjof Schuon's commentary should be read: "Dimensions of the Universe in the Qoranic doctrine of the Divine Names" in *Dimensions of Islam,* trans. P.N. Townsend (London: George Allen and Unwin, 1969) 30–45.

2. This is why the Word Incarnate could leave the world only by *really* ascending along a vertical, which modern theology denies because no longer understood.

Such is the semantic reality of space. It enables us to understand that, according to Ruyer's dictum, ordinary perception or even scientific observation grasp only the 'outside' of the universe, its 'wrong side'; and not its 'right side', which is meaning and awareness: "The universe is only made up of forms conscious of themselves and of the interaction of these forms by mutual information. For consciousness is form and information, but on the *right side* and not the *wrong side*, as object-structure-in-another-consciousness."[1] This semantic reality is not then a mental viewpoint, a way of saying or imagining things that might possibly enable a philosophy of subjective consciousness to be harmonized with the 'mentality' of a traditional symbolism. This has to do with an objective semantism that is the very stuff of things, and which alone can account for their reality. It is not superadded to things from without; they are composed of it intrinsically. Outside this semantism things would have no reality and vanish into a pulverizing and indefinite exteriority, into nothingness.

Two conclusions are the result of these considerations: first, as already stated, the symbolic representation of the universe is not only a possible one but even the only true one. Space not being in space, but every representation being necessarily spatial, we therefore need to depict the ontological finiteness of space as a closed volume, shut up upon itself and surrounded on all sides by the infinity of the divine metacosm. This representation is in conformity with the *nature* of the physical cosmos, which is what is essential, even if not cosmographically exact, in the sense that it does not provide an exact 'photograph' of the cosmic expanse. But, and we are basically just repeating this truth under various forms, *such photography does not exist.* There are no objectively faithful representations of spatial reality. There is no adequate vantage point *overlooking* space where it might be described as it is in itself. There is none, and Einsteinian doctrine should convince us of this. Such is however what the Galilean revolution proposes to do, and anyhow this is what it has succeeded in persuading the vast majority that it does do. Here is why this revolution is *essentially mental.* Indefinite space, which Newton attempted to safeguard by making it the *sensorium Dei*—but he was not understood—exists in truth only 'in the head' of millions of people. As previously noted, it is clearly hard to dispel the fascination that, once produced, this new mental image can exercise over our minds. And yet we imagine that humanity has been unaware of it for millennia;

1. *La gnose de Princeton*, 59; and above all the fundamental chapter in *Néo-finalisme:* "'Surfaces absolues' et domaines de survol," 95–109, as well as the article "Observables et participables," *Revue philosophique*, tome CLVI, 1966, 419–450.

that for millennia mankind has never sought to represent space as such, because every painting in all human civilization *never* represents anything but a spiritual and symbolic space, that is, something transpatial in nature as depicted on a flat surface. To the contrary, the Renaissance also sees the appearance of graphic perspective, or *perspectiva artificialis*,[1] which is the development of procedures aimed at a two-dimensional reproduction of the supposedly objective and universal perception of the three dimensions of space. Whatever the problems raised by this question, it is clear that this perspectivist representation of space introduces the *illusion of depth*.[2] And so this illusory depth realized with the help of a '*trompe l'oeil*' comes to mask the disappearance of the ontological depths of the cosmos. Humanity experiences the need to offer itself pictorial evidence of the indefinite openness of a world nevertheless reduced to its own spatiality: vanishing points are everywhere and are, optically, unhealable wounds out of which the blood and semantic life of the cosmos inexorably gush. Henceforth spiritual space is closed to sacred art, which is degraded into religious art. An entire civilization founders in a theatrical, imitation-antique and lost-paradise decor.[3]

But this conformity of symbolic representation to the nature of space involves not just its object: cosmic space. Also involved are the means for which it is used: space as elementary signifier. In fact, with the traditional depiction of the mythocosm, the indefinite property of space is used to signify the divine Infinity within which the created cosmos is immersed. Now this indefiniteness of space is precisely nothing but, as we have indicated at several points, a reflection, an image, a projection of divine Infinity, that is to say the All-Possibility of the Absolute. It is even only insofar as it is a symbol of universal Possibility that this indefiniteness has a meaning and a reality. Otherwise, as we have shown, it is nothing but indefinite limitation, universal annihilation.

1. The *perspectiva naturalis*, or just *perspectiva*, designates for the ancient and medieval worlds a part of optical science, itself subordinate to mathematics: it applies the rules of geometry to the way in which objects are seen from a particular point of view; astronomy in particular makes use of it (St. Thomas, *Summa Theologiae*, I, q.72, a.9). It is concerned with *things*, while the *perspectiva artificialis* is concerned with *graphic signs* with the help of which it is possible to give the illusion of depth on a flat surface.

2. Cf. Dalai Emiliani's article in the *Encyclopedia Universalis* under the heading "perspective." Anyhow it is certain that prior to the Renaissance humanity was not unaware of a certain perspective; only it did not use it to depict geometric space, but to express spiritual relationships and make the viewer delve into a 'semantic' or 'intelligible space'. This is the case with the 'aerial perspective' of the Taoist painters as well as the 'inverse perspective' of Byzantine painting.

3. On Renaissance perspective, read the scholarly studies by Robert Klein in *La forme et l'intelligible*, (Paris: Gallimard, 1970), 236–338.

Such are the two conclusions to be drawn from these reflections: the adequation of symbolic language to the nature of things involves not only its object. Language itself is also involved, and this is so because the world's reality is of a semantic order, and because there is an objective hermeneutic of the universe that is somehow 'its own interpreter', to the extent that each degree of reality is the expression of those degrees superior to it.[1] The latter constitute the interpretation of the former: descending expression and ascending interpretation criss-cross on the cosmic ladder (this is one of the meanings of Jacob's vision) and establish both traditional symbolism and its hermeneutics.[2]

In conclusion, this is where symbolist metaphysics parts company with Ruyer's 'gnosis'. For Ruyer there is indeed a semantism of the cosmos, there are indeed different strata or regions of the real that philosophical analysis extracts (the here-and-now, the vital, the semantic), but there are no degrees of reality. These strata or regions are not really distinct. They do not exist 'elsewhere'. The other world is already here. Everything is of the present and the semantic or the divine cannot be, and will never be, apprehended by man otherwise than here and now. Never will we know 'God' other than by knowing him in the present. Certainly for Ruyer this is not negligible, for there is in this knowledge something of the immediate, which means that we quite really and directly think 'in God' and start with him. And this is enough for a Ruyerian wisdom that denies the 'beyond' of the soul's immortality. In this perspective human culture, with its symbolism and religions, is only an

1. To be clear about the meaning of 'superior', recall that reality's hierarchy of degrees is of an ontological order, and its conditions of manifestation are imposed on any being under consideration according to that being's degree of limitation. Not a moral evaluation, but an objective observation is involved here. Beings of the subtle world endure less limiting conditions than beings of the corporeal world, and likewise for spiritual beings with respect to subtle ones. This is why, moreover, the 'ontology of individual substance' is no longer appropriate for them: essences, in order to be real, have no need for individual existence; their mode of being is intelligible or semantic. This is precisely what Aristotle did not understand. Basically, it is his ontology of individual substance, that is to say the 'instinctive' sense that he has of the real, which constitutes his chief obstacle to an understanding of Platonism. Hence he cannot 'see' that essence, having no need of being an individual substance to exist, is not limited by this mode of being and therefore *can* be participatively immanent to this mode of being simply by virtue of its transcendence: *only the more can do less.*

2. In reality—and we have already said this—no 'scripture', that of the world no more than of the 'Book', is its own self-interpreter. Hermeneutics (that is to say tradition) 'precedes' and follows the text that it explains. A text is always read *in* a certain tradition that bears it as a most precious possession. For man, the symbolic understanding of the world 'precedes' the world not temporally, but ontologically or essentially. On all of this cf. *Histoire et théorie du symbole*, chap. VI, art. 4.

anthropological function in a psycho-biological or ecological style: culture is mankind's natural habitat, its 'ecological niche'—which we do not deny, of course—but it is not viewed as the sign and proof that mankind's native land is elsewhere. Clearly, it puts us in communication with the Transcendent, the Divine, the Norm, but it would be feeble-mindedness to feel nostalgia for it: "The only solution, if there is one, is to admit that there is an incomprehensible and incommunicable 'God-field'. All life is plunged in this God as in a gravitational field. But it is a hyper-physical gravitational field in which we do not experience impressions of weight but 'ideal' attractions. It governs our conduct by allowing us the freedom to operate and adapt. It leads and dominates us, not by a cord or puppet string, but by having us 'participate'.... Our ties to God, the ruler of all domains, the unity principle of all fields, are much closer (than those of a puppet on its string). Within us God is the organic field itself, the soul of the body. And, at our death, it is God himself who, quitting the material body, leaves only dust behind."[1]

But the metaphysical symbolist, without denying all that is true in the previous viewpoint, also affirms that it is not 'altogether' true, that there is an existential cleavage between what is man, or the human soul, and what he might be insofar as a participant in this 'divine field'. In short, Reality is not only a 'field', but also a 'ladder' with something both of the One *and* the multiple, presence *and* absence. And culture, that is to say symbolism, a *sui generis* reality irreducible to either the human or cosmic pole, as a 'semantic surplus' overflowing with an intelligibility and meaning neither man nor the world can contain or absorb, is precisely the proof of this. And this is why such a metaphysic also gives way to tension, anguish, suffering, and anagogic hope.

COSMIC AND SCRIPTURAL DEMYTHOLOGIZING: FROM SPINOZA TO BULTMANN

There is no better way to close this lengthy reflection on the ontological neutralization of the symbol than by returning to the question that enabled us to broach it, that is, the radical manner in which Bultmann poses this problem.

From the very start we will say that only a Christian could formulate it with precisely this harshness because, in the West, he is the only one for whom it is still meaningful. As we noted earlier, to a basically non-religious mind the symbol's ontological neutralization seems obvious and even trite: such an ontology is quite simply impossible and this is why such a question is never raised. The question is only raised by the

1. *La gnose de Princeton*, 376.

de facto existence of a religious symbolism that must somehow be gotten rid of, at least as to its realist claims. We will have the chance to return to this in the second part of the present work, but this much can be said already: the destruction of the symbolic sign's referent so occupies hermeneutic awareness that it forgets to ask just how this symbolism might have arisen. To tell the truth, this is even quite basic, and yet almost no one has posed it. Surely it is surprising that such 'stuff and nonsense' was believed in, but they do not inquire into its origin: why is such 'stuff and nonsense' just so and not otherwise? And if this question remains so generally unformulated, perhaps this is also because, once clearly stated and considered, it is no longer possible to be rid of it and because it leads Western reasoning to where it does not want to go. Besides, such is the meaning of our entire reflection.

Clearly the existence of religion's sensory forms constitute the *site* of the crisis that is tormenting European culture, precisely because such forms are its most visible possession. They represent a major 'cultural scandal', and for this they are in a state of exclusion. It is this religious portion of itself that the European soul rejects and for three centuries has attempted to eliminate. If it were only a question of theism, the crisis would perhaps not have occurred. If the religious element had been reduced to a purely abstract affirmation of the Divine Being, its exclusion from cultural awareness—inevitable because an outgrowth of its epistemological project—would have passed unnoticed. This is also what Nietzsche asserts: we have killed God and we do not know it. In short, Europeans would have continued to *believe that they believed in God,* all the while doing mathematics, physics and business. But the presence of religious forms of the divine would not let them be spared this crisis. What is left then is an attempt to sever theism's destiny from that of its formal expressions. This is the solution adopted by rationalism, Spinoza's as well as Kant's. True, there is a clear difference between the former's pantheism and the latter's moral theism. But their attitudes with respect to religious symbolism are basically identical. As for the differences, they consist not only in the philosophical infrastructure of their respective systems, but in the fact that Spinoza possessed a certain competence in exegesis, which cannot be said of Kant, who did not know Hebrew. Regarding Scripture, Spinoza's attitude tends to be scientific, Kant's philosophical. Kant interprets Scripture according to a moral religion, in terms of the demands of practical reason: thanks to his philosophical hermeneutics he extracts from the text its true meaning contained beneath an imaged or naïve form, and which is nothing but the contents of practical reason. Spinoza, to the contrary, proceeds with no 'hermeneutic recovery'. What he studied was not so much the meaning but the text itself. It has even been said, with some exaggeration, that he inaugurated the history of textual criticism. Anyhow, he is proposing to study Scripture as one studies nature: "I may sum up the matter by saying that the method of interpreting Scripture does not widely differ from the method of interpreting nature." This is why he "admit[s] no principles for interpreting Scripture, and discussing its contents save such as they find in Scripture itself." Thanks to this "everyone will always advance without danger of error and will be able with equal security to discuss what surpasses our understanding, and what is known by the natural light of reason."[1] However, this "resolution [taken] in total freedom of spirit and sincerity"[2] consists quite simply, as R. Misrahi and M. Francès state, "in establishing the absolute empire of reason to save freedom."[3] When we read that we need "to be freed from the prejudices of the theologians," from "their aberrations," from "human inventions" and the "superstitions" of babblers like Maimonides and some others "who have set about torturing Scripture to draw from it the musings of Aristotle and their own fictions";[4] that the "fervor of the believers, O God, and religion are identified with absurd esoterisms,"[5] that Christ was not God (which would make no more

1. *A Theologico-Political Treatise,* Chap. VII, "On the Interprétation of Scripture", (Elwes trans.). *Œuvres complètes,* (Paris: Gallimard, 1955), 768–769.

2. *Œuvres,* 668.

3. Ibid., 1453.

4. Ibid., 679. We were unable to consult the Latin original of the *Tractacus theologico-politicus* (1670), but it is impossible that Spinoza used the term *esoterismus,* which, we think, did not appear before the beginning of the nineteenth century (in French and English).

5. Ibid., 667. This declaration is instructive as to what Spinoza thinks about the Kabbalah. Besides, he is altogether explicit in chapter IX of his *Treatise* (ibid., 813): "I have also read some Kabbalists and taken cognizance of their nonsense: I never cease to be astonished at their insanity." Must we see in Spinozism a systematic account of the Kabbalah, as Leibniz asserts commenting on the *Elucidarius Cabalisticus* of Wachter (Leibniz, *Textes inédits,* annotated by G. Grua [Paris: PUF, 1949], t. II, 556; and above all G. Friedmann, *Leibniz et Spinoza* [Paris: Gallimard, 1946], 209–210)? The erudite P. Vulliaud refutes this legend at length, a legend nevertheless repeated by many modern historians (*La Kabbale juive* [Paris: Editions d'Aujourd'hui, 1976], 205–277): not only does Spinoza always condemn the Kabbalah, but he never cites the *Zohar,* the *Sepher Yetsirah* or any commentaries on kabbalistic books (213), which obviously does not mean that he was unaware of them. Even without such expertise, reflection is enough to refute the thesis of Spinoza as a dissembling or 'Cartesian' kabbalist according to Kant's expression in a letter to Mendlessohn (Vulliaud, op. cit., 211). We cannot develop this point, but if Spinozism can exhibit a 'family likeness' to the gnosis of the Kabbalah (or oriental metaphysics), this is so as a 'brilliant caricature' out of profound incomprehension. Even more essentially than pantheism (which the Kabbalah rejects), it is the concept of God

sense than imagining a square circle),[1] and therefore that his resurrection is only the materialization of a "wholly spiritual" truth,[2] we begin to understand that in reality Spinoza judges Scripture in terms of his own philosophical rationalism. And he is quite logically in agreement with Kant, who, too, denies the incarnation of the Word in Jesus Christ, seeing in him only the exemplification of an idea of practical reason—the moral ideal.[3]

We should not be surprised that the first systematic rationalist and 'lay' critique of the Bible makes its appearance with Spinoza.[4] It would have been much more surprising if the Galilean and mechanist revolution had not elicited a reaction of this kind (which will be accompanied by a 'Catholic' reaction seeking to vie with non-Christians on this very

that distinguishes them: what Spinoza has misunderstood is that the Absolute is not "a substance *consisting* in an infinity of attributes" (Definition 6, *Ethics Geometrically Demonstrated*). But the Divine Essence transcends its own nature, names or attributes; and this is why It is the supreme Nothing. Perhaps this is what Jacobi meant in retorting to Lessing that Spinoza, "instead of the *En sof* emanating from the Kabbalah, sets up an immanent principle, an inherent, eternal and immutable cause of the universe, one and identical with all of its effects taken together" (*Lettres sur la philosophie de Lessing* [1785], cited by Vulliaud, 211). Recall that *En sof* literally means 'Endless' or 'Nonbeing' or 'Nothing' (Reuchlin, *La Kabbale* [Paris: Aubier, 1973], 248). We write 'Kabbalah', as is customary. But, the term being derived from *qabbala* (= reception, tradition), it should be written with a *qoph* (= Q), and not with a *kaph* (= K).

1. Lettre 73, *Œuvres*, 1339.
2. Lettre 75, ibid., 1343.
3. *Religion within the Limits of Reason Alone*, Greene and Hudson trans., 54–55. Also see Jean-Louis Bruch, *La philosophie religieuse de Kant* [Aubier, 1968], who compares the Kantian attitude with the Spinozan and lists their differences (186, n.82), and describes the commonality of their views (109).
4. It is sometimes held that Spinoza inaugurated the scientific critique. If by that it is meant that he was the first to trace the complete and systematic program of a critique that is *suspicious* of scriptural tradition, claiming to abide by an internal critique and admitting of no other external criterion than that of mechanist rationalism, there is no argument. But if by this is meant the sum total of 'textual sciences' (leaving aside what is concerned with its hermeneutics), then such an assertion is untenable. To determine this it would be necessary to have at one's disposal a *complete* history of biblical exegesis. But this is not to be found "in any language" (Bertrand de Margerie, *Introduction à l'histoire de l'exégèse* [Paris: Cerf, 1980], t. I, 9). Let us set aside St. Augustine, who mastered little Greek and almost no Hebrew (even though his exegetical work is seen as scientifically remarkable). But it cannot be denied that St. Jerome may have possessed an exceptional competence in this realm, and was only barely surpassed by Origen, who, with the multitude of his commentaries and his monumental *Hexaples* or critical comparison of the six chief Hebraic and Greek versions of the Bible, was truly the founder of biblical science—setting aside the Rabbinic science, as well as medieval exegesis, which Gilbert Dahan has shown to be truly critical and scientific: *L'exégèse chrétienne de la Bible en Occident médiéval. XIIe–XIVe siècle* (Paris: Cerf, 1999).

terrain of scientific exegesis: this is above all the case with Richard Simon).[1] What is even more worthy of notice, and which will soon detain us at greater length, is the weakness of the Spinozan doctrine concerning the cause that has produced religious superstitions. In this respect this great philosopher is truly content with little. Fear explains everything, along with the ignorance that is inseparable from it: "Men give in to superstition only as long as their fright lasts; that worship to which they are drawn by an illusory veneration is only addressed to some delirium born of their sad and fearful humor.... From the fact that superstition, as I have just established [sic], is caused by fear ... we observe first that all men are inclined to it ... etc."[2] But how does so modest a cause produce an effect of such magnitude, variety, constant beauty, richness, and semantic productivity, with such striking rigor beneath its apparent expansiveness? This would be an appropriate question, but Spinoza does not seem to have the least awareness of it, because he does not seem in the least aware of either this wealth or beauty.[3] But he is not alone in partaking of this intellectual blindness—an entire society suffers from it and, truly, this betokens something monstrous.

Briefly shown, these were the most notable consequences of the Galilean revolution in the sphere of the religious sciences, and these were the conclusions drawn by one of the seventeen century's greatest minds. It is in fact with Spinoza that, for the first time, biblical criticism is done in the name of the new science, of strict mechanism, of 'natural laws'—or at the very least when it is presented as a complete and coordinated doctrine. No longer is it just irreligion or skepticism at the origin of this critique; first and essentially it is an epistemological demand. And precisely this demand is to be found again in Bultmann, and so we must refer back to the Spinozan critique if we are to discover its primary source and basic theme, that is to say—in the name of scientific truth—the denial of the 'supernatural in nature', the denial of cosmic symbol-

1. The first edition of the *Critical History of the Old Testament* dates from 1678; Spinoza's *Tractatus* from 1670; and Hobbes, in his *Leviathan*, already 'suspects' scriptural tradition in 1651. Nevertheless Richard Simon, a scholar with prodigious erudition, should be considered not so much the founder, but the first to have produced a work of textual criticism in the modern sense of the term.

2. *Œuvres*, 664.

3. This philosopher who has generated almost religious admiration (we are thinking of Romain Rolland) has however ignored one of the major transcendentals of human existence—beauty: "We are not speaking here of beauty and other perfections that men have willed to call perfection through superstition and ignorance" (*Œuvres*, 233). This esthetic insensitivity is, in our eyes, a true intellectual infirmity.

ism and even beauty: "In all that Bultmann writes there is a deep seriousness which comes from his subjective sense of having been seized, in his case, of having been gripped by Christ. But this is a gravity which, alas, is full of anguish because of its total lack of imagery and form."[1]

Nevertheless the differences are considerable and do not consist in differing individual natures alone. Spinoza's cosmology is surely almost the same as Bultmann's, and, we repeat, it is this that draws them to a selfsame rejection of Judeo-Christian 'wonders'. But Bultmann, although a technician of exegesis like Spinoza but obviously even more so, also wanted to be a hermeneut of religious discourse, from which point of view he is much closer to Kant. What remains is that the actual attitude of these three thinkers with respect to Christ's Resurrection, for example, is significantly the same. What then are these differences that do not stem from differing natures and make of Bultmannism, in our opinion, the most telling and developed form of sacred symbolism's ontological neutralization? In Bultmann's eyes they all consist, we think, in the cultural importance assumed by symbolic forms, which Spinoza seems to completely ignore. There are two reasons for this. 1. — Bultmann is a believer, a man of faith 'gripped by Christ', which is not so for Spinoza, who rather makes of Christ 'the greatest of philosophers',[2] a man in whom was realized a perfect knowledge of the third kind. Hence his profound indifference to Gospel mythology. 2. — German Romanticism and its consequences fell between Spinoza and Bultmann, that is, the rediscovery of the importance, not only of the beauty of the Christian past, but also of the universal beauty of the sacred. Spinoza literally saw sacred forms as only aberrations, monstrosities and, in short, superstitions. His 'cultural' or 'ethnological awareness' is almost nil.[3] True, he remarked that we should not judge "the phraseology of an oriental language by European turns of phrase: even though John wrote his Gospel in Greek, he frequently hebraized."[4] Such observations are rare. To the contrary, for more than a century, and especially since the development of the historical sciences and the human sciences in general, the immense presence of the sacred over all the Earth and in every era has

1. Hans Urs von Balthasar, *The Glory of the Lord*, vol. 1 (San Francisco: Ignatius Press: 1982), 52.

2. This is a remark by Spinoza related to Leibniz by Tschirnhaus (G. Friedman, *Leibniz et Spinoza* [Paris: Gallimard, 1975], 102): "He says that Christ may have been the greatest philosopher (*Christum ait fuisse summum philosophum*)."

3. This is not the case with Montaigne (nor therefore Pascal), any more than with Leibniz.

4. *Œuvres*, Letter 75, 1343.

become palpable to us. It has become impossible to mistake man's religious dimension and see in it only a geometrical and abstract reasoning. More and more the Judeo-Christian scriptures are seen to be intertwined with a universal sacred symbolism rendered at the same time more and more improbable by science. Such is the situation of a twentieth-century believer. Has it ever been more difficult to be one?

The believer is in fact placed between two impossibilities: the impossibility of believing in the God of traditional revelation, and the impossibility of believing in the 'God of the philosophers and scholars'. For—and this is physicist rationalism's primary and most definitive victory—believers themselves have accepted the new philosophy just enough to be convinced that the anthropomorphic or cosmic God of the letter of Scripture is no longer credible. To allow for this, precisely another philosophy would be needed, an already renounced metaphysics of the degrees of reality. Henceforth philosophy, that is to say an intelligible and synthetic knowledge, abandoned the sacred and religious order of things, and it should be quite obvious that such a divorce can only be deadly; surely deadly for religion, but also for philosophy as we will demonstrate. But can a believer for all that cleave to the 'God of the philosophers and scholars'? Certainly not. Not, as is too often said, because his faith excludes science: for many centuries the Abrahamic Jewish, Christian and Islamic faith has fully adapted itself to the God of Plato and Aristotle. But the 'God of philosophers and scholars' is a God created by a certain philosophy and a certain science *against* the God of Scripture, Whose impossibility has been demonstrated by scientific reason. Galileo's natural philosophy having demolished the ontological basis for traditional symbolism, all that remained to elaborate in its stead was another God of the cosmos: the Clockmaker God, a celestial Mechanic reduced to the state of first cause. This abstract theism is not contrary to reason. It even presents itself as the only possible solution. But its negation or refutation likewise agrees, although in another way, with the demands of logic. Faith then cannot find in this its needed absolute. This is why this rational god feels profoundly foreign; faith clamors for another God, the God of Abraham, Isaac and Jacob. Doing this faith renounces sacred intellectuality, it endorses the dividing up of the theological field established by the new religious philosophy, and even seems to reclaim for itself the obscurity of its commitment. For the God of Abraham is a God that speaks to our own person, a God of our existence and our life who speaks not to teach the nature of things, but to elicit our freedom. The God of the cosmos has been rejected either as the imagery of a long-vanished mythology, or as an alienating theory derived from false concepts about the divine. To think about God is to subject him to the categories of our understanding and deny his irreducible existential presence.

Such is, we think, the framework within which Bultmann's thought has developed, and the reason for which it has assumed the task of demythologizing Scripture, that is to say of dissociating kerygma from its cultural trappings. But, once again, this task would make no sense if it were not exhibited on the terrain of faith itself, for, for anyone not having this faith, it was high time that this demythologizing took place, or rather that scriptural expressions were definitively transformed into mythology.

As we know, the word *kerygma*, which means 'proclamation, preaching', is used in the New Testament to designate the heralding of Jesus Christ.[1] Bultmann makes it a major concept of his hermeneutics: "Statements of kerygma are not universal truths but are personal address in a concrete situation. Hence they can appear only in a form molded by an individual's understanding of his own existence or by his interpretation of that understanding."[2] But this form reveals not only a certain "understanding of existence" which, in order to be explained without risk of falling back into the alienating theories of discourse and concept, should be situated in the Heideggerian school of philosophy, itself a hermeneutic of existence; it is also tied to a culture in which a certain understanding, or a certain 'image' of the world that has become completely foreign to us, is formulated. "We cannot use electric lights and radios and, in the event of illness, avail ourselves of modern medical and clinical means and at the same time believe in the spirit and wonder world of the New Testament."[3] We cannot just choose a few elements of this mytho-cosmology and reject others.[4] This is because it is totally opposed to our present scientific conception of the universe. Now, even if they wanted to, people today cannot renounce this conception, any more than early Christianity could avoid *spontaneously* imbedding the kerygma in a mythic framework: "A blind acceptance of New Testament mythology would be simply arbitrariness.... [This] would be a forced *sacrificium intellectus*, and any of us who would make it

1. Matt. 12:41, Rom. 16:25, etc.

2. "The Relation between Theology and Proclamation" in *Rudolf Bultmann: Interpreting Faith for the Modern Era*, 239. We will cite from his considerable theological and philosophical work only that which deals more directly with demythologizing, bypassing everything dealing with existential hermeneutics.

3. "New Testament and Mythology," in *New Testament and Mythology and Other Basic Writing*, 4.

4. Ibid., 9.

would be peculiarly split and untruthful. For we would affirm for our faith or religion a world picture that our life otherwise denied."[1]

Besides, the unacceptable character of New Testament mythology is not only bound up with a potential and contingent opposition of two world views, two cosmologies; it is also a direct result of the contradictory nature of this mythology. This contradiction is best summarized by saying that, in Bultmann's view, mythological thought speaks of transcendence in terms of immanence: "That mode of representation is mythology in which what is unworldly and divine appears as what is worldly and human, or what is transcendent appears as what is immanent, as when, for example, God's transcendence is thought of as spatial distance. Mythology is a mode of representation in consequence of which cult is understood as action in which nonmaterial forces are mediated by material means. 'Myth' is not used here, then, in that modern sense in which it means nothing more than ideology."[2]

But, to speak of transcendence in terms of immanence: is not this precisely the function of the symbol? Bultmann does not deny it. Quite the contrary, he even thinks, as we have seen at the beginning of this article, that demythification depends on the recognition of Scripture's symbolic character, while mythology depends on an ignorance of this character: "Mythological thinking . . . naïvely objectifies what is thus beyond the world as though it were something within the world. Against its real intention it represents the transcendent as distant in space and as only quantitatively superior to human power. By contrast, demythologizing seeks to bring out myth's real intention to talk about our own authentic reality. . . . It is often said that neither religion nor Christian faith can dispense with mythological talk. But why not? Such talk does indeed provide pictures and symbols for religious poetry and for cultic and liturgical language in which pious devotion may sense a certain amount of meaning. But the decisive point is that these pictures and symbols conceal a meaning which it is the task of philosophical and theological reflection to make clear. Furthermore, this meaning cannot be re-expressed only in mythological language, for if it is, the meaning of this language also must be interpreted—and so on *ad infinitum*." This is why, Bultmann continues, "[b]ecause God is not a phenomenon within the world that can be objectively established, God's act can be talked about only if we at the same time talk about our own existence as

1. Ibid., 3–4.

2. Ibid., 42, n. 2; likewise René Marlé, *Bultmann et l'interprétation du Nouveau Testament* (Paris: Aubier, 1966), 48.

affected by God's act."[1] Besides, this is the only way to speak of God non-symbolically: "When we speak of God as acting, we mean that we are confronted with God, addressed, asked, judged, or blessed by God. Therefore to speak in this manner is not to speak in symbols or images, but to speak analogically. For when we speak in this manner of God as acting, we conceive God's action as an analogue to the actions taking place between men." Hence "only such statements about God are legitimate as express the existential relation between God and man. Statements which speak of God's actions as cosmic events are illegitimate." This also applies to statements concerned with God's cultic activity (the sacrifice of his Son) or a judicial or political act "unless they are understood in a purely symbolic sense."[2]

As can be seen, we are after all going back to Spinoza and Kant, that is to the moral God. Certainly it is a long way from the pantheism of the former[3] to the transcendental theism of the latter, or to Bultmann's existential theism. But, as already proclaimed by the *Tractatus Theologico-politicus*, what is essential in religion is not knowing God, but living in Him, by acting according to justice and charity. The God of Scripture is moral life, not mathematical truth. If the result is identical, is this not basically because the theodices[4] are similar despite appearances? An affirmation somewhat paradoxical, and ye. . . ! Is not the rejection of the God outside the cosmos with Kant and the post-Kantians (and perhaps already with Pascal) an inverse analogue to the Spinozan identification of Nature with God: to have the geometric cosmos enter into God,[5] or else to decree God's absence from a cosmos reduced to extension? In both cases, with different intellectual temperaments, the same world-concept intervenes and imposes a solution: exclusivist immanentism joins exclusivist transcendentalism in a denial of the degrees of reality and a symbolic cosmos, a cosmos in which sensory forms signify by pre-

1. "On the Problem of Demythologizing," in *New Testament and Mythology and Other Basic Writings*, 161–162.

2. "Jesus Christ and Mythology," in *Rudolf Bultmann: Interpreting Faith for the Modern Era*, 318–319.

3. The word 'pantheism' made its appearance after Kant's death. 'Pantheism' is to be met with for the first time with Toland in 1705, and 'pantheist' with his adversary Fay. It is anachronistic then to apply it to Spinozan doctrine.

4. We are not taking 'theodicy' in the Leibnizian sense, but in the one accepted in nineteenth-century theological treatises: 'natural theology'.

5. To introduce extension into God, Spinoza distinguishes imagined extension, that is finite and divisible, from intelligible extension, that is infinite and indivisible, which can only be in God, since the divine substance is indivisible and infinite (*Ethics*, I, prop. XV, note). Without arguing about the distinction between a conceived space and an imagined space, we will only mention the continuous nature of space, that is to say the

sentification, that is, in which God's presence and absence correspond to each other dialectically because necessarily evoking each other.

Therefore we reject Bultmannism as much from the viewpoint of cosmology as from the viewpoint of symbolism—*from the viewpoint of cosmology* because Bultmann's world-concept rests on classical mechanics, which was already beginning to crumble at the end of the nineteenth century: it is a pre-Einsteinian concept that *completely* ignores that radical calling-into-question of ideas about time, space and matter engaged in by contemporary physics;[1] *from the viewpoint of symbolism*, because the idea of the 'purely symbolic', the equation *demythologization = conscious symbolism*, destroys the symbol's raison d'être. Undoubtedly symbolist thought runs the risk of confusing the referent of the symbolic sign and its signifier, an inevitable danger inherent to the very nature of a symbol since it conjoins the one to the other, and this is called idolatry. But the remedy for this danger does not reside in the rejection of this conjoining which alone establishes the right of the signifier to designate its referent: *if symbolism were purely symbolic, there would be no symbolism.* Either the symbolic sign is one that signifies by virtue of the connaturality of the signifier with its referent—what we have called a signifying by presentification—or it is indistinguishable from either linguistic or inductive signs.[2] Hence the only way to mitigate this risk of idolatry is to reactivate an awareness of the degrees of reality, to show that sensory forms are not only images of intelligible forms in the use made of them by sacred symbolism, but that they are

fact that it is not composed of discrete parts, that it is a property known only by intellectual intuition. This is a given imposed on our minds that cannot be formed axiomatically. In fact Skolem has established that "every axiomatic system having a model has a model of denumerable power" (Léo Apostel, "Syntaxe, sémantique et pragmatique," in *Logique et connaissance scientifique,* [Paris: Gallimard, 1969], 299–300). Now two systems have the same power when a bijective correspondence can be established between their elements. But a continuous agregate, such as a straight line, includes a non-denumerable quantity of elements, since it is indefinitely divisible. Its power is therefore superior to that of the denumerable, which characterizes every axiomatic system (R. Blanché, *L'axiomatique* [Paris: PUF, 1970], 80). In other words, every axiomatic construct having been made with the help of distinct elements, obviously the continuous cannot be composed of these discontinuous elements. As a result, contrary to what Spinoza asserts, an intuitively conceived space is altogether divisible, even indefinitely so. Conceived space is likewise finite (and not infinite) because only the understanding conceptualizes a spatial 'genus', distinguishing it from time or thought, for example. Conversely, space is infinite only for the imagination, because we cannot imagine a spatial 'boundary' to space. That said, space is indeed something in God: it is Universal Possibility, the corporeal expanse of which is only an image or an indefinite limitation.

1. Cf. Wolfgang Smith, *The Wisdom of Ancient Cosmology*, passim.
2. Cf. *Histoire et théorie du symbole,* chap. VI, art. I and II.

likewise so in the order of the real. What restores balance to traditional symbolism is the presence, either explicit or implicit, of a 'Platonic' metaphysic which affirms that things are at once both more and less than they seem: on the one hand they are by no means reduced to the accidentality of a spatial configuration, as Galilean physics would have it, since they lead us to the divine mystery; but on the other, contrary to what they seem to suggest, they do not constitute *all* reality, they do not exhaust the 'being' possibility, they have only a vestigial existence, relatively absolute with respect to nothingness, but relatively nil with respect to the Absolute. When awareness of this metaphysic is obscured in traditional symbolism, it is almost inevitably degraded into superstition and magic. But, to maintain metaphysical awareness, it is obviously necessary to renounce the—in reality impossible—cosmology that the geometrism of the physicist has constructed.[1]

We do not think then that Bultmann's demythologizing restores the truth about symbolism. Actually the mythocosm, as conceived of by him, is still seen from the Galilean *episteme,* it has been marginalized from the outset, *a priori* rejected; it has been posited as the insane '*alter ego*' of the rational universe; in short, its truth is simply not understood. This idea turns out to be basically identical, then, to the way in which idolatry and magic conceive of symbolism. Doubtless the *episteme* of magic is opposed to the *episteme* of physics, but they can be in mutual opposition only because both are situated on the same plane, that plane formed by the rejection or ignorance of *scalar ontology,* that is to say the *metaphysics of the degrees of reality.*[2]

1. We hardly need to specify that this rejection by no means implies a rejection of the numerical results to which the physicist's geometrism gives access. We see here a philosophic thesis on the nature of the universe, not the putting of the movement of falling bodies into an equation.

2. It is up to the metaphysics of sacred symbolism to show that what is incontestable in the Heideggerian concept of being is only truly understood from a superontological point of view: Being can only be 'posited' starting from Beyond-Being. And it is only out of That which surpasses Being that a critique of the confusion between Being [*Être*] and being [*l'étant*] becomes fully cogent, for only More-than-Being can be the basis for least-being; only the More can do less. This axiom is one of metaphysics' major keys. From this viewpoint Being is itself already part of *Maya,* but by a divine and supreme magic, of which It is the 'effect' insofar as the original ontological determination. Being can be conceived of as a radical upwelling or affirmation only out of the Depthless Deep of the Absolute. As for the Heideggerian critique of Platonic essences, it basically betrays the same incomprehension as Aristotle's. It reifies the Ideas, ignoring that they are still part of the cosmic theater, since they are portrayed by the puppets, and since the Good is beyond the ontological degree, that is to say it is not the First of a series; cf. *Penser l'analogie,* 136–192.

Conclusion: On Culturological Reduction

And so we have come to the end of our reflections about the ontological neutralization of sacred symbolism. But have we for all that come to the end of the crisis of religious symbolism? Certainly not, for once we notice we have reached an end we have already gone beyond it. Now, far from having exhausted its power, the critique of the symbol seems triumphant. By amputating the symbolic sign from its referent, this critique encloses it within the sphere of its jurisdiction. It allows only two of the three poles of its triadic structure to remain, signifier and meaning, both of which seem able to appeal to nothing but the human. All real transcendence is set aside; nothing is left but the illusory transcendence of a consciousness taken in the snare of its own productions.

Also, the cosmological analyses just undertaken will surely not disturb the critique in its self-assurance, or compel it to reintroduce a referent into the definition of a symbol, for the following question is unavoidable: do we need to define the symbol such as the traditional civilizations have conceived of it, or such as it is in reality? Or again, can we rightly attribute to the symbol what pre-scientific culture in fact attributes to it? "We will not deny," some will object, "that traditional symbolism implies a scalar ontology, since it rests on the capacity for one thing to designate another of superior reality." Even an Aristotelian like St. Thomas agrees: "It would be wrong to believe," he says, "that the multiplicity of these senses produces equivocation or any other kind of multiplicity, seeing that these senses are not multiplied because one word signifies several things, but because the things signified by the words can be themselves signs of other things."[1] However, it does not follow that this ontology is actually true. It is itself only a *cultural artifact*. After all, Platonic metaphysics is as unbelievable as sacred symbolism, and it would be a sophistry to claim to base religious superstitions on a philosophical one. And also the one lends support so well to the other only because they are basically identical and are in themselves two translations, the first in religious mode, the second in intellectual mode, of the same definitively surpassed, definitively historical vision of things. As A. Vergote says, never again can we situate ourselves in the original light of myth. We must therefore resign ourselves to the inevitable: the founding of sacred symbolism by a scalar ontology is only a vicious circle that in no way removes us from the cultural order. On the other hand, the definition of the symbol to which its ontological neutralization leads is altogether sufficient for covering the field of symbol-

1. *Summa theologiae* I, q. 1, a. 10. This is an article on the four senses of Scripture.

ism. Besides, in the best-case scenario, this only means going back to the traditional definition without changing anything, except to consider it as a cultural thesis, that is, except to proceed with what might be called a *culturological reduction*, a reduction basic to every phenomenology of culture. We can therefore retain the definition of symbolism given by Paul Ricoeur "as any structure of signification in which a direct, primary, literal meaning designates, in addition, another meaning which is indirect, secondary, and figurative and which can be apprehended only through the first."[1] This definition, reiterated as well by many authors,[2] brings in only two terms, 'signifier' and 'meaning', distinguishing only a dividing of the second into a direct and indirect meaning. Basically, this parallels what we have just given from St. Thomas and is essentially presented as a description of the real functioning of symbolic signification. Neutral with respect to every ontological thesis, it does however enable this symbolic structure to be distinguished from every other signifying structure. Can philosophy ask for anything more?

We do not deny that this 'phenomenological' definition is in fact true. And this is the one that underlies the works of the most remarkable specialists of the history of religions, van der Leeuw for example and, above all, Mircea Eliade. Our concern, like our proposal, is not to contest this. We would only like to draw attention to the culturological reduction on which it is based, because it seems so self-evident that it ends up going unperceived. Moreover, is it not the symbol itself that seems to invite us to see within itself only a cultural entity? To speak of a symbol: is this not to speak, by definition, of something situated from the start outside of the real? And, under these conditions, does the concept of the ontological neutralization of the symbol still make any sense? Can one neutralize an ontological dimension which, by definition, is excluded from the symbolic sign? Clearly, we come up against a major difficulty here for every philosophy of the symbol summed up in the equation: symbolism = unreal. We have already encountered this many times. It functions in an almost automatic manner and claims the title of primary definition of the symbol; this is what bedevils our reflection and impedes every metaphysics of the symbol. Ever reborn, ever ready to serve, with such a banality that it eludes the grasp of critical analysis, it is then right to constantly keep an eye on it and avert its incessant threat. For symbolism clearly has a relationship with reality just as directly as it has a relation-

1. *The Conflict of Interpretations* (Evanston, IL: Northwestern University Press, 1974), 12–13.

2. There is, among others, the remarkable book by Clémence Hélou, *Symbole et langage dans lest écrits johanniques* (Paris: Mame, 1981), 10–11.

ship with knowledge; the very fact that it can be defined as the non-real plainly proves this relationship, since it is thus posed as that which has for its function *not to be* what it is. The *non-real* does not in fact designate a logical class of a negative sort that excludes a quality, in the way that *non-white* encompasses all entities in which this determination is lacking without saying what their colors are. But *non-real* here is an almost positive determination that defines a dialectical relationship with the real, and which is itself defined by this relationship.

What are the possible meanings attributed to this non-real, it being understood that it cannot signify pure and simple nonexistence? The first, most widespread meaning today is that the non-real designates something imaginary. If the non-real were to purely and simply signify the non-existent, the nothing as such, we could not even speak of it, it would not exist in any manner, and neither would symbolism. The non-real must be real then, must exist, after a certain manner. The answer is immediate: it exists 'in our head', in 'imagination', but not like things existing before us in the real world. This answer is incontestably true. Symbolism is bound up with the imaginary. Beyond its onto-cosmological neutralization, the culturological reduction refers the symbol back to the reality of the imaginary. But are we sufficiently cautious about what this thesis implies? Can what is imaginary be the ultimate explanatory instance? Surely it will be admitted that this thesis accounts for the *existence* of symbolism, that is, culture as the sum total of non-realist signs. Again, the producing of these signs seems to require an awareness of the non-real, that is to say of the degrees of being, surpassing the order of the imaginary as a representative function because it presupposes a direct *experience* of the non-real, or rather the 'less real', which is intellective in nature. But it cannot account for the *qualitative essence* of symbols. To refer symbolism back to the imaginary is to defer the problem, not resolve it. It is not enough to endow the imaginary with a kind of natural 'insanity' which would dispense us from asking why it produces a particular symbolic form, seeing that, for its part, 'we might expect anything' (and besides we will see precisely in the two following parts that this attitude is in fact impossible). We are also constrained to ask what is the raison d'être for these forms. Man is condemned to meaning, to the intelligible, to reason. And nothing is more insane than to use insanity for an explanation. Everything has a raison d'être. And so the non-real or symbolic requires an objective meaning founded on the nature of things, that is, on a scalar onto-cosmology; for the multiplicity of the degrees of being implies that none of them are fully real and that all are arranged in a hierarchical realization, an anagogical dynamic that is the very basis for all symbolism. The symbol therefore affirms by itself an onto-cosmological thesis; it speaks of transcendence in terms of immanence, it makes known what it is not and what no eye can see, because it is its sensible presentification. Clearly then there is an *ontological* neutralization of the symbol, because there is a rejection of the onto-cosmological thesis native to the symbol's essence. And this is why the culturological reduction is unfaithful to the symbol. What is allowed to subsist, in terms of its *epoche*, is not altogether the same entity as the one received directly from cultural data.

At least this much should be said. However, apart from Paul Ricœur, who shows that the notion of 'metaphoric truth' calls into question the very idea of reality,[1] this is a stipulation never, so to say, encountered in contemporary philosophy, and even in the most remarkable works on the symbol. Neither Jung, Gilbert Durand,[2] Mircea Eliade, nor René Alleau seems to view the symbolic other than as a product of the imaginary. Questioning himself on the nature of his discipline, Eliade writes: "In the broad sense of the term, the science of religions embraces the phenomenology as well as the philosophy of religion. But the historian of religions . . . applies himself to deciphering in the temporally and his-

1. *The Rule of Metaphor*, trans. R. Czerny (Toronto and Buffalo: University of Toronto Press, 1977), eighth study, especially 305–306: "It is first as a *critical* instance, directed against our conventional concept of reality, that speculative discourse in its own sphere of articulation takes up the notion of split reference. The following question has arisen repeatedly: do we know what is meant by world, truth, reality?"; likewise 247ff: "Towards the Concept of 'Metaphorical Truth.'"

2. In a work nevertheless quite interesting, the Jesuit Charles A. Bernard declares: "To suppose that the imagination possesses no ontological value, the value of the symbol would vanish into the fictitious. But why should we deny all value to the image? We refer here to works which, like those of Gilbert Durand, have declared the transcendental value of the imaginary" (*Théologie symbolique* [Paris: Téqui, 1978], 101). But whatever may otherwise be the depth, the wealth and the pertinence of Gilbert Durand's analyses, as much in his now-classic 1963 treatise *The Anthropological Structures of the Imaginary* (trans. M. Sankey and J. Hatten [Brisbane: Boombana Publications, 1999]) as in his later works, *L'imagination symbolique* (1964), *Science de l'homme et tradition* (1975), etc., works that we often agree with fully, we must note that the 'transcendental fantastic' that he has worked out (*Anthropological Structures*, 363–406) does not at all correspond to our idea of a symbolic onto-cosmology. What we have here is a Kantian transcendental, a component of objectivity, in other words of the universality of the symbolic imagination as a structuring capacity of human knowing, feeling and acting: "...as well as 'objective truths'—products of repression and of the blind adjustment of the ego to its objective milieu—there are also 'subjective truths' that are more basic to the functioning of thought than are external phenomena.... Is a lie still a lie when it can be called 'vital'?" (ibid., 381). And later: "In this area 'vital lies' are truer and more valid than mortal truths" (ibid., 408). Now that is a proposition that seems unacceptable (which in no way means that we would deny the reality of the structural patterning of the imaginary): a lie is always a lie or indeed has never been so.

torically concrete the destined course of experiences that arise from an irresistible human desire to transcend time and history. All authentic religious experience implies a desperate effort to disclose the foundation of things, the ultimate reality. But all expression or conceptual formulation of such religious experience is imbedded in an historical context. Consequently, these expressions and formulations become 'historical documents,' comparable to all other cultural data."[1] Such is the concept of 'religious experience' formed by one of the greatest historians of religion, someone quite open to the 'mystical' and one of the greatest lovers of symbolism. But, as we see, the so-called objectivity of the culturological reduction is accompanied in fact by a thesis on the *psychological* nature of the religious as a 'desperate effort,' which offers no speculative consistency, and which begins by denying, in the name of a disputable but undisputed etiological pantheism, the most fundamental affirmation of religious discourse.

Incontestably, this generally prevalent attitude is the result of the epistemological revolution achieved in the seventeenth century. We have described this revolution—and the evolution that preceded it—in three stages illustrated by the names Plato, Aristotle, and Galileo. Plato's onto-cosmology is open 'toward the heights' and is completed in the super-ontological Good. Thus it institutes a vertical disequilibrium between the intelligible and the sensible in which the corporeal world 'aspires to fall upwards', towards the spiritual world. This disequilibrium, this dynamic tension corresponds most exactly to the nature of the symbolic sign, and to the anagogical movement that a hermeneutic understanding perceives in it. With Aristotle, conversely, there appears the first attempt to create a closed onto-cosmology or, perhaps better said, an ontologically equilibrated cosmology. No longer is there a dis-equilibrating tension between the intelligible and sensible, but an agreement and almost an identification. This is why Aristotle is the West's first physicist, the first to intend to create a true science of *physis*. In fact, however, the same reason for intelligibility persists from Plato to Aristotle: it is the form, which is also *eidos* ("we other Platonists," as Aristotle occasionally states).[2] But the object to which this intelligibility is applied has changed, and by that Aristotle is a distant herald of Galileo. In this respect we could speak of a 'logical' closure of the concept which likewise prepares for its epistemological closure under the effect of a mathematizing of scientific dis-

1. *The History of Religions: Essays in Methodology*, ed. Mircea Eliade and Joseph M. Kitagawa (Chicago: University of Chicago Press, 1970), 88–89.

2. Aristotelianism is basically a truncated Platonism (with all the refitting demanded by this amputation).

course. All these factors are correlative. The Platonic openness of the concept, the only viewpoint in conformity with the philosophical intent, basically signifies that there is no sensible object to which the intelligible capacity of the concept can be truly applied, and by which it can be adequately enclosed. The intelligible burden of our thinking is basically stronger than anything offered by the sensible. The human intellect, desirous of finding rest, is in quest of the being that could support this intelligible weight, and finds it only beyond the sensible, and even beyond the immutable forms. This is what maintains the openness of a genuinely philosophical concept, that is to say what the wisdom of Reality truly loves. Its incompleteness is a sign of its surplus of intelligibility with respect to the sensible objects to which it can be applied.[1] For Aristotle, to the contrary, with his setting of the sensible and the intelligible into separate blocks, the concept is adequate for being applied to things. This is still not an epistemic closure, one by which a concept forms itself in an autonomous way and becomes in some manner '*causa sui*', but this is a 'logical' closure. With Aristotle in fact, for the first time in the West, philosophy intends to gain perfect mastery over its own discourse. Aristotle is the first to produce a technically elaborated philosophical language, one in which the terms and the relationships sustained among them are defined in such a univocal manner (at least in principle) that the knowledge of things can be transformed into operations on discourse. The concept closes itself around being and being closes itself around the concept. One is adjusted exactly to the other. Logic is thus the science of philosophy and bestows on it the quality of a science. This is why "it is necessary to know the Analytics before broaching any science."[2] We are far from Plato, for whom the primary condition for gaining knowledge of wisdom is to become aware of our ignorance.

But lastly Aristotelianism perhaps did not truly have the means to accomplish its hopes. Mathematical formality would have to be substituted for logical formality for this already-emphasized reason: mathematical concepts alone realize necessarily and by definition the most radical closure. It is Galileo who then, as we have seen, brought the scientific intent that orients all Aristotelianism to successful issue, but by

1. By that we see how this surplus of intelligibility of the concept relative to sensible being can equally lead to Hegelian idealism, that degraded and perverted Platonism: instead of submitting the intellect to super-intelligible Being, and by that accomplishing the most deep-seated hope of intellection, infra-intelligible being is reduced to a concept. It is Galilean science that has come up with this as if inevitable 'solution' by eliminating intelligible and transcendent being from the field of what is possible, all the while maintaining, naturally, the intellective need. Naturally, for the intellect itself *cannot* be eliminated.

2. *Metaphysics*, IV, 3, 1005b.

breaking with its qualitativist realism. Concepts, having become mathematical, are applied to the real only to the extent that it is itself mechanistically reduced to the geometrical. For Plato concepts are mental symbols of the eternal Ideas and therefore participate in them. For Aristotle they are abstracted and even emanate from things. For Galileo the cognitive trajectory is reversed (even if Galileo himself was unaware of this), as Cartesianism has proven; the physically real is constructed in the image of its concept. Succeeding the static equilibrium of Peripatism is a disequilibrium towards the bottom: the intelligible, as with Aristotle, is indeed 'lodged' in the sensible, but the second is no longer an expression of the first, it is its lie and betrayal. The only upshot of this divorce instituted by Galileism at the very core of physical reality is the technical possession of the world that, at each moment, 'proves' that things are indeed so and that geometry is right. The mythocosm in that case utterly disappears. The meaning of anagogical tension is reversed and becomes, so to say, categorical. Symbolism is rendered impossible. By "driving away everything that Aristotle had left to draw souls into things," in the words of Brunschvicg, everything that would draw things into souls is also driven away. The vital and nurturing immersion of culture in nature is scientifically forbidden. Henceforth culture drifts, rootless, with its universe of forms, colors, and rites which bespeak a reality from which they are forever exiled. Ontological neutralization indeed allows for the subsistence of symbolic signifiers along with their meanings, but a meaning that empties into the void. Here now is humanity, cut asunder in the very depths of its cultural soul. The Galilean revolution has not only introduced a divorce into the very heart of appearances, it also separates human beings from their symbolic representations. Within this epistemic cleft the European soul discovers that what it took to be an inexhaustible semantic well-spring, a perpetual stream of vital, affective, and cognitive meanings, is only monstrous imaginings, a thinking disorder of the insane, a blind stare that will never be satisfied with the light of reality.

Appendix 1

Very Elementary Notions of Ancient Cosmography (cf. 19)

1. — If we look at the sky on a beautiful August night, we perceive a multitude of luminous points. For the most part these points are motionless with respect to each other: these are the stars; they depict unchanging figures: the constellations. They all seem fixed to the celestial vault like golden nails on a spherical enclosure called the sphere of the fixed stars.

However, if we observe closely, we will notice that the stars shift all together without changing their respective positions, as if the sphere of the fixed stars pivoted in a single block around an axis called the 'axis of the world' (a complete turn every twenty-four hours). A single star is immobile, the *alpha* star of the Little Dipper, called the pole star because it defines (almost) the north pole of the sphere of the fixed stars' rotation. The axis of the world passes then through this star and crosses the celestial sphere to a diametrically opposite point, the south pole (where no star is to be found). On the other hand, if the observer is between the equator and the terrestrial north pole—the case for Europeans or North Americans—the pole star is not above his head (the local zenith). Therefore the axis of the world is not perpendicular to the plane of the local horizon, but oblique with respect to it.

2. — Now, among all the stars in the sky, we see some (with the naked eye) that do not exactly obey the regular rotation of the sphere of the fixed stars. These are the 'wandering' stars that the ancients called *planets* (*planes* in Greek = wanderer, vagabond). This is the case with the Moon, Mercury, Venus, Mars, Jupiter, Saturn, and also the Sun. The Greeks imagined then that each of these wandering stars (the Sun included) possessed its own sphere (a crystalline, ætheric or invisible sphere) to which it was attached and which would draw it along in its movement of concentric rotation around the Earth. The universe for them was then made up of a system of concentric spheres nested one inside the other and all turning in the same direction. Their number and the order in which they were arranged varied according to authors and traditions. In general ten spheres were counted and arranged in this way starting from the lowest point (the Earth): 1st Moon, 2nd Mercury, 3rd Venus, 4th Sun, 5th Mars, 6th Jupiter, 7th Saturn, 8th the Heaven or sphere of the Fixed Stars, or 'firmament', 9th the Crystalline Heaven or sphere (the starless heaven), and 10th the Heaven or sphere of the First Mover (the movement of which drew along all the others). Beyond and therefore outside the world was 'situated' the Empyrean Heaven where God abides with the elect. This is the doctrine of Ptolemy, St. Thomas Aquinas, and Muhyi 'd-Din ibn al-'Arabi. The elementary spheres of water, air, and fire (non-astronomical, infra-lunar spheres not constituting 'heavens') are often intercalated between the Earth and the sphere of the Moon.

However, this general agreement masks important divergences, summed up in the opposition between Aristotelian and Ptolemaic astronomies (homocentric or eccentric spheres). These divergences are the result in part of the observed facts. On the one hand, the movements described by the planets sometimes follow the rotation of the sphere of the fixed stars and sometimes seem to go retrograde and travel in the opposite direction. On the other, precise readings of the orbital positions of the heavenly bodies show that the Earth is not always exactly at their center (which anticipates the elliptical orbits of Kepler). This is why the preceding diagram was modified, chiefly by Hipparchus of Nicea (second century before Christ), who conjectured about planet-bearing secondary circles (the 'epicycles'), where a center of rotation 'A' is itself fixed upon a principal circle (the 'deferent') turning around a point 'C' situated at a certain distance from the center of the Earth (cf. figure 1). It is this system that Ptolemy, in the second century after Christ, corrected and brought to a point of admirable perfection.

3.— Among all the wandering stars, the Sun seems animated by a more regular motion. It rises in the East and sets in the West. But it does not rise—or set—at the same time every day, nor does it travel exactly the same path in the sky. If, during a year, one marks each day in the sphere of the fixed stars the Sun's place at noon, when all the points obtained are joined together these points will be seen to draw a great circle in the celestial sphere, a circle to which the name *ecliptic* has been given, because it is in the plane of this circle that eclipses are produced whenever the ecliptic meets the lunar orbit. In Copernican astronomy, with the Earth turning around the Sun, the circle of the ecliptic is the one according to which the plane of the Earth's orbit cuts across the celestial sphere (figure 2).

Now this ecliptic is oblique with respect to the celestial (and terrestrial) equator, which means that the axis around which the rotation of the Earth upon itself occurs is inclined with respect to the plane in which it moves about the Sun, and not perpendicular to this plane (hence, for any given point on the Earth's surface, the inequalities of days and nights and the different seasons). This inclination (which varies) is around 23° 27′. Or again, in the geocentric system, the great circle annually traveled by the Sun is inclined 23° 27′ with respect to the celestial (or terrestrial) equator taken as the horizontal plane.

The circle of the ecliptic cuts across the circle of the celestial equator at two points, called equinoctial points, for there the duration of the days equals that of the nights. When the Sun at its rising occupies one of these points, it is the spring (vernal point) or autumn equinox. Two other points of the circle mark the moment when the Sun reaches its greatest distance from the equatorial plane (more or less 23° 27′ above or below): these are the solstice points (from *sol*, Sun, and *stare*, to stand still).

Lastly we observe that wandering stars other than the Sun (our actual planets) are likewise moving—if their trajectories are inscribed on the sphere of the fixed stars—in a band that goes around the sky and from which they never leave; this band extends 8° 5′ on either side of the ecliptic. It has received the name *zodiac* (or cycle of 'life', from the Greek *zoon*, 'to be living'), and was divided (no doubt by the Babylonians) into twelve sections or 'signs', the names of which are sometimes oddly analogous in different cultures, and refer to the figures formed by the constellations in the sphere of the fixed stars that lie along the zodiac seen from the Earth (figure 3).

Figure 1: PTOLEMAIC SYSTEM

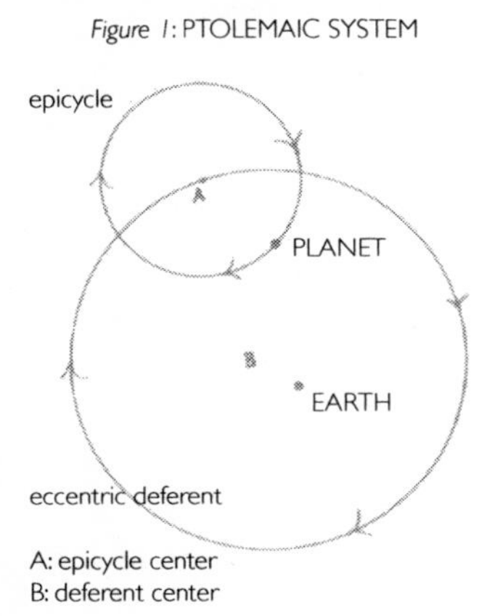

The planet's motion remains bound up with that of the deferent, even though it moves at times in the opposite direction.

Figure 2: HELIOCENTRIC REPRESENTATION

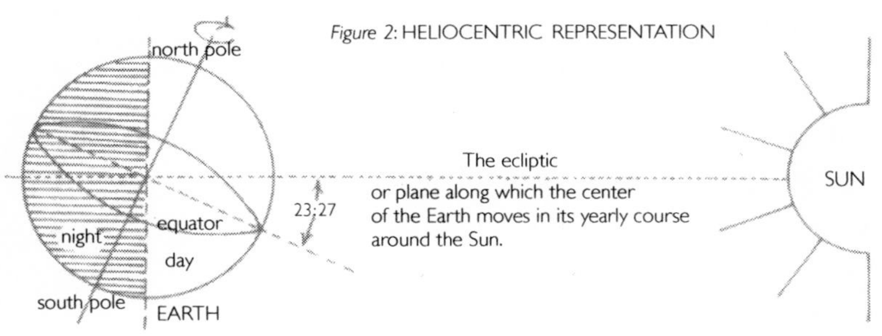

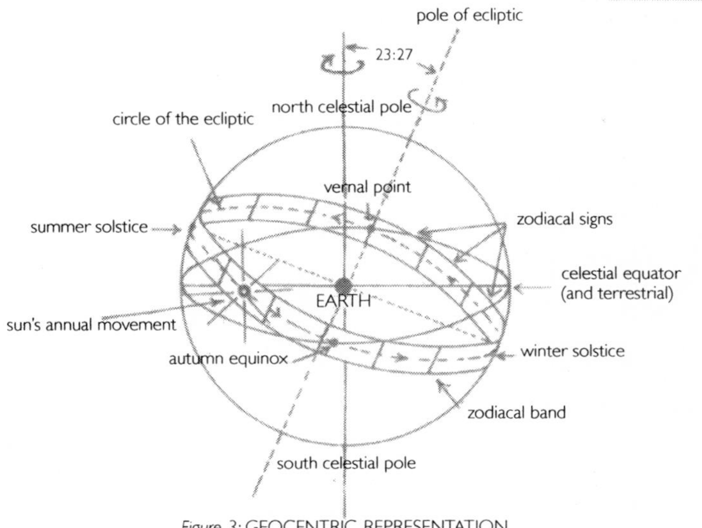

Figure 3: GEOCENTRIC REPRESENTATION

Appendix 2

Ptolemy and the Reality of the Celestial Spheres (cf. 22, n3)

Ptolemy is regarded as a mathematician-astronomer for whom the physical reality of the constructions elaborated in order to 'save the appearances' (epicycles and eccentric deferents) hardly matters. This thesis, considered as certain by Duhem (*Système du monde*, t. II, 85) and taken up again by Koyré (cf. supra 22, n 1) does not seem to be confirmed by the texts. It is true that the *Almagest*, where the Ptolemaic system is entirely worked out, never poses the question of its reality. But, ten years after the *Almagest*, Ptolemy wrote the *Planetary Hypotheses* (two books, the second of which has reached us in Arabic [*Ptolemaei opera minora*, ed. Teubner]; 'hypotheses' does not mean 'simple suppositions', but 'starting principles', 'necessary presuppositions'). In this work he expressly deals with the solidity of the celestial spheres, even regarding the epicycles as made up of little spheres (rings perhaps) set into the substance of the eccentrics (cf. Theodor Litt, *Les corps célestes dans l'univers de S. Thomas d'Aquin* [Louvain: Nauwelaerts, 1963], 336–341).

Appendix 3

St. Thomas Aquinas, the Precursor of Copernicus? (cf. 31, n3)

This question was used as the title of a talk given not long ago by the Dominican Father Tomo Vereˇs, a professor at the University of Zagreb, at a conference devoted to St. Thomas (pages 247–253 of the collection of the proceedings, the reference to which we have lost). His conclusion was clearly in the negative: "We think that it is impossible to consider St. Thomas as a precursor of Copernicus, even though he too called the geocentric system into question. But St. Thomas, even though knowing of the heliocentric system, is not a defender of it. In a certain sense, he went further than Copernicus . . . because he called into question the apodeictic value of all astronomical systems in general." Father Vereˇs in fact accords the greatest importance to a text of the *Summa Theologiae* (I, q.32, 1, 2m) where St. Thomas declares (with respect to the impossibility of our rationally proving the Trinity): "In astronomy the theory of eccentrics and epicycles is considered as established, because thereby the sensible appearances of the heavenly movements can be explained; not, however, as if this proof were sufficient, forasmuch as some other theory might explain them."

This text is generally interpreted by saying that St. Thomas refuses to decide between the astronomy of Aristotle and Eudoxus of Cnidus (fourth–third century before Christ), where all the celestial spheres are homocentric to the Earth's center, and that of Hipparchus of Nicea (second century before Christ) and Ptolemy for whom the spheres are more or less eccentric. The fact that St. Thomas presents this uncertainty as analogous to the impotence of reason faced with the trinitarian mystery invites us, however, to see there something other than the indifference of a metaphysician to physics, and tends to *epistemologically* establish the idea that, in physics, knowledge is always hypothetical, which, it must be stressed—we have done so on page 47—is in disagreement with Aristotle's quite confident physicism, otherwise seemingly shared by St. Thomas.

But, although St. Thomas is by no means a pre-Copernican, are there reasons to suppose that he would have accepted the Copernican system if he knew of it? Father Joseph de Tondéquec leans in this direction: *Questions de cosmologie et de physique chez Aristote et S. Thomas* (Vrin, 1959, 22). This is also our opinion. There is in fact at least one express declaration of St. Thomas in his commentary on Aristotle's *De Caelo* (L. II, l. 17, n. 451), where he states that one cannot *a priori* discard the possibility of a third hypothesis that is neither Aristotelian nor Ptolemaic: "These hypotheses discovered by Aristotle and Ptolemy are not necessarily true; although the constructing of such hypotheses does help to explain phenomena, it need not be said that they are true; for it might be that astronomical phenomena will be explained by some other means of which men until now have had no idea (*alium modum nondum hominibus comprehensum*)." Which, St. Thomas adds, is not the case with Aristotle who "has recourse to these hypotheses as if to truths (*utitur suppositionibus . . . tamquam veris*)." As we see, St. Thomas does not on this fundamental point hesitate to break with Aristotle, indicating by this his epistemology's basic openness.

Appendix 4

St. Thomas Aquinas: a Theophanic Cosmos? (cf. 47, n1)

One knowledgeable theologian, after reading our chapter, thought it too severe towards Aristotle. "If it is true," he writes, "that the essences are present in things themselves, the latter constitute a set of entelechies hierarchized by the finalizing motion of the Prime Mover, the Thought of Thought, the perfection of which undoubtedly prohibits it from knowing the finite, but which nevertheless assures the order of the world and its intelligibility by finalizing it and by truly forming it into a cosmos. There is then a presentification of transcendence in immanence and the first is signified by the order of the world." Next, recalling the Thomist doctrine of the hierarchy of causes, degrees of universality, and perfection, he concludes: "Supported as it is by a scalar metaphysics, is this cosmology ambiguous?"

These remarks have made us pause to reflect, for we do not wish to be unjust to anyone, insofar as we hold Aristotle to be one of the two greatest philosophers of the West, and St. Thomas Aquinas the greatest theologian of Latin Christianity. However, we do not on the whole think that we should change our analyses. But they should be brought to completion and nuanced on certain points. The following considerations are only a brief outline.

That Aristotle's cosmology is in agreement with the requirements of symbolism by its *structure*, we readily acknowledge, and have even emphasized this expressly (42). Likewise we find with him a hierarchy of forms up to the Pure Form or Pure Act of divine Thought, upon which everything else *depends* for its actuality, a hierarchy which represents "the basis and source of this immense analogy composed of all dependent substances" (Léon Robin, *Aristote*, 92). But it seems equally certain that Aristotle's physics, or even physicalism, in many respects compromises the structural symbol-making power of the world because it *reifies* everything: an individual substance (a tree, a cat, a man, water) is an absolute being existing by and in itself, and having reference to nothing but itself. And since there is only the reality of individual substances, ontologically this universe is peopled with solitary existences. There lies, it seems, the tension if not the contradiction of his philosophy. Lastly, what we will certainly not find in Aristotle is a doctrine of God's epiphany in the world. Now this doctrine is necessary for a metaphysical foundation of symbolism. But is it to be found in St. Thomas?

Without hesitation we reply yes, and this is where our analyses demand clarification. We should stress first the importance of the *theory of celestial bodies in the thought of Saint Thomas Aquinas*, an importance revealed by Theodor Litt's dossier-book bearing the same title [in French] (Béatrice-Nauwelaerts: Paris and Louvain, 1963), which the greatest historians of Thomism (a Gilson for example) have entirely neglected. The concept of a hierarchical universe of concentric spheres constitutes the general and determining framework for all metaphysics and even Thomas' theology, as proven by the multitude of texts collected in all of the Master's works, from first to last. This universe includes four regions, two corporeal and two incorporeal: 1. — corruptible and sublunary bodies composed of the four elements and subject to becoming; 2. — incorruptible and supra-lunary bodies, made of a matter entirely actuated by the form received (and which therefore can no longer become something else), bodies animated by a circular motion (enabling them to be moved without changing place); 3. — the angels, themselves hierarchized and some of whom are put in charge of the movements of the celestial spheres; 4. — the divine region (cf. *Compendium theologiae*, c.74). By affirming the heterogeneity of terrestrial and celestial matters, St. Thomas expressly rejects (*Summa theologiae*, I, q. 66, 2) Plato's thesis in the *Timaeus* (31b–32b). These cosmic (and metacosmic) degrees are also degrees of nobility and beauty, and constitute the various degrees of a universal analogy. As for the raison d'être of this hierarchical ordering and these 'immutable' movements, in the last analysis it is the deification of the elect (*De spiritualibus creaturis*, 6, c); the celestial movements will cease when the number of the elect is complete (*Summa theologiae*, I, q. 66, 3, and numerous other texts).

On the other hand, it is not only the structure of the cosmos that accounts for symbolism; it is also Thomas' ontology of creatures and Creator. Father Faucon de Boylesve, O.P. strives to show this in *Être et Savoir—Etude du fondement de intelligibilité dans la pensée médiévale* (Vrin, 1985). The author affirms that "the medieval interpretation of Aristotelian semantics allows for the expression of an authentically neoplatonic content" (100) and does not hesitate to speak, in connection with St. Thomas, of an "epistemology in compliance with theophanic manifestation" (93), as this declaration testifies: "all knowing beings implicitly know God in every being known... nothing is knowable if not by a similarity to the first Truth [= God]" (*De veritate*, q. 22 a. 2, ad 1).

To this exemplarist epistemology is added an ontology of participation that establishes the possibility for the Divine Being to manifest Itself in things, an ontology that St. Thomas borrows from St. Dionysius the Areopagite. Commenting on the fundamental text of St. Paul in the *Epistle to the Romans* (cap. 1, lect. 6) on the invisible God known by visible things, St. Thomas speaks of the 'book of creation' (*sicut in quodam libro*). In this way the possibility of a theology, a discourse on God, is established, as this very beautiful text from the *Summa Theologiae* (I, q. 13, a. 5) shows: "Whatever is said of God and creatures, is said according to the relation of a creature to God as its principle and cause, wherein all perfections of things pre-exist excellently."

However Father Faucon—who in reality confines himself pretty much to St. Thomas, despite some (inexact) allusions to Meister Eckhart—acknowledges that this discourse "derives from a theological intention *foreign* to the system of Aristotle" (119), which seems to us self-evident. It must therefore be asked in what measure it was possible to express without contradiction a Neoplatonic vision of the cosmos in the language of Aristotelian *physics*. Father Faucon's book does not answer this question. Was St. Thomas aware of this difficulty? Already Father Theodor Litt, in connection with the centuries-long conflict between Aristotle's homocentrism and Ptolemy's geocentrism, concludes that St. Thomas "did not understand the importance" of the dilemma (*La théorie des corps célestes*..., 365). He did not see "at what point the system of Ptolemy shattered the philosophical physics of Aristotle" (364).

The same question is posed in connection with the Neoplatonism of Dionysius, which shattered peripatetic physics no less, a question so much more embroiled since the Middle Ages readily attributed texts of Platonic inspiration to Aristotle. Leaving the task of answering to someone more expert, we will just say that St. Thomas is a more mysterious genius than he appears. Besides, it is clearly not only the Dionysian influence that impels Thomas towards a little-acknowledged Neoplatonism; it is first the doctrine of creation and the complete dependence of the creature, which is hardly in agreement with Aristotle. True, as Father Faucon declares, Aristotelianism seems to confer on created nature a *consistency* "removed from it by Platonism" (146)—which is, in our eyes, nonsense: to the contrary, Plato intended to establish the visible world's consistency, but on its participative relationship with the world of the Ideas; what he rightly denied is the self-consistency, the self-sufficiency of the created. Far from preserving its reality, a reductive Aristotelianism consigns it to the unreality of its pure contingency. *It is not the world that can preserve the symbol, it is the symbol that preserves the world.*

The whole question is reduced to the following point: which has primacy, astronomy or physics? Does *order* determine the cosmic situation of things, or indeed is this order only the consequence of the nature of various cosmic substances? For Plato it is the world inasmuch as world (cosmos) is an order. For Aristotle physics shapes astronomy, and things are autonomous substances before being ordered relationships. (A remark: It is precisely this ordered relationship, viewed in itself, that Plato calls an Idea.) With St. Thomas the two perspectives coexist and are variously combined. When he speaks of being, he is an Aristotelian; when he speaks of order, he is (without knowing it?) a Platonist: "the universe of St. Thomas is comprised of series of causes. Let us immediately add that, in this universe, not only are there series, *but that is all there is*!" (Litt, op. cit., 370). Here substances have almost disappeared.

We see how complex the question is, and how difficult the synthesis realized by Thomas' genius: the elements, at least for what concerns natural philosophy, could not fail to become dissociated after a brief time. We repeat, Aristotle's universe is not completely opposed to the symbolic, but neither does it impose it. To the contrary, by his epistemology, his plan to establish a sure physics (not probable or availing itself of fables, as with Plato), and by his ontology of individual substance, he brought to light and legitimized physicist preoccupations in the field of medieval culture at its height, which could not help but reify the cosmos and make it opaque, causing it to lose its 'metaphysical transparency', or give rise to the emergence of a physics that would then be truly 'scientific'.

PART II

The Subversion of Meaning or the Neutralization of Religious Consciousness

3
Nature and Culture

Linguistic Referent and Symbolic Referent

Here we have then the symbolic sign deprived of its referent, that is deprived of the capacity to be related to a transcendent reality, a capacity that implies, in its own order of things, use of the sacred sign. The Galilean revolution has made this referential capacity ontologically impossible, first by denying that an invisible and transcendent 'world' exists, and next by correlatively denying its possible immanence in sensible forms. This effect of a cosmological revolution on the symbolic sign is inconceivable if one does not accept, like ourselves, that the symbolic referent, although differing from something simply denoted, from an objective referent, is nonetheless inseparable from the symbol, and is even much more intimately conjoined with it than in the case of a linguistic sign. As we have shown in *Histoire et théorie du symbole*, the reduction that Saussure carries out on the linguistic sign, by amputating it from its referent (the sign is a unity with two faces: signifier and signified), raises many difficulties, but it is not impossible. It corresponds to what we call the epistemic closure of the concept, one by which linguistics transforms language into a scientific object; this is naturally in agreement then with the Galilean geometrization of the physically real. After all, neither a straight line nor a plane without thickness exists, and yet Galileo has reduced bodies to these abstract forms. But such an operation is unacceptable for the symbol, because, if it were amputated from its referent, what would subsist would be no longer a symbol at all. There is a symbolic sign only if the reality it signifies is presentified within itself by virtue of a participative analogy.

This presentification—or presence of the referent *in* the symbol—is verified whenever the symbol is composed of only linguistic signs, for as

St. Thomas says, "*things* signified by the words have themselves also a signification,"[1] and not the words themselves. This is the whole difference between the metaphorical and the symbolic. These two categories can obviously be applied to one and the same expression; but the first envisages only the rhetorical process, the other the profound essence of things. To call the lion 'king of the animals' is a metaphor because one thereby 'trans-ports' (the etymological sense of meta-phor) the meaning of the word king outside its natural political context. To say that the lion symbolizes royalty in the animal world is to say that the essence of royalty is presentified by it.

Clearly then the referent must be taken into account, or rather a whole ontology of reference, which brings into play signifier and meaning. Seeing that the symbol signifies by presentification, *both* because the reality signified is of another order than the symbol, *and* because the former yet can be presentified within it. There must be at once existential discontinuity and essential continuity. Without the first, no *sign*; without the second, no *symbolism*. What mechanistic physics has destroyed is the referential ontology of a cosmos open to the spiritual and, vice versa, of a spiritual and a divine that do not refuse to become incarnate in cosmic forms. Not only did it destroy the object to which the symbolic sign is related, but even more it rendered impossible its manner of relating. This is clearly the most important, for this mode of reference specifically characterizes the symbolic sign, and, what is more, the destruction of the object is only the consequence of the impossibility of that referential mode proper to the symbol. And this is an inevitable outcome: the negation of the Transcendent follows on the disappearance of the signs of transcendence, since, precisely, It can only be affirmed in symbolic mode, that is, by presentification. In other words, 'science' does not deny God; it denies the signs of the divine or, even more exactly, it denies that there might be signs of the divine. As already mentioned, the symbolic referent is not *a* denoted object, in the manner of a linguistic referent, but this has to do with a synthetic referential potentiality which alone actualizes and defines, provisionally, an explicit hermeneutic. Provisionally for, being an act, that is to say an event, hermeneutics is always about recapturing, beginning again: each person must understand on his or her own behalf. In a certain manner language is always speaking 'about itself', while the symbol only speaks thanks to the hermeneutic that its word causes to resound and be heard

1. *Summa theologiae*, I, q. 1, a. 10.

(which does not at all mean that this word invents the hermeneutic).[1] By itself the linguistic sign is almost nothing; its naturalness is as if extenuated, everything in it is a signifying function. Conversely, the symbolic sign is everything, or nearly everything. In the naturalness and very substance of the signifier there resides a signifying function that speaks only to the one who hears and understands it.

The Traditional Notion of Culture

The conclusion is clear: the symbol is the pre-eminent locus where nature and culture meet, join together, and intersect. It is the mediation in which nature becomes culture and culture nature. Without repeating the analyses we have made elsewhere,[2] we will recall however the essential thesis: man is by nature a cultural being; his nature is only realized in culture, that is through the ordered complex of symbolic representations of the world, man, and the divine conveyed by tradition. As a function this complex enables man to situate himself with respect to the three orders of reality (macrocosm, microcosm and metacosm), understand them, and behave accordingly: existing (living), knowing, and acting. That culture exercises this function is doubly required by man's very nature. First, because he is endowed with conscious thought, man must *learn* everything that he is, knows, and does: to be human is to be in apprenticeship. And this implies that what he lives, knows, and accomplishes is always, more or less, according to a semantic, intelligible, essentialist mode, since to think about his life, knowledge, and action is always to perceive a primordial and universal significance in it, and therefore a value to be realized, hence the moral character of his entire existence. On the other hand, the representations assisting culture in actualizing and determining the virtualities of human nature ought to have themselves the obviousness, immediacy, unity, and eternity of a natural fact. They should not seem to be contingent products of history, nor abstract constructions, as are the teachings of the human sciences. Men learn how to live and act (to be a man or woman, to love, suffer, die, rejoice, celebrate, etc.) only through teachings that have themselves the natural, all-encompassing, and 'just like that' character

1. A symbol without hermeneutics is like a musical note played by a bow on a violin deprived of its sounding box. The note is indeed emitted, it exists, but it is not heard. Or again it might be compared: to a sound wave that would be deprived of the air that carries and transmits it, which in this case represents hermeneutics; to an unlighted painting; etc. Hermeneutics is a resonance, an atmosphere, a light, a medium for manifestation and propagation: it is the function of the *Pneuma* with respect to the *Logos*.

2. *Amour et vérité*, 42 and 53–69.

of life, and are, as such, the only ones that man can assimilate to his own nature. And this is why they assume the form of symbols. They give form to his being, then, in the manner of a real instinct, culturally acquired no doubt, but for all that lived like an innate aptitude. Man has no instinct, they say? Yes, he has cultural instincts. But every culture is not suited to form these instincts in man. For this it must convey its teachings through the naturalness of symbols.

Such is, briefly recalled, the concept of culture that seems indispensable if we have any concern for real human happiness. There is nothing very original here, and yet it is surprising to see what little space it occupies in philosophic doctrines for the last two centuries.[1] Not that the culture-nature debate is to be ignored. Quite the contrary, we even think this is what gives organization and structure to the philosophic field of the modern and contemporary West; this is the major problem of European thought since the middle of the eighteenth century, because the Galilean revolution, by overthrowing the world of Christian culture, has posed it in a quite inevitable fashion: it has definitively separated culture and nature. And humanity poses itself problems it no longer knows how to resolve. But philosophy and the social sciences generally deal with this question only in order to oppose culture to nature, to separate and isolate them from each other, or to see in this duality, following ethnological structuralism, the very imprint by which man signifies, posits, and defines himself, because he is himself the difference between nature and culture. As we see, the stakes of a philosophy of the symbol if true is not

1. Pascal is one of the rare Classical thinkers to clearly situate the politico-social importance of symbolic relationships. He understood that the scientific critique which denounced the 'lie' of cosmic and religious forms would likewise bring its destructive—one might even say revolutionary—power to bear on social forms. The signs by which the social order and hierarchical relationships are expressed are no more truthful than the sensible appearances of the universe: the habit does not make the monk, a doctoral hat does not produce medical knowledge, and four lackeys behind the carriage of important people are not enough to prove their real nobility and their right to be respected. This is why a class of tough-minded libertines has come into being, called the 'half-educated' by Pascal, who with gleeful ferocity strive to 'destroy the idols' and 'wise up' the naïve, who believe that appearances are simply truthful. But these half-educated people have not understood that we should not be bound only by 'cords of necessity' based on strength or even reason; 'cords of the imagination' (B. 304, L. 828) are also needed. The truly learned know that the half-educated are not wrong, but they speak like naïve people, harboring all the while their 'private thoughts', because they know the 'reasons for effects', and have understood that no social life is possible without order and hierarchy, and no hierarchy subsists if not signified symbolically, men not being completely reasonable. In short, political and social life require what Ruyer has so well termed 'psychic nutrition'. Such is in our opinion the basic meaning of Pascal's thinking.

paltry, for, as we have tried to show, the symbol is the site of conjunction and exchange for nature and culture. All the same, to view the nature-culture relationship as *the* problem of modernity is not to deny the existence of other problems; this is only to designate it as the problematic horizon of all *modern* philosophy. This is also to admit that prior philosophy more or less ignores this problematic, and therefore that culture is not considered as an independent and separate reality unto itself. This in no way signifies that human thought is ignorant of the distinction. But culture is, then, only regarded, or preferably regarded, as the extension and development of nature. This concept—the only normal one after all—appears under various guises: that of Thoth-Hermes for the Egyptian and Greek traditions, Enoch in the Bible, Ganesha among the Hindus, or Lugh among the Gauls. "He it was," says Plato, "that invented number and calculation, geometry and astronomy, not to speak of draughts and dice, and above all writing."[1] Secretary to Osiris, master of knowledge, revealer of language, music, and all kinds of art, Thoth's magic dominates nature since he can give life to the dispersed members of the husband of Isis. And this operation defines the very essence of the symbol and culture: manifestation or creation being conceived of as the sacrifice of the Logos who disperses his cosmic body, and culture, cult, or symbol as that which restores the lost unity, gathers what is scattered, restores his body to God-the-Creator.[2] All these figures, and the ones corresponding to them in the various religious traditions, clearly express the distinction of culture and nature and the function of the first with respect to the second. But we must refrain from seeing here only a mythological (and therefore unconscious) translation of this dialectic understood in the modern sense. The modern concept of the nature-culture duality is not the truth of these symbolic figures.

1. *Phaedrus*, 274 c–d; Hackforth trans., *Plato, The Collected Dialogues* (op. cit.), 520.

2. This is also the theme of the "sacrifice of Purusha" (*Rig Veda*, X, 90; *Sacred Writings, Hinduism: The Rig Veda*, trans. R.T.H. Griffith [New York: Book-of-the-Month Club, 1992], 602–603). On the creating sacrifice of Purusha, the divine man, or of Prajapati, the 'Progenitor', read A.K. Coomaraswamy, "*Atmayajna*: Self-Sacrifice," in *The Door in the Sky: Coomaraswamy on Myth and Meaning* (Princeton, NJ: Princeton University Press, 1997), especially 73–74. Ganesha, the elephant-headed god, son of Shiva and Parvati, is the "Lord (*isha*) of Categories (*gana*)," the patron of letters and schools, the scribe that transcribes the *Mahabharata*. He symbolizes the subjacent and essential identity of the Principle and manifestation realized by the knowledge of cosmic numbers and the sacred arts (A. Daniélou, *Hindu Polytheism*, 291). His enormous belly signifies that he is a "fullness of knowledge," according to the expression by which Dionysius the Areopagite translates *Cherubim* (*The Celestial Hierarchy*, chap. 7, §1). Enoch (who appears in Genesis 5:21), of whom the Ecclesiast (44:16, Hebrew text) states that he was

They have their own truth which is totally opposed to the modern vision of this problematic; they affirm, in particular, that 'culture' is at origin divine—either God himself has revealed and taught it, or an angel (or a 'god') has gone in quest of it from heaven—which seems pure and simple 'fantasy', while it should be rationally evident however that only a 'divine' genesis can, for example, account for the origin of language.[1] Also affirmed, which is no less foreign to modern thinking, is that the function of culture, essentially sacred by nature, has for an end the restoring of the multiple exteriority of the relative to the unifying interiority of the Principle.[2] With these two theses on the origin and end of culture, traditional civilizations sufficiently show that under this term they saw something altogether different, and has so to say nothing in common with current nature-culture ideas.

"an example of knowledge to all generations," is presented in the *Book of Enoch* (translated from the Ethiopian text by E. Isaac, in *The Old Testament Pseudoepigrapha*, vol. 1 [Garden City, NY: Doubleday & Company, 1983], 13–89) as mediator between God and man of the sciences and arts, having been allowed to contemplate the divine secrets (chap. 71, 49–50). Rabbinic Judaism interprets his name *Hanok* as 'learned' and identifies him with the mysterious *Metatron*. He should return with Elijah at the end of time. The *Koran* (19, 57–58) substitutes the name *Idris* (= savant) for him. Finally, let us recall that the 'hermetic' or hermeneutic function is expressly attributed to St. Paul by the New Testament (Acts 14:12): at the sight of a miracle, the inhabitants of Lystra actually identify him with Hermes insofar as "master of speech," which would have no meaning other than as an anecdote if St. Paul was not the pre-eminent hermeneut of revelation, the one who has given its definitive and guiding interpretation.

1. What is true of language is also true for many other arts and bodies of knowledge, especially agriculture. Where do the cultivated plants, wheat, barley, corn and carrots come from? How did wheat first become bread? "The first homeland of the vegetables most useful to man, and which have been his companions since remotest times, is a secret as impenetrable as the abode of the domestic animals" (Humboldt, *Essay on the Geography of Plants*, cited by Jean Servier, *L'homme et l'invisible* [Paris: Laffont, 1965], 214). This is why all traditions assign to these edible plants a celestial origin.

2. This does not mean an annihilation of creation as such, but to the contrary its perfect realization, that is the access of the created to a perfection of being compatible with its relative nature. This is then the "new heaven" and the "new earth" heralded by the Apocalypse (21:1), a perfection that we interpret as designating the 'divine' state of creation about which the first verse of Scripture speaks: "In the Principle Elohim created heaven and earth." We are following here in particular the interpretation of John the Scot, for whom this verse "signifies nothing other than the creation of primordial causes, prior to their effects. The Father created these causes in his only Son who is called 'principle'; the term 'heaven' designates the primordial causes of the intelligible and celestial essences, and the term 'earth' designates the causes of the sensible beings of our corporeal universe" (René Roques, *Libres sentiers vers l'érigénisme* [Roma: Edizioni de l'Ateneo, 1975], 137).

Pure Culture and Pure Nature

Clearly, in this perspective we see that culture is not posited for itself, but always with respect to its origin and the end that it ought to fulfill.[1] We are therefore led to ask under what conditions such a position is possible, a question that puts us on the track of an important remark. Actually, to posit culture for itself implies first that it is conceived of in itself, in a separate state, and as entering into a relationship with nature only 'from without'. This means that there can be a 'pure culture' only if a 'pure nature' is correlatively posited. And what then is this 'pure nature'? Is it, as is occasionally said, only the result of an abstractive operation, something that remains when everything having to do with culture is suppressed? Here we have, in a way, the conception imposed by the Rousseauist method: naked man, such as he is produced by nature, appears at the end of a reductive process, a veritable mental experiment by which he is stripped of everything added to him by the arts and sciences. After this, the concept of pure nature (or the natural state) thus obtained will serve as a criterion to discriminate between what is natural and what is not in the historical and concrete man. Nature here is non-culture, and its concept seems then to derive from this. But, in reality, the abstractive process is itself guided and determined by the idea that culture is non-nature. If, to the contrary, Rousseau had envisioned culture in the traditional manner as an orderly development of nature, which is itself therefore seen as a virtuality, as a reality open to becoming, calling for its own completion by itself and, hence, implying a *possible culture* within itself, just as culture is in some way only the making explicit of nature, if he had viewed matters in this way, he would have understood that it is impossible to actually separate what is natural from what is cultural in a real person except from a 'log-

1. It might be objected that we are posing here a false problem, the word *culture* having acquired its current sense only lately. But this is not altogether exact. "*Colere* and *cultura* are rightly understood as care for the fields, orchard, and livestock. These are words of the rustic tongue, but the meaning has been extended through metaphor. Some have thus come to speak of *cultura animi* or *ingenii*; one text from Cicero underscores the image with all necessary precision: 'Every field that we cultivate does not bear fruit . . . just as every mind that we cultivate does not bear fruit' (*Tusculanes*, 2, 5)" (H.I. Marrou, *Saint Augustine et la fin de la culture antique*, 550). Although for a long time the term was only applied to the cultivating of the mind by literature and philosophy, we understand that it was in some manner destined to designate, in the unrestricted sense, the sum of means by which a society educates and forms the whole person. And, as this sum is proper to each society, a comparison of cultures becomes possible in which the concept of culture acquires its present meaning.

ical' point of view.[1] In other words, and to employ the language of Scholastic philosophy, the distinction between nature and culture is neither real, in the sense that they would be two separately existing realities, nor simply ideal or by reason (as between a ring and a circle, for example). But this is a formal or even virtual distinction, because if the form, essence, or quiddity of nature and that of culture are indeed distinct, they do not exist however in a separated state: there is neither pure nature, nor pure culture, but both constitute virtualities of the real man, corresponding analogically to animality and rationality. But, even more profoundly and in a general way, it must be said that, in non-Platonic philosophy, the identification (and therefore the identity) of nature, as well as culture and every other reality, can be assured only by the 'horizontal' exclusion of everything that is not itself and is yet found on the same plane of existence, and therefore by oppositional relationships. For a Platonist the identity of anything is its transcendent essence, an essence independent and of itself indifferent to a particular state of existence. This is the 'vertical' relationship of a being with its essence, a relationship traversing all the degrees of existence, whatever this being's modalities, and assuring and maintaining its identity. The encounters, combinations of all kinds, changes and transformations which come to it from its 'exterior' relationships are not to be feared. To the contrary—and this is already the case with Aristotle, but even more so with Galileo and Descartes—when this relationship with the essence disappears the identity of anything can no longer be imagined except 'from without',

1. In a general way Rousseau's political philosophy crosses constantly and unduly from the order of logical concepts to that of concrete realities. Insofar as Rousseau did not study the relationship of the two orders in itself, we will not be able to truly judge about this philosophy. Now, with him, concepts are obtained at the end of an altogether novel process of abstraction. Aristotelian abstraction starts with things such as they are given in concrete experience. It is then always standardized by this concrete given, and this is what it should make intelligible, or rather concepts should express its intelligibility. To the contrary, Rousseau begins "by setting all facts aside" (*Discourse on the Origin and Foundations of Inequality among Men*, in *The Collected Works of Rousseau*, vol. 3 [Dartmouth, NH: Dartmouth College, 1992], 19). Directly and axiomatically, reason builds up the concepts that it needs in such a way that these concepts designate something "which no longer exists, which perhaps never existed, which probably never will exist, and about which it is nevertheless necessary to have precise Notions in order to judge our present state correctly" (ibid., 13). Rousseau is generally praised for having dared this rationalist coup d'état. We see in this the original sin of political philosophy. The result is that 'hypothetical history' (ibid., 16) becomes the norm for the real: "logical antecedence compulsorily entails historical antecedence" (J. Starobinski, *Rousseau, Œuvres complètes*, tome III [Paris: Gallimard, 1964], notes, 1285). This is to say that rationality, intoxicated with itself, has lost the sense of the real (which is deemed confused, deformed, and indiscernible), and knows nothing but its own rights.

that is by distinguishing it horizontally from everything that is not itself, by separating it from, isolating it from, and opposing it to all the rest. *The state of nature becomes then the war of all against all*, in a sense that Hobbes had not foreseen, but which only lays bare the speculative underpinnings of his political thought. In other words, Hobbes' theory expresses, on the particular plane of politics, the general speculative situation of his time. A being, a thing is maintained in its own nature only on condition of being contraposed to every being and every thing: a situation that we discover again in pure philosophy, with the Cartesian subject contraposed in its solitude to the totality of the cosmic expanse. However, we think that the first to have realized this consequence of a complete rejection of the Platonism of essences was Galileo; and it is the speculative fruitfulness of this model that has brought in its train all other intellectual developments in every domain.

Rousseau's nature is basically then the world of Galilean physics,[1] a thesis that Rousseau shares moreover with the whole eighteenth century: "[I]n the order of the world," he declares, "I find a never-failing system,"[2] a mechanical and yet finished order. And so, faced with this 'opened watch' that is the world, "I admire the workman in the details of

1. Obviously the word *nature* has, for Rousseau, several meanings: nature, in the psychological sense, is the spontaneity of the human being, the nature Rousseau discovered in himself, in his own heart; it is also the authentic, the primitive willed by God, opposed by the artificial, the work of men; there is lastly nature as model, the universe regulated by strict laws. We think that this third and last sense is the most important in Rousseau's thinking, for, either for what belongs to the hypothetical state of nature at origin or for the legitimate civil state described in his *Social Contract*, nature is the idea of a norm, an order that men should strive to imitate: "If the laws of nations could have the inflexibility of the laws of nature that no human force could overcome, then the dependence of men would become once again a dependence on things. Thus one would reunite in the republic all the advantages of the natural state with those of the civil state; one could bring together the freedom that keeps man exempt from vice with the morality that raises him to virtue" (*Émile*, Book 2, §237; Foxley/Roosevelt trans.). What remains is that Rousseau never dealt with nature as such, and that this idea interested him more for the critical use he could make of it than in itself by its philosophical content. Paradoxically, there is no Rousseauist philosophy of nature.

2. *Émile*, Book 4, §1011. This is basically the very definition given by Descartes in the Sixth Meditation (§11): "...by 'nature regarded generally' I now understand nothing other than God himself—or the coordination of created things established by God" (*Meditations on First Philosophy*, 199–201). And 150 years later Kant likewise declares: "Nature is the existence (*Dasein*) of things, insofar as that existence is determined according to universal laws" (*Prolegomena to Any Future Metaphysics*, trans. G. Hatfield [Cambridge UK: Cambridge University Press, 1997], 46); cf. equally the texts cited by Jean Ehrard, *L'idée de nature en France à l'aube des lumières* (Paris: Flammarion, 1970), 41, 47, 87, etc.

his work, and I am quite certain that all these wheels only work together in this fashion for some common end which I cannot perceive."[1] He even presses mechanism so far as to adopt the Cartesian thesis on animal-machines: "In every animal I see only an ingenious machine to which nature has given senses in order to revitalize itself.... I perceive precisely the same things in the human machine, with the difference that Nature alone does everything in the operations of a Beast, whereas man contributes to his operations by being a free agent."[2] But in all this Rousseau is just repeating, more or less, the most common theses of his time. In reality nature is only scenery, and nothing else is asked of it than what is asked of any scenery, which is to isolate the stage from the rest of the world, trace its boundaries, and shelter it from 'outside' influence. And what, then, is this influence outside nature that might intervene in its order to change or transform it if not 'super-nature'? The basic function of a globally mechanical nature is quite clear: it is to *shelter man from the supernatural.* Rousseau plainly declares: "Supernatural! What do you mean by the word? I do not understand it."[3]

The Pure Subject Confronting Pure Nature and Pure Culture

Compared to this radically posited nature that encloses existence within the rigorous circle of its determinism, culture appears as necessarily unnatural, or even as anti-nature. This is only when it is posited in itself and seen as pure culture. However, we still do not possess all the details of our problematic. Clearly the nature-culture polarity delimits the speculative field of European philosophy, but this polarity implies a third term precisely with respect to which this formative opposition makes its appearance: the cognitive subject. This third term is doubly implied, both by the position of nature and that of culture. By the position of nature: this is what Cartesian philosophy shows exemplarily. The ontological closure of cosmic reality, which is as a whole enclosed within extended substance (as if extension could be a substance!), which is identified with exteriority as such, gives rise to the extremely contradictory position of a thinking subject reduced to pure interiority faced with pure exteriority. But this position is not only contradictory (the cognitive subject is non-exteriority), it is also required more or less consciously for Descartes by the strict logic of mechanistic theory. For it is

1. *Émile*, Book 4, §988.
2. *Discourse on the Origin and Foundations of Inequality among Men*, 25.
3. *Émile*, Book 4, §1069.

hard to posit an absolute object, a purely physical *en-soi* [in-itself], purely and exclusively 'there'. To do this the thinking that posits such an *en-soi*, and to which everything else is necessarily relative, must think of itself with an absolute regard, not relative to what it posits, and identify itself with a universal reason, radically 'exterior' to nature. In this way Descartes thinks he can save the 'spirituality of the soul', and this is in some manner his alibi, his religious guarantee. But at what price? Everything must be eliminated from the cognitive subject that gives it roots in the world, everything within it that results from its situation of being worldly, historically and geographically determined. Now, from where do these 'cosmic' traces come that dwell in our spirit and tie it to the world in which it lives? Basically from education, that is from culture, since the one is inseparable from the other. As H. Gouhier says, what Descartes wants to kill in himself is the child, that is, the child 'informed' by the received culture: "there is indeed a scandal with the human condition: in Plato's eyes it is that humanity is the exile of the soul; in Descartes' eyes it is that man begins by being a child."[1] This involves expelling from the adult soul everything that remains of received education; this involves a shock treatment: "this is not so much a child-birth (in the manner of psychoanalysis) as an infanticide."[2] This is why Descartes proposes in the first part of *Discourse on Method* the first model of a cultural revolution.

But, to tell the truth, this revolution is individual. It occurs in the interior of one and the same person,[3] and this is why putting the thinking subject at a distance with respect to his own cultural determinations does not lead Descartes to posit culture in itself, the latter being always viewed as a state of the subject, a bias or a prejudice. Wherever this revolution—for which Descartes is only a spokesman—is spread and propagated in European society, it will be necessarily expressed by the

1. *La pensée métaphysique de Descartes* (Paris: Vrin, 1962), 52.

2. Ibid., 58. Cardinal de Bérulle, who strongly encouraged the Cartesian enterprise (Jacques Chevalier, *Histoire de la pensée*, tome III [Paris: Flammarion, 1961], 110), wrote: "the state of childhood is, in nature, the state most opposed to true wisdom . . . , the vilest and most abject state of human nature" (*Opuscules de piété*, éd. Rotureau [Paris: Aubier, 1944], 251, 254).

3. It is even here, in our opinion, that the key to *all* the difficulties of Cartesianism reside. But this is not the place to explain this thesis. Let us only say that, breaking with tradition, Cartesian philosophy is necessarily an individual, solitary adventure, an experience really lived, the first philosophy to really say 'I'. But at the same time and no less necessarily—otherwise it would be devoid of legitimacy—in this adventure it is universal reason that speaks, the pure *logos*: logic, the ordering of reason, becomes a psychological adventure. The *ego cogito* is inseparably the pure understanding of Guéroult and the existential man of Alquié.

opposition of universal man, the common man, or even the one that Rousseau calls 'abstract man',[1] to the diversity of particular cultures. Philosophical with Descartes, this theme becomes sociological and ethnographic from the moment that men are thought about as a whole along with the diversity of civilizations and societies. Universal reason makes way for universal man. "Having compared men of every class and every nation," states Rousseau, "which I have been able to observe in the course of a life spent in this pursuit, I have discarded as artificial what belonged to one nation and not to another, to one rank and not to another; and I have regarded as proper to mankind what was common to all, at any age, in any station, and in any nation whatsoever."[2] Surely there is some difference between the thinking Cartesian subject and the abstract man of Rousseau, since this last seems devoid of reason. However, in reality they play the same role: the first is what remains when methodical doubt has banished all existence and every essence, the second is what remains when a methodological hypothesis has banished from concrete man everything added by the diversity of time and place. On the other hand, what the natural man is deprived of, according to Rousseau, is not so much reason, which exists in a virtual state,[3] but just the use he makes of it. As soon as human relationships require it, it appears, and then that which speaks within it is clearly the voice of nature such as God has ordained: "to tell me to submit my reason is to insult the giver of reason."[4]

It should not be otherwise. To posit culture in itself, by elaborating a closed concept that will surface later in the formation of the human sciences, is only possible if disconnected from the one that is its natural support and through whom it exists: man. For culture does not exist all alone. To the very extent that it 'informs', a material in which this information becomes visible and manifest is needed: man himself. Hence, to abstract culture and posit it in itself is to isolate it from its human bearer and, as a result, to abstract the bearer in turn.

Conclusion: Religious Consciousness as Illusion (Kant)

We can conclude now and more clearly untangle the situation that occurs with sacred symbolism at the moment when European scientific rationalism elaborates and sets in place the basic elements of its own

1. "We must therefore look at the general rather than the particular, and consider our pupil as man in the abstract" (*Émile*, Book 1, §40).

2. Ibid., Book 4, §901.

3. Ibid., Book 2, §254.

4. Ibid., Book 4, §1067.

epistemic system. This situation is the result of the transforming of an 'ideological' rejection of the sacred into its 'structural' imprisonment. By this we mean that the rejection of the sacred with which the seventeenth century proceeded was at first only a reaction of a rather emotional nature, a reaction provoked in the European soul by the rise of the scientific revolution. When science left the very restricted realm of the scholars who exchanged letters throughout Europe, from Great Britain to Poland, to burst upon the 'public at large'—and this is what the Galilean crisis and the resounding condemnation of its hero realized exemplarily—it necessarily ceased being the object of a purely speculative consideration, and could not help but provoke feelings and passions in the mass of men, men more responsive to the religious or moral effects of the scientific revolution than to its physical or mathematical consequences. These emotional reactions, these infatuations or hatreds, contributed to developing what might be called an 'ideology', that is, the more-or-less well-coordinated body of imperative themes by which a social, political, or cultural movement expresses its 'ideal'. As Jules Monnerot has shown, ideology is a substitute for myth.[1] But, like myth, ideology also ends up organizing knowledge, structuring the epistemic field. It then ceases being only ideology, or rather being felt to be such, to become the "unsurpassable intellectual horizon" of an age, as Sartre says that Marxism is for the twentieth century. For reason thinks about and sets in order only what is given it to think about and set in order. Only, once this operation is accomplished, the fluctuating themes of an ideology acquire, beneath their rational disguise, the solidity of primary data, meaning that which is connected to a problem's data. And yet these are criticizable facts if one notices that, in the problem-data dialectic, the data receives a kind of justification and guarantee by the very fact of its being introduced into the problematic: attention is diverted from the terms to focus exclusively on the problematic relationship that they support; that the problem is, perhaps, badly posed is lost sight of.[2]

It seems that this is indeed what happened with the advent of scientific rationalism and the appearance of 'Enlightenment' ideology. The shaping and rational structuring of this ideology was basically the work of Kant. He is the one who systematically constructed the triangular sit-

1. Jules Monnerot, *Sociology and Psychology of Communism*, trans. J. Degras and R. Rees (Boston: The Beacon Press, 1953), 146 ff.

2. Many 'logical trap' games are set up around this principle: Should the name of the United States capital be pronounced 'nou york' or 'niou york'? A ladder fixed to the hull of a ship is submerged to a length of a half meter; the tide rises one meter. What length will be submerged? etc.

uation of the epistemic field polarized by the mutual exclusion of cognitive subject, nature, and culture, and who drew all of its consequences from it. Nature is identified with existence (*Dasein*) insofar as determined by mechanistic laws, that is, with the being (entirely)-there of physics. Its complete heterogeneity with respect to the knowing subject, to whom however this being is related, means that, first, this being (entirely)-there totally eludes, in its noumenal reality, the cognitive subject, and therefore that, in a certain manner, it is 'elsewhere', although nevertheless belonging to the order of nature[1]—the 'other side' of the world is in the world; second, it means that all that the subject may know of the object is nothing but the form of its own cognitive structures. These *a priori* cognitive structures define *a priori* man, the universal man, and are contraposed to the pathological, the empirical, and the cultural, in short to everything *a posteriori* of which the practical reason must be purified in its subjective patterns as well as its objective representations: the symbol stems from a pathology of the imagination, as do feelings from a pathology of the will.

Thus the Kantian critique of religious symbolism leaves consciousness confronted with its own representations, as it leaves reason confronted with the transcendental illusion.[2] And so we would like to get to the bottom of this idea, because it is this idea that is laden with all the later forms assumed by the critique of the symbol. Having definitively lost all relationship to a referent for which it would be a messenger, the symbol no longer has, in the proper sense, any other place to exist than in the human mind. Just as, in the philosophic theater that is the Kantian system, the struggle of ideas transpires only on reason's interior stage, in thrall to its own representations and caught in the trap of their pseudo-objectivity, so religious reason is obliged to recognize that it is itself the source and 'stage' producer of all the imageries, so many examples of which are presented by the historical religions. "The one true religion," Kant declares, "comprises nothing but laws, that is, those practical principles of whose unconditioned necessity we can becomes aware, and which we therefore recognize as revealed through pure rea-

1. This is the previously mentioned 'faked' world of Descartes, the other-world of Platonism inexplicably imbedded within the 'density' of this one, the reverse side—forever invisible—of the stage scenery; "whoever would see Nature as she is," Fontenelle asserts, "would see only the backstage at the Opera" (cited by J. Ehrard, op. cit., 47).

2. As we know, the transcendental illusion designates for Kant the appearance of objectivity that the ideas of God, the self, and the world assume for the reason that thinks about them, whereas they are only the objectification of the rational subject's needs for unification; cf. Appendix 5 at the end of this chapter: "Brief Remarks on Symbolism in Kant."

son (not empirically). Only for the sake of the church . . . can there be statutes, i.e., ordinances held to be divine, which are arbitrary and contingent as viewed by our pure moral judgment. To deem this statutory faith . . . as essential to the service of God generally, and to make it the highest condition of the divine approval of man, is religious illusion [*Religionswahn*]"; that is, Kant explains in a note, "illusion is the deception [*Täuschung*] of regarding the mere representation of a thing as equivalent to the thing itself."[1] Note that this definition of religious illusion (or insanity) corresponds exactly to that of the sacred symbol, which is quite really equivalent, by participative presentification, to the 'thing' symbolized. Now, just as Kant explains, with this insanity there is (which rightly involves the *ritual behavior* of religious man) no other basis than a 'subjective' one.[2] This was a consequence of the critique of the sacred symbol still not being posed so clearly. But surely the critique was there from the outset, seeing that the symbol had been amputated from its ontological reference. As we have already indicated many times, the great 'shutting away' was not at first the one for insanity, which is, in the Classical age, structurally framed as anti-reason; above all it was a 'shutting away' of reason itself, which had been 'scientifically' isolated from its religious soul.[3] And perhaps—but we have no statistic in this respect—the seventeenth century witnessed an increase in the number of the insane, itself resulting from the progressive disappearance of the medieval mythocosm and religion's cultural universe. The frantic and permanent exclusion of insanity, and the never-satisfied desire for a totally pure reason, are themselves only a kind of desperate wish to conjure away the threat of a 'sacred madness' ever reborn at the very heart of the human spirit. However, at the first moment of the sacred's exclusion, the very act by which it is held at a distance prevents any thorough inquiry into its true 'place' and profound origin. Wholly absorbed by the prodigious mutation that it was in the process of knowing, the mind did

1. *Religion Within the Limits of Reason Alone*, Book 4, Part 2, trans. T. Greene and H.H. Hudson (LaSalle, IL: Open Court Publishing, 1960), 156. The translation of *Religionswahn* as 'religious insanity' [in the French edition] is a little strong, as Bruch underscores (*La philosophie religieuse de Kant* [Paris: Aubier, 1968], 191), but 'illusion' is a little weak. In any case, this is clearly a question of religion as neurosis.

2. *Religion Within the Limits of Reason Alone*, Book 4, Part 2, §1, 156–158. This subjective ground is the illusion, deadly for reason, that we can act upon God through sham activities that exempt us from the only required action: good conduct. Thus the hermit, the fakir and the monk are equally insane and chimerical.

3. It was in 1656, as Foucault recalls (*Madness and Civilization* [New York: Random House Books, 1965], 39) that the founding of the first 'general hospital' for the insane was ordered.

not perceive that the impurities of which it was ridding itself, all the symbolism that it expelled, are in reality a part of itself, or, at the very least, were within itself. And if the truth of man and even his reality are his totally pure reason, then *how was such an insanity possible*? That is the most redoubtable question when attempting to get to the bottom of this.

As a matter of fact, Kant did not reply to this question in any very explicit way. Without doubt he was not content, like Spinoza, to conjure up a superstition-producing fear, which truly falls somewhat short. He makes of it an illusion of the practical reason, and here we have a decisive step; for the cause of 'religious folly' is then the spontaneity of a subjectivity that wants to gain divine favor by acting on surrogates, whereas fear is a reaction (a passion) to an external stimulus. However, the question of the exclusion of sacred culture would not assume any major importance for him, quite simply because in great part he ignored it. His knowledge in this respect, compared to Hegel's or Schelling's, seems extremely scant. Kant's religious philosophy is often mentioned, but the religion with which it deals, its real object, is nothing but Christianity, and a Christianity reduced to the idea that an eighteenth century pietist philosopher has of it, that is, amputated of fifteen hundred years of patristics, theology and sacred art. Nevertheless, the whole movement of Kantian philosophy (speculative as well as practical) inclines it to take signs of the Transcendent—which are symbols first and foremost—as illusions of consciousness, the prototype of which is the transcendental illusion. We will come back to this in-all-respects basic notion, but we can already stress how exactly it corresponds to the present situation of the critique of symbolism. Deprived of its referent, the symbol seems like a meaning (without object), that is, a pure meaning. Now a pure meaning has no other reality (unless one is a Platonist) than that of an act of human consciousness. Focus on the signifier is still almost non-existent. It will only make its appearance when the critique will be striving to 'reduce' symbolic meaning in its turn. For the moment the signifier remains in some manner hidden by the meaning. Symbolism is basically seen as an attitude of consciousness, a human behavior, and the concrete positivity of sacred signifiers is as if blotted out and submerged in the general notion of the 'religious'. Here we have then the result of the whole evolution and the terms in which the problem of the symbol will be posed henceforth. The immense realm of traditional culture is completely absorbed by the sphere of religious consciousness. But, contradictorily, this religious dimension of human consciousness is posited as a foreign body in this very consciousness, an aberration, a disorder, a madness. And since symbolism is no longer anything but an affair of meaning, but with a meaning that seems nonsensical, then it is clearly necessary to account for this foreign semantic body by unveiling its 'true' meaning. Illusion is never pure: it is the delusive sign of hidden truth. Symbols are no longer signs of the Transcendent, they are symptoms of a drama immanent to consciousness itself. It was Hegel, we think, who was first to explicitly propose making of philosophy a hermeneutic of religious consciousness, as proven by *Phenomenology of Mind*, and who strove to resolve the 'triangular' contradiction of cognitive subject, nature and culture in which Kant had left philosophy, or rather, by which Kant had wanted to think about philosophy. However, we should point out, to the very extent that, according to us, Hegelianism subverted the true meaning of the sacred symbol, the metaphysics of the symbol will have to revert, this side of Hegel, to the Kantian position of the problem. Surely this position is untenable, but at least it has the merit of constraining us to seek something else, whereas the Hegelian pseudo-resolution has totally falsified the very nature of speculative activity. Only the 'Structuralist reaction' has 'delivered' us from the fascination for the Hegelian mode of thinking by affirming the formative primacy of the signifier over consciousness. But this deliverance is another shutting-away that will bring us to the conclusion of the symbol's critique.

Appendix 5

Brief Remarks on Symbolism in Kant (cf. 158, n2)

In *Histoire et théorie du symbole* (71), we have presented the Kantian conception of the symbol in a few lines. This conception is expressed in the *Critique of the Power of Judgment*. Credit is generally given to Kant for having given back to the word 'symbol' the significance of a '*sensible* representation' withdrawn from it by Leibniz, who reserved this term for logical or mathematical symbols only: "The use of the word *symbolic*...," writes Kant, "in contrast to the intuitive kind of representation has been accepted by recent logicians, but this is a distorted and incorrect use of the word."[1] In fact it does not seem that the Leibnizian terminology prevailed everywhere (nor does Leibniz always adhere to it).[2] "In general usage, one can say that, in the second half of the eighteenth century, the terms *Symbol*, *Sinnbild* (literally: sense-image), and Hieroglyph are only

1. Cf. AK V, § 59; trans. P. Guyer and E. Matthews (Cambridge, UK: Cambridge University Press, 2000), 226. 'Intuitive' signifies perceived by the senses.

2. Cf. *Monadology*, § 61.

separated by nuances."[1] *Symbol* (and *Sinnbild*, which has the same meaning) is often employed then to designate the figurative components of an allegory (for example the hour-glass and scythe in allegories of Time or Death). It also happens, with Herder especially,[2] that the term 'allegory' is reserved to designate a *plastic* representation as a whole, and that *Sinnbild* is itself applied to figures of *discourse*.

However that may be, Kant's doctrine aims chiefly at describing the thought process utilized in the symbol understood as the sensible (that is to say: figured) presentation of an idea not directly representable in itself. The idea of a circle can be directly presented to the sensible intuition by drawing a ring, but the idea of a despotic state could not be directly represented by a figure. One can only offer an indirect figuration of it, a 'hand-mill' for example, by making use of the analogy between the *schema* of a despotic state's functioning (where a single person moves all the wheels of state) and that of the functioning of a hand-mill (all the wheels of which obey the movement of a single hand). The hand-mill is not a sensible representation of a despotic state (which is not directly representable), only its *symbol*.

This doctrine makes use of the notion of schematicism, a difficult and variously interpreted notion.[3] It presents a certain interest, but we would not think of enlarging on it here. The essential and, we think, the right idea is that there is a 'hidden art' of the imaginative faculty that offers our thinking the means of linking what is intelligible in a concept with what is sensible in a perception. As for the Kantian doctrine of the symbol, the essential idea that Schiller retained of it and that led Goethe to his own doctrine is that the symbol is not an indicator, a mark, or a 'character', but a sensible reality in which is represented, but indirectly, something that can be the occasion—directly—for a sensible figuration.

However, as Marache (117) stresses, Goethe's conception of the symbol in reality "differs profoundly" from Kant's. It was actually Goethe who very robustly formulated the opposition between symbol and allegory: what is said *in* the symbol can be said in no other way, whereas the allegory only restates in figured terms what in itself is perfectly conceivable and sayable. To the contrary, and this seems the least contestable point, what Kant describes under the name of symbol is applied exclusively by him to allegory in the most rhetorical and most abstract sense

1. Cf. M. Marache, *Le symbole dans la pensée et l'œuvre de Goethe* (Paris: A.G. Nizet, 1960), 20, note 1.

2. Ibid., 38.

3. Cf. Roger Daval, *La Métaphysique de Kant* (Paris: PUF, 1951), 97–100; Alexis Philonenko, *L'œuvre de Kant* (Paris: Vrin, 1975), tome 1, 170–190.

of the term. A brief examination of the practical conception that Kant had of *symbolism* will confirm this.

Although it is true that there is no 'sensible thing' that can be identified as the sensible presentation of the despotic state, it can be conceded that it is no less certain that the notion of such a state is perfectly conceivable in itself and that sufficiently identifiable realizations of it exist, even if none of these realizations is altogether in conformity with its concept (but neither does there exist a perfectly circular ring). This is why we do not see why it is necessary to either represent such a concept or, to use the language of Kant, realize an indirect hypotyposis of it. At the most we can see a need to illustrate the notion in a brief and 'telling' manner. This is exactly what is done in allegory: an abstraction is clothed in a sensible form by virtue of a certain analogy between a guiding scheme of this form and that of an abstraction. There is not a single allegory that Kant's definition verifies, and without doubt a Kœnigsbergian and transcendental imagination was needed to transform the 'hand-mill' into a symbol for generations of philosophical commentaries.

The doctrine formulated in the *Critique of the Power of Judgment* (1789) is taken up again and used in later works, especially in *Anthropology from a pragmatic point of view*, published in 1798, but which collects university lectures delivered "during more than thirty years." At §38 of this work,[1] Kant recalls his definition of the symbol: "Symbols are merely means that understanding uses to provide the concept with meaning through the presentation of an object for it. But they are only indirect means, owing to an *analogy* with certain intuitions to which the concept can be applied." However, in Kant's eyes, recourse to symbolism, such as we find in the majority of traditional cultures, including Judaism and patristic and medieval Christianity, is only the result of a lack of conceptual intelligence: "He who can only express himself symbolically still has only a few concepts of understanding, and the lively presentation so often admired in the speeches presented by savages (and sometimes also the alleged wise men among a still uncultivated people) is nothing but poverty in concepts and, therefore, also in the words to express them. For example, when the American savage says: 'We want to bury the hatchet,' this means: 'We want to make peace,' and in fact the ancient songs, from Homer to Ossian or from Orpheus to the prophets, owe their bright eloquence merely to the lack of means for expressing their concepts" (ibid.). As for cosmic symbolism in the manner of Swe-

1. AK VII; trans. R.B. Louden (Cambridge, UK: Cambridge University Press, 2006), 84–85.

denborg, where "the real appearances of the world present to the senses are merely a symbol of an intelligible world hidden in reserve," Kant states that this "is *enthusiasm* [*Schwärmerei*]," which could just as well be termed 'mysticism'.

We need to recall then in what esteem Kant held 'mystical enthusiasm'. In a work dating from his pre-critical period, *Dreams of a Spirit-seer, Illustrated by Dreams of Metaphysics* (1766), wholly dedicated to refuting Swedenborg, the austere professor of Kœnigsberg did not hesitate to make the crudest 'jokes' his own when this involved overwhelming the mystic with his enlightened man's scorn. "The clever Hudibras alone," he writes, "could have solved the riddle for us; according to his opinion, when a hypochondriac wind rattles through the intestines, it all depends on the direction it takes: if down, it becomes an f..., if up, it turns into an apparition or a holy inspiration."[1] Sir Hudbras is the grotesque, pot-bellied (eponymous) hero of a vast satiric poem in three parts directed against Puritanism by Samuel Butler (1612–1680). Although sometimes tedious, this Rabelaisian imitation of Don Quixote was highly appreciated in the eighteenth century: Henry Fielding cites some verses from it in his famous *History of Tom Jones, A Foundling*.[2]

Basically, Kant does not know what a symbol truly is and envisages under this term what will be called, starting with Goethe, an allegory, that is, an imaged figuration of something abstract. Surely his thinking, viewed in its full sweep, is more complex than the impression given by our brief account. François Marty, in an important study,[3] discerns hesitations if not regrets, and shows that, in the thread of Kant's reflections, a more positive regard for symbolism is outlined in the discourse on God. And yet the philosopher remains a prisoner to the ideology of his own time, an ideology that was not only scornful of the cultures being discovered beyond Europe, but also rather arrogantly ignored seventeen hundred years of Christian symbolism. As Jean-Louis Bruch writes: "What has prevented Kant from elaborating a true theory of religious schematicism is the idea, often exploited by the Enlightenment, according to which revealed religion constitutes a fabrication proper for aiding the development of humanity in its childhood, but destined to be removed with its coming to maturity."[4]

1. AK II, 3; *Dreams of a Spirit-seer and Other Related Writings*, trans. J. Manolesco (New York: Vintage Press, 1969), 66.

2. Book IV, chap. i.

3. *Symbole et discours théologique chez Kant: Le travail d'une pensée*, in *Le Mythe et le Symbole* (Paris: Institut Catholique de Paris, Beauchesne, 1977), 55–92.

4. *La philosophie religieuse de Kant* (Paris: Aubier, 1968), 99.

4

From Hegel to Freud: The Agony of the Sacred

Introduction: Religious Alienation in Hegel

With the neutralization of religious consciousness, the crisis and critique of religious symbolism reached their essential moment. Hegel, then Feuerbach, Marx, and Freud characterize the major phases at the same time that they summarize the history of modern European philosophy—which proves, if there were still need, that this history is that of a long and systematic destruction of our religious soul. The 'essential moment' because rational consciousness finds itself faced with its own contradiction: by withdrawing its objective basis from sacred symbolism, the scientific revolution obliges consciousness to consider it to be its involuntary and monstrous creation. That the human soul could bring to birth such a bastard, and for so long a time, urgently requires that its genesis be made intelligible, under pain of seeing the nocturnal power of the imaginary subvert clear rationality from within. Such is, we think, the threat that all philosophies intend to avert, and such is the desire that drives them.

But, seeing that religious symbolism is carried by a culture with which it is most often identified, thinking that undertakes a genetic critique of it is necessarily transformed into a philosophy of culture. No longer is it just the world or consciousness that must be explained, it is also the history of human works. Man produces from culture and these cultural productions form a particular order of realities about which it is important to create a theory, to the very extent that this order seems to bear witness against its producer. In short, culture enters into the philosophic field only by means of its suddenly obvious problematic nature. Now here we have—at least in appearance—a new object and an unusual task. The human mind is discovered to be not only the site of a timeless and transparent rationality; it is also, in its very substance, history and contingency—which would mask the former state of things in which cultural forms would claim to be productions of the divine *Logos*.

Theology, it must be acknowledged, was hardly aware of this task. A few great minds, however, like Joseph de Maistre and Bonald, immediately perceived the urgency and strove to respond by a philosophy of tradition (an effort taken up again and systematized in a whole other perspective by Auguste Comte). But they were misunderstood and condemned, whereas their attempt (and those of some others) should have been rectified by the working-out of a true theology of culture. For lack of this German philosophy, and especially Hegel's, imposed itself and still imposes itself as the model for all modern speculative theory, because it explicitly poses the question of the relationship of the spirit to its works. At the same time it provides the general principle for a critique of religious consciousness, that of alienation. This is what we will reiterate briefly.[1]

Hegel's basic intention was to cure the state of separation in which Kantianism had left philosophy: on one side the knowing subject, on the other the knowable being of the object. The solution is provided by the doctrine of absolute idealism: nature, in its substantial reality, is nothing other than mind, but an 'objectified' mind or spirit that is opposed to spiritual subjectivity only by reason of a kind of blinding of consciousness, which does not succeed in recognizing its own *identity* in the *other* that is the world of things. It suffers from this because this alterity threatens it and calls into question its own reality: what exists in its eyes, what is real, is the world's material fullness with respect to which the spirit experiences itself as unreal. In short, the mind is in an *alienated* situation: it necessarily perceives itself as other than material substance, yet this substance seems to hold the key to its being since it presents itself as what alone is real. This is why it will strive to recover its being by seizing upon this world so to transform it and put its stamp upon it. This is the whole history of human culture (art, science, religion, philosophy), which, in its evolution, is an immense process of disalienation. The succession of cultural works is the temporal mirror in which the mind becomes visible to itself and becomes what it is ('phenomenology of the spirit'). But this is also an endeavor that fails to the very extent that it progresses until the moment when it attains full consciousness of self in Hegelian philosophy, which is precisely nothing but a thinking about the process itself. In other words, all figures of the

1. What follows summarizes in a few lines a forty-page explanation that can be read in *Problèmes de gnose*, 75–112.

mind (all cultural products) are forms of alienation that should be surpassed. Only in this surpassing is the truth value of alienation itself revealed and achieved.

Hegelian philosophy is thus pervaded by an uncontrolled tension, the result of the dual nature of cultural forms: they are revelations about the mind, so that all Hegelianism is as if a philosophy of symbolism, but they are also figures of the mind, revelatory symbols of its truth only in the light of pure philosophic reason, a light by which alone, thanks to its hermeneutic work of disalienation, they lose their opaqueness and set free their essence. In certain respects, this is the situation of every hermeneutic, and therefore of all philosophy, but a situation that requires being posited for itself (which we will do in the last part of this work) and that can only be resolved by having recourse to the 'absolute surpassing'—eternally accomplished—of an authentic Transcendence. Otherwise, in this dialectic of symbol and hermeneutic, or again, in Hegelian language, of representation and concept, it is the concept that will prevail. All things considered, the Hegelian 'surpassing' (the famous *Aufhebung*) therefore suppresses more than it preserves. Out of that tension spoken about above, it is the angel Reason that triumphs: this System integrates only transparent cadavers with the pure noonday light of thought.

This is why religion is only the penultimate form assumed by the spirit, preceding that of philosophy, in which it becomes truly absolute and free. This is also, as a consequence, the form in which alienation is at its height, or rather summarizes in itself all possible alienations. Actually, if, in religion, God at last realizes the infinite identity of being and knowledge, what remains is that this identity is still posited outside of human consciousness, as an object. In short, a *figure* of the mind is involved here, a *representation* of consciousness. All sacred symbolisms of all religions express, according to various modes, the same alienation and require the same surpassing. Such is the secret source of religious productions. In God and in all signs of the Transcendent, consciousness is retelling, without knowing it, its own truth: "The religions and what they contain of mythology are the productions of man in which he has deposited what there is of the most profound and sublime in himself.... Religions and mythologies are productions of the reason in the act of becoming conscious." This is true of Greek mythology: "the spirit is enlightened with respect to itself in a sensible existence; this is still more the case for Christian mythology."[1] Incarnation, Resurrection,

1. *Leçons sur l'histoire de la philosophie*, trans. J. Gibelin, tome I (Paris: Gallimard, 1954), 231–232.

Ascension, and Pentecost—so many mythic figures in which is revealed, still alienated, the truth of the spirit. Therefore all transcendence has disappeared. The ontological referent is definitely neutralized: symbols are no more than immanent signs of consciousness, which only surpasses itself towards the horizon of its own becoming.

And so is realized the ultimate shutting-away of human culture within a mental and rational cavern. For Kant, 'simple reason' still had 'frontiers' that preserved it from 'religious madness'. For Hegel, it apparently no longer has any, since he extends reason to the totality of the real and becoming, accommodating even what seems the most anti-rational. But in reality, faced with the scandal of religious consciousness and its representations, Hegel has reduced everything to the limits of reason, integrating sacred symbols with its substance as a momentary alienation in the course of its historical realization. We have therefore shifted from culturological reduction to 'consciousness' reduction, and from consciousness reduction to rational reduction: clearly, a reason must be found for what seems devoid of it. Henceforth a general model is perfected. Modern and contemporary thought will not change this, even though it has led to giving various interpretations of it. But there will always be a question, after the mental (or psychic) shutting-away of the sacred—because it cannot, in all decency, be left at the door of the human spirit as triumphant scientism would have it—of finding out why such a spirit, forgetful of its true nature, could have produced such monsters, through which, however, it can gain access to a certain self-recognition.

Feuerbach and the Demystification of the Sacred

IT IS NECESSARY TO QUIT HEGEL TO PROVE HIM RIGHT

Feuerbach's philosophy does not enjoy a prestige comparable to that of Hegel's.[1] It has, moreover, neither the breadth nor the speculative power of Hegel's, even though it claims to be the only true interpretation of Hegel's thought. Its greatest 'claim to fame' is to have worked out the general form of every possible critique of religion, a form from which later critiques will retain only a working concept to apply to various

1. Wedged between Hegel, whom he wishes to overturn, and Marx, whom he introduces, Feuerbach is nearly absent from the histories of philosophy; that of the *Encyclopédie de la Pléiade* (Paris: Gallimard, 1974) tome III, 237–238) accords him only a brief mention while on the subject of Marx. Jacques Chevalier, in his monumental *Histoire de la pensée* (tome IV [Paris: Flammarion, 1966], 150–151) is a little more explicit. Is Feuerbach a 'hidden' philosopher?

contents. It is then his very success that has made him commonplace and risks making us inattentive to his importance. Besides, major philosophies always somehow overlook the sole 'truth' of their historical significance, and Hegelianism is no exception. This is why this 'truth' is more obvious among the—even unfaithful—successors; to these we should turn then to look for it if we are to understand it in all its clarity. And so it is with Feuerbach, who directly confronts the essential task of European thought: to account for the sacred and its signs in such a way that religion is at once justified and abolished.

For such is the chief obligation imposed on philosophers, so that by inaugurating the systematic analysis of the crisis of religious symbolism here we are dealing, not with a peculiar or marginal object, but with the very one basically called into question in the history of the modern West. Of course, with speculative metaphysics and religious symbolism so intertwined, the Kantian critique, robbing metaphysical ideas of their claim to being intellectual signs of the Transcendent, is only the speculative translation of that which attacked sacred symbols in their claim to theophany. Deprived of their ontological referent by the Galilean revolution, ideas and symbols are necessarily reduced to productions of consciousness. Kantianism simply registers this consequence, all while casting the sacred outside rational consciousness, thus dealing with it under the heading of insanity. But Hegel, understanding that one cannot be satisfied with a consciousness split in two under pain of seeing the rejected half threaten, from within, the rejecting half (reason), strove to transform this insanity into wisdom by annexing it to reason, discovering therein an alienated figure of itself. The sacred is clearly then shut away in the consciousness, no longer at risk of escaping: it is a necessary moment of its history.

However—and this is where Feuerbach intervened, by opening reason infinitely to the religious—is there not a risk of seeing reason converted into its exact opposite having become its 'contents'? Is not the enemy, introduced into the stronghold for want of an ability to expel him, going to prevail? Can the Hegelian *logos totally* integrate *mythos* without perishing from it? No doubt religion is transformed for Hegel into philosophy, which alone is triumphant, and this *logos* reduces everything within it to itself: Hegelianism is a 'panlogism', an absolute rationalism.[1]

1. This is what a sustained analysis of the Hegelian conception of the symbol *proprio sensu* would confirm as explained in the lectures on *Aesthetics* (trans. Jankélévitch, tome I, 134–138, and tome II, 1–142 [Paris: Aubier, 1944]). Clearly, if symbolism is an attempt to clothe an idea in a sensible form, this attempt nevertheless always fails because the idea is conceived of in a confused and indeterminate way ('vague generalities') and because

But does not this triumph rightly transform reason into a 'myth'? And does not this mythic (or mystical—the expression is Feuerbach's) rationalism, this religion of philosophy, betray the secret triumph of the conquered over the conqueror? Even more, if the phenomenology of the spirit is a philosophy of culture, does not this lead to *reducing spirit to the series of its historical apparitions*? Actually there is nothing transcendent to history in the 'Hegelian' spirit: "there is only one reason, there is no superhuman second one; it is the divine in the human."[1] God himself becomes conscious of self only in the human intellect: "God is accessible only in pure speculative knowing, and is only in this very knowing, and is only this very knowing."[2] Hence the spirit should not be distinguished from the temporal deployment of its own figures: philosophy identifies itself with religion and its symbols (which are its contents), just as the *logos* identifies itself with the discourse that manifests it; content has devoured form, philosophy completely exhausts itself in thinking about theology.

And so Hegelian thought seems to oscillate from the universal transparency of a pure idea dissolving all forms to the supersaturated opaqueness of formal coagulations: absolute idealism is converted into absolute naturalism and *vice versa*. Hence the dual Hegelian posterity: one on the 'right' and one on the 'left', either of which can only avoid this oscillation by 'quitting' Hegelianism in order to posit, *outside of it*, an objective referent in connection with which its true significance will become apparent. For to know what a text means is to tell *about what* it is speaking. Hegel wanted to escape from this extraneousness by making discourse (in its historical manifestation) the very object of discourse.

the natural forms assumed are clumsy and artificial (Hegel thinking about oriental art). "Thus, instead of a perfect identification," one obtains only "a purely abstract agreement of meaning and form which appear so much more exterior and foreign to each other." Symbolism in sacred art therefore expresses the battle between spiritual and sensible, a battle that ends with the definitive victory of the spirit. In the last analysis, it is art as such that should give way to the realization of the absolute spirit in religion, then, ultimately, in philosophy; this conception reveals the deep-seated angelism of Hegel's system.

1. *Leçons sur l'histoire de la philosophie* (Introduction), tome I, 146.

2. Cited by J. Hyppolite, *Logique et existence* (Paris: PUF, 1961), 70. Can we speak here of pantheism? Hegel rebuffed the accusation many times (for example: *Encyclopédie des sciences philosophiques*, trans. Gandillac [Paris: Gallimard, 1970], 55–57, 116–117, 490–492). His underlying intent aimed not at identifying, from without, God and the world—besides, the supposed pantheism of Spinoza seemed to him rather, and rightly so, an 'acosmism'—but at making it understood that the absolute Spirit is the very 'place' where, in the words of André Breton, the divine and the human "cease being contradictorily perceived."

But his heirs, in order to understand this text—and barring repeating it literally—can only return to it afresh and direct it to an exterior referent, in short can only transform absolute discourse into a discourse relative to something. For Feuerbach, this referent is none other than man as a natural species. Only with respect to man as a natural species does alienation makes sense; this alone prevents Hegelianism from falling back into theology and saves us from religion. We understand that the realization of this project, at the time of the publication of *The Essence of Christianity*,[1] must have provoked a wave of enthusiasm in the 'Hegelian left', to which Marx and Engels bear witness. "It is necessary to cast oneself into this 'river of fire',[2] into this Purgatory of modern times, so to leave it cleansed of all the myths, and tempered for all the battles."[3]

THE GENETICO-CRITICAL METHOD

The judgment that Feuerbach pronounced against Hegel appeared clearly in the text by which he publicly broke with his master in 1839: *Contribution to a Critique of the Philosophy of Hegel*.[4] Until then Feuerbach, after having renounced the study of Protestant theology to which he had wanted to dedicate himself, had at first followed the teaching of Schleiermacher, then that of Hegel. "His conversion to Hegelianism was swift. He renounced theology for the study of philosophy."[5] However, as his correspondence with the master and various writings, including his university thesis, prove, Feuerbach already had a tendency to interpret the supremacy of the Spirit as that of the human species over individual subjectivity, that is, to endow it with a concrete and natural content. He was thus led to perceive more clearly what he rejected of Hegelianism until the moment when the rupture was confirmed. "Hegelian philosophy is rational mysticism," he declares;[6] "... this unity [of the mystical and the rational] is for philosophy a principle as sterile as it is deadly, because it suppresses right down to the particular the distinction between objective and subjective, and compromises genetico-critical and conditional thinking, and the problem of truth. Thus Hegel too has

1. 1841. *Das Wesen des Christentums*, neuherausgegeben von Wilhelm Bolin (Stuttgart-Bad Cannstadt: Fromman, 1960); Eng. trans. George Eliot (New York: Harper Torchbooks, 1957). A good general idea is given by *Manifestes philosophiques, textes choisis, (*1839–1845*)*, trans. Louis Althusser (Paris: PUF, 1960).
2. Allusion to the German meaning of *Feuer-Bach*.
3. *Manifestes philosophiques* (Althusser), 3.
4. Ibid., 11–56.
5. Jacqueline Russ, *Précurseurs de Marx* (Paris-Bruxelles-Montréal: Bordas, 1973), 213.
6. *Manifestes philosophiques*, 47.

truly conceived as objective truth representations that express only subjective needs, and, in order not to go back to the source, back to the *need* that would engender these representations, he has taken them at face value."[1]

This text is important because it contains what is essential to Feuerbach's methodological principles. In particular we will draw attention to the term 'genetico-critical thought' that is 'compromised' by the rational mysticism of Hegelianism. If it is *compromised* that indeed signifies that this method, at the same time that it constitutes Hegel's great discovery, is deformed and altered by the bad use made of it. In other words, the Hegelian method is more genetic that critical for want of finding what was the true cause of the genesis that it implemented. Now, as we have stated, from the point of view of the critique of sacred symbolism, religious forms can *only* be seen as productions of consciousness. By realizing the ontological neutralization of the symbol, scientific rationalism is also involved, whether wishing to or not, in accounting for these productions, in other words, in describing their genesis starting with consciousness itself. The genetic method is inevitably, then, imbedded in the rejection of the symbolic referent. Every post-Galilean philosophy consistent with itself will encounter it sooner or later. Human consciousness has a history; this is what Hegelianism has shown. And if consciousness has a history, if it engenders itself, this is because there is within it something non-conscious with which it illusorily identifies itself or from which it believes itself, no less falsely, forever distinct. In short, if consciousness evolves, this is because it is not altogether realized.

But this genesis should be equally a critique; otherwise it would amount to justifying whatever is taken into account, so to neutralize it as completely as possible. Now a mere genesis of consciousness lacks the required critique. Consciousness being a becoming, every subsequent consciousness is more truly conscious than every prior consciousness: the different moments of this genetic history are all separated from each other according to temporal successiveness. Prospectively, the present moment has its truth only in the one to come. Retrospectively, the new consciousness rejects the old one into the unconscious. But then it becomes impossible to discriminate (critical function) between true and false. Everything is true and everything is false. Feuerbach thus reproaches Hegel for viewing everything according to the exclusion of temporal succession, and for forgetting spatial "coordination and coexistence."[2] Permanent principles must be found then which will enable

1. Ibid., 48.
2. Ibid., 12.

us to discriminate what is true and what illusory in this genesis, what is objective reality and what subjective need.

We are not at all surprised then when we find in Hegel many formulations anticipating those of Feuerbach. When he declares in *The Phenomenology of Mind* that "the Absolute Reality professed by belief is a being that comes from belief's own consciousness, is its own thought, something produced from and by consciousness,"[1] or when he reveals that "the religions . . . are productions of man wherein he has deposited what there is in him of the most sublime and most profound . . . ,"[2] we are evidently already in the presence, at least formally speaking, of the first principle of Feuerbach's hermeneutics. But, as to contents, that is a different matter. For Hegel, and according to his own terms, what is expressed unconsciously in religion is reason and ultimately the absolute spirit which, through it, becomes conscious of itself; whereas, for Feuerbach, it is man himself who expresses the sublimity of his generic essence in this way. He therefore realizes a veritable *naturalization* of Hegelian doctrine. Man is a being of nature, who himself possesses a nature, and who functions according to natural laws, starting with his natural needs. By this we have at our disposal an objective and enduring *criterion* that will enable us to decipher the true significance of religious alienation, without needing to expect from it the revelation of the absolute spirit's hypothetical realization: ". . . Hegel has rejected the *natural* principles and causes, the foundations of the genetico-critical method. . . . Without doubt the former types of naturalist and psychological explanations were superficial, but they were so for this reason alone, that there was no recognition of logic in psychology, metaphysics in physics, and reason in nature. If, to the contrary, nature is truly apprehended, apprehended as concrete reason, it is then the unique canon for philosophy as well as for art. The peak of art is the human *form* . . . the peak of philosophy is the human being."[3] And he concludes: "Philosophy is the science of reality in its truth and its totality; but the essence of reality is *nature* (nature in the most universal sense of the word)."[4] In a certain manner this is a return to Hume, but to a Hume who would have read Hegel and posited the identity of nature and reason. We thus rediscover the terms of the triangle: rational subject, physical nature, and religious culture, which we isolated in the previous chapter and which effectively dominates the entire thought of the nineteenth century.

1. Trans. J.B. Baillie, rev. 2nd ed. (London: George Allen and Unwin, 1949), 567.
2. *Leçons sur l'histoire de la philosophie*, tome I. 231–232.
3. Feuerbach, *Manifestes*. . ., 55.
4. Ibid., 57.

Such is the method that Feuerbach means to follow, and he is well aware of being its creator. This is what one of his historians, S. Rawidowicz, remarks: "Without having undertaken to isolate the logical structure of his 'own method,' he declares he nevertheless thought that he had created a new method. In a letter to his editor Otto Wiegand[1] (1841), Feuerbach characterizes the method of his *Essence of Christianity* as being speculativo-empirical, speculativo-rational, or again genetico-critical."[2] In any case, this method will be universally imposed each time the transcendent and the sacred are taken into account so as to effect their suppression. One can even see in him, Rawidowicz states, "the discoverer of psychological analysis, an idea which is without doubt highly exaggerated."[3] 'Highly exaggerated' if one rests content with the application that Feuerbach made of his method, but altogether pertinent to the contrary if its principle is considered. To tell the truth, we already encounter it with Hegel and even Kant in the notion of transcendental illusion,[4] at least in a certain fashion. It is found again in Marx and Freud. This is inevitable to the very extent that—the Transcendent presenting itself as that which, by definition, comes from 'above'—every genetic explanation of transcendental appearances is equivalent, not only to its negation or rejection, but also and more profoundly to its pure and simple disappearance. It constitutes therefore pre-eminently *the* critique of sacred symbolism. Here, with Feuerbach, this critique comes to full self-awareness. In a decisive manner this is at last what we have called "the subversion of meaning or the neutralization of the religious consciousness" and what defines the contents of this book's second part. And this subversion of meaning is clearly governed by the disappearance of the referent. But, as we have stated several times, this disappearance leaves unchanged religious consciousness, which can, in a certain manner and by a kind of cultural schizoidism, continue to believe in the sacred; whereas the genetico-critical method penetrates to the very heart of religious consciousness and subverts it from within. There is then nothing more effective. Even the critique of the signifier, the third pole of the symbolic sign triad, does not escape the genetico-critical method; it only prolongs and extends it to an at-first undreamt-

1. This involves the first edition, *Sämtliche Werke* (Leipzig, 1846–1890), 10 volumes.

2. *Ludwig Feuerbach, Philosophie-Ursprung und Schicksal*, 2e éd., (Berlin: Walter de Gruyter & Co., 1964), 101.

3. Ibid., 102. We have not read the work of H. Weser, *Sigmund Freud und L. Feuerbachs Religionskritik* (Bottrop: Buch- und Kunstdruckerei W. Postberg, 1936).

4. That is, the illusion in which we find a reasoning that sees in metaphysical ideas representative signs of the Transcendent (the idea of God, for example). These ideas have a transcendent value in appearance only; cf. supra, chap. 3, 158, n1.

of realm, absorbed as it was by its task of occupying our consciousness. This is also its limit, and we shall see why the critique of the symbol should necessarily and last of all come to the signifier itself, which was in some respects ignored for so long.

MAN IS THE GOD OF MAN[1]

We can now briefly review what Feuerbach's hermeneutics entails and the ontological reversal it presupposes.

Feuerbach was perfectly aware that a hermeneutic was involved. And, on this point too, he registered his differences with respect to Hegel: "Speculation makes religion say only what it has *itself* thought, and expressed far better than religion; it assigns a meaning to religion without any reference to the actual meaning of religion; it does not *look beyond itself.* I, on the contrary, let religion *itself speak*; I constitute myself only its listener and interpreter, not its prompter. Not to invent, but to discover, 'to unveil existence,' has been my sole endeavour."[2] These words, drawn from the preface to the second edition of *The Essence of Christianity* (1843),[3] should not mislead us. Contrary to what is sometimes claimed, it is actually a weapon of war against religion, destined to permanently free us from its hold. Feuerbach undoubtedly gives religion ample room, but this is to better subvert its meaning completely. Marx and Engels are at times reproached today for having 'pigeonholed' Feuerbach's philosophy, which would have no other interest in their eyes than to permanently rid us of religion, enabling us, having accomplished this work, to go on to something else; whereas Feuerbach's atheism would be ambiguous, and, paradoxically, the function that he assigns to religion as an unconscious expression of human greatness would entirely restore its importance.[4] But we must be serious: this importance can only satisfy an atheism anxious to preserve certain forms of human

1. This formula which summarizes all of Feuerbach's hermeneutics is not his; we already encounter it in Francis Bacon (*De augmentis scientiarum*, t. VI, c. 3, Antitheta, 20): "*Justitiae debetur quod homo homini deus sit, non lupus.*" It is likewise found, as is acknowledged, in the dedication of Hobbes' *De cive* to the Earl of Devonshire: "To speak impartially, both sayings are very true; That *Man to Man is a kind of God*; and that *Man to Man is an arrant Wolfe.*" It seems it was Plautus who affirmed: "*homo homini lupus.*"

2. *The Essence of Christianity*, xxxv–xxxvi.

3. Feuerbach's most famous work, published in 1841, was a great success from the start. This success was due not only to the attraction of a revolutionary yet on the whole rather simple thesis, but also to its great explanatory power. The book equally owes much to a very clear and not very technical language, a sustained and urgent rhythm, and a dynamic tone filled with emotion. Feuerbach was speaking to the general public.

4. Significant in this respect is the article by H. Arvon ("Anarchisme") in the *Encyclopaedia Universalis* (Paris: Club français du livre,1968), 988–991.

culture in the name of esthetics or a 'broad humanism' for which the formula would be: "How interesting all that is!"[1] Now Feuerbach is altogether explicit about his intentions: "It is not I, but religion that worships man, although religion, or rather theology, denies this.... I have only found the key to the cipher of the Christian religion, only extricated its true meaning from the *web of contradictions and delusions called theology*;—but in so doing I have certainly committed a sacrilege. If therefore my work is negative, irreligious, atheistic, let it be remembered that atheism—at least in the sense of this work—is the secret of religion itself, that religion itself, not indeed on the surface, but fundamentally, not in intention or according to its own supposition, but in its heart, in its essence, believes in nothing else than the truth and divinity of human nature."[2] In other words, there are two things: the religious form and its contents. Insofar as form "religion is nothing, is an absurdity.... But I by no means say (that were an easy task!): God is *nothing*, the Trinity is *nothing*...." These mysteries are simply "native mysteries, the mysteries of human nature."[3] And this is why he can call himself a "natural philosopher in the domain of the mind."[4]

What then is the philosophic basis for this hermeneutic that intends to realize a 'Critique of pure irrationality'?[5] We will briefly touch on this. What distinguishes man from animal is consciousness. But, for man, to have self-consciousness is to have self-consciousness insofar as 'man', that is: to be conscious of the human species within oneself. In consciousness (individual) man is speaking to (species) man, the individual enters into a relationship with the essence of the species, which is reason, will and heart. However, this essence being an object for the consciousness, the individual man does not recognize it as his own subjective truth. And yet the object of consciousness is the revelation of the subject. "In the object which he contemplates, therefore, man becomes acquainted with himself; consciousness of the objective is the *self-consciousness of man*. We know the man by the object ... this object is his manifested nature, *his true objective ego*."[6] Thus the objective pole, the absolute that orients individual consciousness, and into which it projects itself, is the generic essence or nature: "The *absolute* to man is

1. Alain and Valéry are excellent at this kind of exercise.

2. *The Essence of Christianity*, xxxvi.

3. Ibid., xxxviii.

4. Ibid., xxxiv.

5. This was the first title that Feuerbach thought of to designate his book, or rather the first subtitle, the title being *Know Thyself*.

6. *The Essence of Christianity*, 5.

his own nature [*Wesen*]. The power of the object over him is therefore the power of his own nature. Thus the power of the object of feeling is the power of feeling itself; the power of the object of the intellect is the power of the intellect itself; the power of the object of the will is the power of the will itself."[1] Moreover, this generic essence is not only absolute as an object; it is also infinite and unlimited as essence of the species, and this is why, in this very consciousness, individual man discovers by contrast his own finiteness: "It is true that the human being, as an individual, can and must—herein consists his distinction from the brute—feel and recognize himself to be limited; but he can become conscious of his limits, his finiteness, only because the perfection, the infinitude of his species, is perceived by him, whether as an object of feeling, of conscience, or of the thinking consciousness."[2] Such is the essence of man. There is then in man an opposition between himself and his own essence which really constitutes the absolute object of his consciousness.

It is not at all surprising then when we turn to the essence of religion and notice that it constantly opposes God to man, divine perfection to human imperfection. In this opposition we recognize the reality of the opposition we have established by a simple analysis of human consciousness: ". . . it is our task to show that the antithesis of divine and human is altogether illusory, that it is nothing else than the antithesis between the human nature in general and the human individual; that, consequently, the object and contents of the Christian religion are altogether human."[3] Religion is then an especially precious thing, since it reveals to man, without knowing it and therefore without altering it, the truth about himself. Lacking this revelatory mirror, man would not know who he is: "so much worth as a man has, so much and no more has his God. Consciousness of God is self-consciousness, knowledge of God is self-knowledge. By his God thou knowest the man, and by the man his God; the two are identical. Whatever is God to a man, that is his heart and soul; and conversely, God is the manifested inward nature, the expressed self of a man,—religion the solemn unveiling of man's hidden treasures, the revelation of his intimate thoughts, the open confession of his love-secrets."[4]

We have just seen the principle of Feuerbach's hermeneutics. Now what of the ontological reversal that its application to the realm of the

1. Ibid., 5.
2. Ibid., 7.
3. Ibid., 13–14.
4. Ibid., 12–13.

sacred entails? One of the merits of this thinking, besides the clarity and simplicity which contributed so much to its success, is its radical character. Feuerbach goes right to the end of his (speculative) theses. In particular, he does not hesitate to draw all the consequences that the hermeneutic reversal entails and that are concerned with being itself. In other words, the hermeneutic reversal is correlative to an ontological reversal, which should not surprise us, since to the contrary we have already noticed the solidarity that ties these two orders together. If we actually take Feuerbach's naturalism seriously, we cannot help but conclude that the generic essence of man is much more important and real than individual humanity. We have already pointed out this thesis by remarking that the young Hegelian Feuerbach interprets the supremacy of the Spirit over the individual as being that of the species. The certainty of individual death is accompanied by a feeling of the nonexistence of isolated subjectivity. The individual only exists in his relationship to the species, especially in love: "In death your being-for-self steps forth on its own ground. But the nothingness, the death of the self at the moment of isolation, at the moment when it wishes to exist without the object, is the revelation of love, is the revelation that you can exist only with and in the object."[1] And moreover, by dying, the individual enables the species to exist and continue. This non-reality of the individual, or at least this lesser reality, provokes Max Stirner's criticism. Feuerbach, he says "presents us with 'the species, Man, an abstraction, an idea, as our true essence,'" unlike the individual and real ego that looks upon itself as essential. And Feuerbach, who cites this criticism to defend himself, strives to show that he also divinizes the individual.[2] This is true, but to the extent that the individual identifies himself with the species. The real world of Feuerbach is therefore a world of natures, existing by themselves, and not a world of subjective beings or individual substances. It is a world of determinations, qualities, and physical essences: these are what constitute the truth and reality of all that is. And yet the propositions by the help of which we express our thinking about man or God seem to imply the contrary, since they are constituted from a subject to which a predicate is attributed; this supposes that the subject is real and exists by itself. But, in truth, "*qualities* are the fire, the vital breath, the oxygen, the *salt* of existence. An existence *in general*, an existence without qualities, is *an insi-*

1. *Thoughts on Death and Immortality*, trans. J.A. Massey (Berkeley, Los Angeles, London: University of California Press, 1980), 126.

2. "*The Essence Of Christianity*" in Relation to "*The Ego And Its Own*" (1845), trans. F.M. Gordon, *The Philosophical Forum*, vol. viii, nos. 2–3–4 (1978), 83–88.

pidity, an absurdity."[1] The classical ontological order must then be reversed, which impels us to deny the predicates for the sake of the subject and to see in the mere existence of man or God the basis of all reality. For, states Feuerbach, "the necessity of the subject lies only in the necessity of the predicate. Thou art a *subject* only in so far as thou art a *human* subject; the certainty and reality of thy existence lie only in the certainty and reality of thy human attributes. What the subject is lies only in the predicate; the predicate is the *truth* of the subject—the subject only the personified, existing predicate."[2] And this is obviously the same for God. Without his predicates, God would not be God. Without his greatness, his absoluteness, his infinity, his wisdom, God would not be divine, and therefore would not exist. His *existence in himself* is not then something *other* than his predicates: it is only the personification of the existence of predicates. We rediscover here the Spinozan conception of a God consisting of an infinity of infinite attributes, but without substance.[3] An idea so basic to the purely metaphysical perspective as a 'transcendence' of God with respect to His own determinations (the superessential Good of Plato, the distinction in God of pure Essence, *adh-Dhat*, and the qualities, *as-Sifat*, in Sufism,[4] the Eckhartian distinction of Godhead and God,[5] already misunderstood by Spinoza) is here completely rejected: "What theology and philosophy have held to be God, the Absolute, the Infinite, is not God; but that which they have held not to be God is God: namely, *the attribute, the quality, whatever has reality*."[6] This is why, although the suppression of the predicates entails the suppression of the subject, the converse is not true. To sup-

1. *The Essence of Christianity*, 15.

2. Ibid., 18–19. This is why it seems at least ambiguous to summarize the thought of Feuerbach, as the Communist philosopher Jacques Milhau has done, by saying: "God is not subject, but predicate; man is not predicate, but subject" (*L'athéisme, idéologie et religion*, published in *Philosophie et religion* [Paris: Éditions Sociales, 1974], 58). Is there not to the contrary an attempt with Feuerbach to break away from the ontology of the subject?

3. Moreover, Feuerbach expressly refers to this; ibid., 24. In his *Provisional Theses for the Reformation of Philosophy* (1842), he writes (thesis 3): "*Pantheism* is the *necessary consequence* of theology (or theism), it is *consistent* theology; *atheism* is the *necessary consequence* of 'pantheism,' *consistent* 'pantheism'" (*Manifestes*..., 104).

4. "He is beyond what they attribute to him" (*Qur'an*, XLIII, 82). L. Schaya, *La doctrine soufique de l'unité* (Paris: A. Maisonneuve, 1962), 42–47.

5. *Meister Eckhart: Sermons & Treatises*, trans. M. O'C. Walshe (London & Dulverton: Watkins, 1981), 81. If Feuerbach (and before him Spinoza) had understood, as Eckhart says, that "when all creatures say 'God'—then God comes to be," he would only be able to say that pantheism is the consequence of theism, and atheism of pantheism.

6. *The Essence of Christianity*, 21.

press the subject God is not to suppress the predicates of the divine; to the contrary, this is to accept them for what they are—the truth of the human essence—and, at the same stroke, to definitively liberate man from all alienations and all illusions. Actually, to attribute all these predicates to the subject 'God', "to enrich God, man must become poor; that God may be all, man must be nothing."[1] Conversely, by reading the truth of man into the illusion of religion, the new hermeneutic in some manner verifies its ontological 'humanism' and restores to human beings what in reality always belonged to them. Feuerbach's whole endeavor thus leads to a dual reductive identification: of the divine subject to its predicates and of its predicates to the predicates of the human essence, itself seen as a mosaic of natural determinations.

We will not expand further on Feuerbach's hermeneutics. Its development presents itself as a meticulous exegesis of all sacred forms, its dogmas, rites, and sacraments: Creator-God, God-the-Trinity, God incarnate, the Virgin Mary, etc. Everything is translated into the language of philosophic truth, but also with an often quite lyrical tone. Of greater interest, though, are the consequences to be drawn from this in what concerns the critique of religious symbolism.

THE ILLUSIONS OF A HERMENEUTIC OF ILLUSION

First of all and chiefly the superiority of Feuerbach's hermeneutic over all of its predecessors is that, as we have said already, it is a true hermeneutic. Surely, in our opinion, it was only developing a possibility already inherent to Hegelian philosophy: the shutting-away of the sacred inside consciousness as an alienated figure of this consciousness. But, with Hegel, the figure is crushed or rather drowned beneath the extent of his interpretation in such a way that it rightly stops being a symbol; whereas, with Feuerbach, we rediscover a true conception of sacred forms as so many symbols in the full sense of the term. Religion's *text* is posited in all its concrete determination, and a complete translation is offered, that is, we are shown its referent. Religious forms thus rediscover their nature as symbolic signs which they had more or less lost since Spinoza and Kant; they had become mere witnesses to superstition or human folly.

But these signs were radically subverted in their express significance. In the realm of sacred traditions, sacred signs are indicators of transcendence. They have meaning only to the extent that they testify to a supernatural or metaphysical other-world, and it is even because this function seems self-evident that the adversaries of symbolism reject it.

1. Ibid., 26.

In the realm of 'new philosophy', to the contrary, signs are accepted as such because the referent towards which they point, far from being situated in the 'other-world', is to be found entirely 'here-below'. This idea could only enter the mind of a classical thinker, for whom the symbol, always having a natural element, belongs to the same plane of reality as man, and therefore can only designate a human 'other-world', since a visible sign is necessarily a sign of something invisible. And yet the solution was, in appearance, rather simple. It was enough to consider that the true man—authentic human nature—is invisible or not yet disclosed, so that visible signs might point towards this nature. And how is it veiled if not by the signs themselves? Or rather it is veiled by the form in which they present themselves, that is to say religion or, if one prefers, culture, for "the progressive development of religion . . . is identical with the progressive development of human culture."[1] We rediscover then the components of our triangle: cognitive subject–nature–culture, with this idea that culture hides nature, but in a sense quite different from Rousseau or Kant. Culture hides nature because it is its unperceived revelation, which obliges us to ask if Feuerbach's philosophy, rather than an anthropological *naturalism*, might not be a *culturalism*.

Feuerbach is thus seen introducing the distinction of visible and invisible within the sole natural order, following Galileism, for which the 'beyond' of the world resides within the world itself. Here, in the same way, the 'beyond' of man resides within man: this is also what '*homo homini deus*' signifies—man is to himself his own 'beyond'. The negation of transcendence, or 'trans-ascendance', does not lead to a true immanence, but to an inverted transcendence, to a 'trans-descendance'. Is it a certainty then that the *humanity that was adored in God will continue to be adored in man*? Is not the religious form necessary to the sublimity of our generic essence? We indeed understand that Feuerbach wished to realize the truth of religion and not purely and simply destroy it. Yet there is one thing of religion that disappeared, or otherwise Feuerbach's hermeneutics would be purely and simply Christian theology itself: this is the illusion, that is, the affirmation that the propositions of faith and, in a general manner, all sacred signs, possess a rightly transcendent, eternal, absolute, and infinite referent—which constitutes the primary referent of all religion. Feuerbach can indeed recover all of religion except religion in its essence, that is in its 'illusion'. And that is valid for all doctrines that introduce a relationship of illusion to reality *within a single, self-same order of existence*. The whole question basically

1. Ibid., 20; or again: "The eras of humanity are only distinguished among themselves by religious transformations" (*Manifestes*. . ., 97).

comes down to knowing if there is for Feuerbach, yes or no, a religious *illusion*. The answer is clear. Now then, by positing religion as an illusion, one is forever forbidden to realize a hermeneutic of it, and therefore a truly integral science, since this hermeneutic begins by denying, for this religion, its essential affirmation by denouncing it precisely as illusion. It would be quite another question if it were to denounce religion as an error. That would entail no serious consequence for the one who rejects it, unless to affirm the presence in man of an inexplicably non-human attitude. But from the very instant that we see the source and cause of this non-human (not in conformity with human nature) element in man himself (and how do otherwise?), we are inevitably led to define it as an illusion of man about himself, that is, as a humanity ignorant of itself. At this same instant illusion vanishes as illusion, even if its contents remain, and, as a result, the illusion as illusion is not explained. Either illusion is something, or it is nothing. If nothing, we cannot speak of it, it does not exist, and there is no religious illusion. If something, if it exists, its explanation—to the extent possible—would be equivalent to its disappearance, in other words its non-existence, for then this explanation would make what it had to explain unintelligible. And this is what Feuerbach does, he who is truly a victim of the illusion he thinks he is denouncing. For the illusion of an illusion is to have itself pass for illusory, that is to say unreal.[1] In other words, there is no *mere* illusion, or again: illusion should not be a mere appearance, *simply an effect of consciousness*; it is necessarily a 'lesser reality', and therefore it calls for a greater reality. Sacred forms are indeed 'relatively illusory' insofar as forms, that is, less real than their contents, but this is also why they call for a perfect and absolute Reality, and that is a symbol. But when they are interpreted as human signs they, by losing their transcendence, also lose their sublimity, their *value*, which can be preserved only

1. It will be objected that there is, in Feuerbach, a basis for illusion: this is the consciousness that man has of himself as a generic essence and, therefore, as an apparently transcendent object. Hence his illusory determination as God. But this explanation specifically makes religious illusion unintelligible. Why would individual man become conscious of self as a generic essence only under a religious form? Why must he perceive his own essence, his 'humanity', as 'God' and not for what it is? What is it that makes this mistake necessary? The best evidence eludes him precisely because Feuerbach is not a victim. Why him and no one else since men have existed? An exception for such a basic and universal illusion is quite simply unintelligible—especially since men seem to have had—forever—an awareness of their humanity and have not confused it with God, since they have defined it as the image and likeness of God. We clearly see that the explanation of illusion actually makes illusion impossible. If it were true, it would be incomprehensible to the very victims themselves that they had succumbed to it; they could

on condition that the 'religious illusion' be preserved; that is to say that religion be not exactly an illusion, but rather an allusion, a symbol. When God is man, man can no longer be God.

On the other hand—and this is only a consequence of the loss of transcendence—this 'disalienating' hermeneutic in reality only reinforces man's alienation definitively, and it is in this sense that we have spoken of culturalism or culturalist determinism. Culture becomes *cultura naturans*. Of course it will at once be asserted that religion only expresses the truth of nature which, without it, would remain in some manner unmanifested. But to the very extent that religion is the sole revelation of nature, Feuerbach is forever denied the ability to demonstrate that this is indeed so and that nature actually includes many determinations that appear in the illusory mirror of religion. He is condemned to rely on what religion says about man. Assuredly all sacred cultures, as we stated, present themselves as modes of expression about the nature of things, intelligible and symbolic expressions, messages of the eternal archetypes addressed to the human form and destined to teach it about the nature of things and itself. But this is a function of a non-human origin of these cultural messages. Transcending all modes of existence and all degrees of reality, the divine Word, by revealing itself as such, reveals itself at the same time as the immanent *Logos* of things and beings, and therefore as their own intelligible unity. The sacred forms, the culture-defining orderly whole of symbolic signs, are, as we have shown, like rays emanating from the eternal Word, crossing the ocean of the worlds with a single lightning-flash. Feuerbach's hermeneutics *a priori* forbids such a *fundamentum in re* for the determinations

only remark it as a fact, the key to it forever eluding them. A true explanation of an illusion does not suppress it; to the contrary, it permanently establishes its existence (thus a stick partially plunged in water will always appear broken). One can therefore be aware of an illusion only by gaining access to a higher degree of reality, and not just by a mutation from within one and the same consciousness. It is from the vantage point of a higher reality that the (relatively) illusory nature of the lower degree becomes manifest: far from disappearing, it can only be seen in this way. Knowledge about illusion implies the doctrine of the degrees of reality or multiple states of the being. This is precisely what the stick partially plunged in water symbolizes, the distinction between aquatic and aerial environments representing the distinction in degrees of reality. Only from the air would the stick appear broken in the water. But seen from the water it is straight. It is the laws of refraction in an aquatic environment that deform it. Religion might well be compared to a deforming environment, but then it is itself an objective reality, a cause of the deformity and not produced by it. And it must be considered afresh. Feuerbach—and every hermeneutic derived from his—posits at once and contradictorily that religion is the cause and effect of an illusion. We will return to this objection in the following article with respect to the Marxist theory of 'reflection'.

and qualities revealed in the religious mirror. It can only grasp the trajectory that goes from religion to nature, never the one that, we are told, sets the precedent and goes from nature to religion. Of course this second trajectory is everywhere presupposed and, since we cannot get away from hermeneutics, this will never be deniable or verifiable. And there is nothing to enable an 'outside' observer to decide if it might not rather be nature that reflects religion, nothing except Feuerbach's atheist 'wager'. This proves in an irrefutable way that atheism is an extra-hermeneutic proposition that determines, from without, the atheist sense of this hermeneutic; in other words no one can avoid such a 'wager', that is to say a judgment dealing solely with the *existence or non-existence* of God. Surely Feuerbach posits *a priori* nature and individual man as two extra-hermeneutic realities. But these realities, by very virtue of the ontological reversal of predicate and subject, cannot be viewed as substances. Their whole reality is a function of their qualities and determinations. Now, as it happens, these determinations are known only in religion. As a result there is no longer anything outside hermeneutics. Everything is taken into this interpretive discourse, but without a set, immutable, and transcendent referent that would allow the meaning of the hermeneutic to be *determined*. To the contrary, it is hermeneutics or rather the (cultural) natures that it reveals that are determining. These cultural natures exist by themselves, in an autonomous manner: "The predicates . . . have an *intrinsic, independent reality*; they force their recognition upon man by their very nature; they are self-evident truths to him; they *prove*, they attest themselves."[1] Existing by themselves, these natural determinations have total power over man: "Religion is that conception of the nature of the world and of man which is essential to, *i.e.* identical with, a man's nature. But man does not stand above this his necessary conception; on the contrary, it stands above him; it animates, determines, governs him. The necessity of a proof, of a middle term to unite qualities with existence, the possibility of a doubt, is abolished."[2]

Feuerbach's materialism is so much more a cultural materialism than a naturalist one. His contradiction, we think, lies there. If it went unnoticed, this was because people were above all responsive to the antireligious prophetism that seemed to animate the least of his lines, and because little heed was paid to the choice and value of arguments when they went in the desired direction. Nevertheless this philosophy functions with a very lively feeling of having finally discovered a solid footing on which to set Hegelian religious alienation, which it has thus

1. *The Essence of Christianity*, 21.

2. Ibid., 20.

brought down from the heaven of speculative philosophy onto the earth of men—whereas it is itself, however, and by virtue of its own principles, only a hermeneutic that deals just with meanings and never with realities.

Feuerbach's hermeneutics clearly seems to be the model for all antireligious philosophy. Neither before nor after it, to our knowledge, do we encounter a reflection more closely ordered towards the demystification of the sacred. There are certainly more violently atheistic or antireligious philosophies, but none more attentive to or meticulous about overturning the illusion of the sacred. This is why every philosophy that intends, as an objective, to expel religious mystification from human consciousness is in some manner Feuerbachian, even if the *cause* of the alienation, that is the basis of the hermeneutic, varies considerably from one to another. Lastly, as we shall see, it is not until cultural positivism that the determinism of the signifier is anticipated in a certain manner, which structuralist surfacialism will bring to light, shaped by a critique of the third and last pole of the symbolic sign. And this is, we repeat, a necessary and inevitable consequence already quite clearly shown by the example of Feuerbach's philosophy. Every demystifying hermeneutic, that is, any relying on an interior illusion of the consciousness, being ultimately unable to leave the hermeneutic sphere within which it has methodologically enclosed itself, finds *determinations* only in cultural forms the true meanings of which it claims to reveal. Everything else is nothing but an indefinitely reciprocal game of meanings. But this extreme consequence only shows the extreme absurdity to which the anti-symbolic critique has led western thought.

Marxism and the Abolition of the Sacred

THE MARXIST THEORY OF ALIENATION

In certain respects the Marxist critique of religion and the sacred only repeat Feuerbach's, which it considers to be well established and to which it constantly refers. In that case, what is the benefit of dealing with it in particular? Especially since the texts that Marx, Engels, and Lenin[1] devoted to these questions represent only a miniscule part of

1. Karl Marx and Friedrich Engels, *On Religion* (New York: Schocken Books, 1964); Lenin, *Philosophical Notebooks*, in *Collected Works*, vol. 38 (Moscow: Foreign Languages Publishing House, 1961). The best overall account of the Marxist conception of religion (but which includes, and for a very good reason, no study on Marx himself) is *Philosophie et religion*, a collective work (Paris: Éd. Sociales, 1974). Michèle Bertrand, *Le statut de la religion chez Marx et Engels* (Paris: Ed. Sociales, 1979) raises some interesting problems.

their works: for Marx nothing or nearly so;[1] for Engels some detailed explanations of religious movements in their relationship to social history, and only once, in a few pages, a general account of the materialist idea of religion[2]; for Lenin about a hundred pages among a total of twenty-six thousand, and which, moreover, tackle religion only indirectly or are spread about in sarcasms and invectives. Marx is known to have experienced a pathological hatred of the religious: Christianity preaches "cowardice, self-contempt, abasement, submission, dejection, in a word all the qualities of the *canaille*";[3] concerning the idea of a heavenly survival, he has this to say: "the hair of my head stands on end in fright, my soul is a prey to anguish."[4] Lenin's invectives are no different: "All worship of a divinity . . . be it the cleanest, most ideal . . . is ideological necrophily. . . . Any religious idea, any idea of any god at all . . . is the most inexpressible foulness."[5] Or again: "The materialist exalts the knowledge of matter, of nature, consigning God, and the philosophical rabble that defends God, to the rubbish heap."[6]

The founders of Communism are, in fact, torn between two contradictory attitudes: on the one hand, as Marx and Lenin repeated *ad nauseam*, this does not involve thinking about religion but destroying it: "That is the ABC of all materialism, and consequently of Marxism,"[7] and the efficacy of communist regimes in this matter is well known; on the other, they would like to convince themselves that religion no longer exists: "For the mass of men, i.e., the proletariat, these theoretical notions do not exist and hence do not require to be dissolved, and if this mass ever had any theoretical notions, e.g., religion, these have now

1. To our knowledge, the longest text that Marx devoted to religion is found at the beginning of *Contribution to the Critique of Hegel's Philosophy of Right* (end of 1843–January 1844) and includes a little less than two pages. The remainder only consists of allusions in a few lines or a paragraph. It is in this *Contribution* that is to be found the famous declaration: "Religious distress is at the same time the expression of real distress and the protest against real distress. Religion is the sigh of the oppressed creature, the heart of a heartless world, just as it is the spirit of a spiritless situation. It is the opium of the people" (42). Comparing religion to opium was quite commonplace, as C. Wackenheim has shown in *La faillite de la religion d'après K. Marx* (Paris: PUF, 1963), 187: we find it in Heine, Feuerbach, Bruno Bauer, Goethe, etc.

2. *Anti-Dühring* (1878); *On Religion*, 145–151.

3. *The Communism of the Paper* Rheinischer Beobachter, 1847); *On Religion*, 84.

4. Poem of 1836, cited by J. Natanson, *La mort de Dieu. Essai sur l'athéisme* (Paris: PUF, 1975), 163.

5. Letter to Maxim Gorky, mid-Nov. 1913; *Collected Works*, vol. 35, 121–122.

6. "Philosophical Notebooks," *Collected Works*, vol. 38, 171.

7. "The Attitude of the Workers Party to Religion," *Collected Works*, vol. 15, 405.

long been dissolved by circumstances."[1] To tell the truth these are wild assertions, but are nevertheless in close accord with the Marxist attitude as a whole, an attitude that consists in denying reality in the name of doctrine. This would lead Communism in Russia and elsewhere to what one contemporary philosopher, André Conrad, has rightly named "police surrealism." Religion is theoretically dead; so, since it should no longer exist, let us kill it in practice.

However, as negligible as it might be, the Marxist[2] philosophy of religion is not satisfied with simply repeating Feuerbach; it also wants to correct him.

And from the very start, should we concede, along with Althusser, that Marx has abandoned the still Hegelian vocabulary of alienation for a purely economic one of 'fetishism of commodities'? Certainly not. Until the end of his life Marx spoke of the "alienation of work."[3] What changes with him is not the *form* of the alienating process, but its nature. With Feuerbach, individual man becomes aware of himself as a generic transcendent essence clothed with the attributes of divinity. For this 'ideal' duplication in consciousness Marx at first substitutes an actual duplication in the productive process of labor: man, who is a worker *essentially*, a transformer of nature, realizes his essence in object-producing activities: "The object of labor is, therefore, the *objectification of man's species-life*: for he duplicates himself not only, as in consciousness, intellectually, but also actively, in reality, and therefore he contemplates himself in a world that he has created."[4] Labor's product then "confronts it [labor] as *something alien*, as a *power independent* of the producer. . . . The worker puts his life into the object; but now his life no longer belongs to him but to the object."[5] Subsequently, especially in *The German Ideology* (1846), Marx specified that it is not the product of labor as such that is alienating, but the economic production process:

1. Karl Marx and Frederich Engels, *The German Ideology* (Amherst NY: Prometheus Books, 1998), 64.

2. Marx, if not insightful, was at least a quite learned economist, and this at the cost of an intense labor that soon made him entirely neglect philosophical speculation, the impress of which remained deep-rooted in him. As for materialisms, whether dialectical or historical—terms unknown to Marx—among his successors they have never achieved the dignity of a true philosophy, and indeed evince an extreme conceptual coarseness. One can always, obviously, extract a philosophy thought to be implicit in a work and suggest a reconstruction of it, as Althusser or Michel Henry have done.

3. *Un chapitre inédit du "Capital"* (Paris: Union Générale d'Éditions, 1971), 10/18, 142; cf. Lucien Sève, "Analyses marxistes de l'aliénation," in *Philosophie et religion*, 209.

4. *Economic and Philosophic Manuscripts of 1844*, trans. M. Milligan (New York: International Publishers, 1964), 114.

5. Ibid., 108.

the division of labor obliges the worker to manufacture a product that does not belong to him: "as long as a cleavage exists between the particular and the common interest, as long, therefore, as activity is not voluntarily, but naturally, divided, man's own deed becomes an alien power opposed to him, which enslaves him."[1] This economic alienation involves an ideological illusion under the form of a 'reflection in the brain': "The phantoms formed in the human brain are also, necessarily, sublimates of their material life-process. . . . Morality, religion, metaphysics, all the rest of ideology and their corresponding forms of consciousness, thus no longer retain the semblance of independence."[2] And finally the last interpretation of alienation: 'fetishism of commodities'. The product of labor, inserted into the merchandising circuit, loses all concrete individuality to become identified with, in an almost substantial manner, an anonymous mercantile value, a reified abstraction imposed as a pseudo-reality; "Christianity with its *cultus* of abstract man" is "the most fitting form of religion" for a society that practices this fetishism, this kind of fantastic worship of 'commodity value'.[3]

EVERY ALIENATION IS RELIGIOUS IN FORM

Such is the doctrine of Marx. And apparently religion is seen as hardly having any place in it. It is one ideological form among others, engendered by economic alienation as a reflection of this alienation 'in the human brain'. This is both not much and vague. *All the same, no demonstration has ever shown the mechanism by which we actually go from economic alienation to its ideological religious reflection.* This is simply said to be so, and religion thereby loses all proper reality, all cultural autonomy, since it is an effect, the inverse reflection of an economic structure; a determinist dependence (between a particular economic structure and a particular positive content of dogmas, symbols, or rites) is defined by this, and for it clearly no one is able to provide the least beginning of a proof. With the same stroke we gauge the extraordinary insignificance of this so-called doctrine, which is in reality only a philosophical imposture.

But we must go further than appearances. Far from being one ideological form among others, religion is, for Marx himself, the primary and general *form* of *every* ideology, and therefore of every alienation, even economic, the one that all the others presuppose and that survives,

1. *The German Ideology*, 53.

2. Ibid., 47.

3. *Capital*, ed. David McLellan (Oxford and New York: Oxford University Press, 1995), 49.

implicitly or not, in all the others; a thesis that will seem obvious once we notice that the idea of religion is confused with that of 'dominating transcendence', which is presupposed in the very notion of ideology. Religion is therefore the 'primary site' in which every alienation is *signified*, the site without which alienation would only be a relationship between things, and would ultimately not exist as such, even disguised or inverted. This is a fundamental but often misunderstood point, and so it would be fitting to prove it.

"Religion," notes Marx, "is from the outset [*von vornherein*] consciousness of the *transcendental* arising from *actually existing* forces."[1] In other words, every real domination, that is one that man is actually obliged to obey, is perceived first, and essentially, as a transcendence. Consciousness of this transcendence *is* religion. This is why, each time Marx recalls alienating situations, even of a purely economic nature like the fetishism of commodities, he refers "to the mechanism of religious consciousness, considered to be a primary analogue."[2] But then, whether many specialists like it or not, we must conclude with Lucien Sève that "the basis of religion is by no means summed up in the fetishism of commodities, but is identical to the sum total of barriers that individuals encounter in their relationships among themselves and with nature."[3] The root of alienation is not economic but natural, and basically consists in our insufficient technical mastery of the material world (still seen by Marx as an adversary). Our thesis is confirmed by the most explicit text that Engels devoted to religion; it begins in this way: "All religion, however, is nothing but the fantastic reflection in men's minds of those external forces which control their daily life, a reflection in which the terrestrial forces assume the form of supernatural forces."[4]

THE REFLECTION OF MYSTERY OR THE MYSTERY OF REFLECTION

We are now ready to go to the heart of the contradiction that the Marxist explanation of religion conceals. We have already encountered this contradiction in Feuerbach, we will rediscover it in Freud, and we will be led, *per absurdem*, to the certainty of the semantic principle, that is, to the founding and regulating principle of the metaphysical field.

The basic concept used by Marxist methodology is that of reflection:

1. *The German Ideology*, 102; the manuscript—left unfinished and never published by its authors—is flawed: the subject of 'arising' (*hervorgeht*) is uncertain (consciousness or the transcendental?). This has to do with an isolated note.

2. Jacques Bidet, "Engels et la religion," in *Philosophie et religion*, 171.

3. "Analyses marxistes de l'aliénation," ibid., 242 (penetrating and well-informed).

4. *Anti-Dühring*; cf. *On Religion*, 147.

in Marx, Engels, and still more in Lenin, this is what is at work and explains everything but remains inexplicable. It is only a translation, under the form of an image (sometimes ascribed to the experience of the *camera obscura*),[1] of Feuerbach's notion of the genetico-critical method, a translation that accentuates its ambiguity and contradiction. This is why it would be good to undertake now a more rigorous analysis, even though similar to the one introduced for Feuerbach.

In the ordinary sense of the terms, 'genesis' and 'critical' belong to two different realms: respectively, to the realms of being and knowing; they should not then be joined. Genesis designates the entire formative process, from its initial state, of a completed reality. It explains its becoming in such a way that every subsequent state finds its raison d'être in a prior state, which implies an intelligible continuity between cause and effect, origin and term, an intelligibility conveyed by the necessary order of succession for the various phases. As for criticism, it is a reflexive judgment or mental act by which someone reconsiders the (sensible or intellectual) knowledge they have of something and the implicit judgment of the reality encompassed to appreciate its truth-value, and therefore distinguish true from false, that is, from the appearance of the true or illusion. Only indirectly and secondarily does criticism encounter genesis when, having exposed possible illusions, it has to take it into account: namely, to explain why the appearance does not correspond to the reality (which presupposes moreover that what is apparent should express what is hidden). Genesis is required then to explain the distortion or deformity that an appearance has suffered and confers precisely such an appearance on it. Illusion does not then reside *in* the apparent itself (its appearance is what it ought to be), or *in* the knowing subject (which cannot see what is apparent other than as it actually offers itself to be seen), but in the judgment that the mind thinks it can infer from appearances, and therefore in the relationship of one to the other. Neither object nor cognitive awareness is false. But precisely because they are distinct from each other, the relationship that knowledge implicitly or explicitly posits between perceived object and its reality is always hazardous. With respect to things known, this is the ransom for the freedom of the mind that judges. The eye is not free to

1. "If in all ideology men and their circumstances appear upside-down as in a *camera obscura*, this phenomenon arises just as much from their historical life-process as the inversion of objects on the retina does from their physical life-process" (*The German Ideology*, 47). In such a 'darkened room' the image obtained is, as we know, always inverted.

not see the perceived broken stick, but the mind is free to judge whether or not this knowledge corresponds to reality. And so the possibility of illusion is only relative to the 'freedom' that the mind has surrendered to the suggestions of appearance—freedom of the yes which is only the other side of the critical freedom of the no. Once its genesis is explained, the potential illusion is transformed into certainty, and the laws of refraction tell us why the eye can only see a broken stick when this is not so in reality.

As for its possibility, criticism presupposes then a mind that knows (for which there are appearances) and a knowledge including an implicit judgment of reality. As for its operation, it consists in a reflexive judgment that suspends and then confirms or refutes the implicit judgment. This judgment in the second degree rightly expresses the mind's freedom as such. Independent, it is not and should not be the product of any genesis, no more than a judgment in the first degree or even the simple cognitive apprehension of the mind. If critical judgment is exercised this is always, in the last analysis, in the name of purely intuitive evidence, or of an already recognized principle, or of a previously established factual truth (for example: by touch we know that the stick is not broken), but this is not in the name of a genetic explanation of a deceptive appearance, which is always *a posteriori*.

Now the concept of a genetico-critical method alters this model radically. To be able to unite genesis and criticism, that is for genesis to be *by itself* a critique, it must necessarily deal no longer with the object, but with the mind's faulty knowledge of it. And in fact, since criticism can only be a mental act, a genesis that would be by itself a critique (that is, such that it effects a discernment between true mind and false mind) can likewise only be concerned with the mind. It supposes then that illusion resides *in* cognitive consciousness, and its whole effort is focused on explaining why this is so. A deceptive appearance will no longer result then from the property of an object or the objective laws of its perception acting on the mind—as falsely believed by the 'old philosophy'—but from a property of the knowing subject, namely a falsified cognitive consciousness, and genesis will consist in explaining the presence of this property in consciousness through the history of its formation.

The entire issue then comes down to the following question: does this involve genetically explaining the appearance of a deformity affecting an originally healthy consciousness, or else genetically explaining the appearance of consciousness as such, the faultiness of which would be in that case essential or congenital? There is no doubt about the answer. The first instance in reality refers us back to the classical model: consciousness is only accidentally affected by illusion because it yields to the suggestion of appearances. Therefore only the second eventuality remains. In other words, if genesis has a bearing on the production of a faulty consciousness, it also necessarily has a bearing on the production of consciousness as such, the faultiness of which is therefore intrinsic: to be conscious, to be cognizant is to speculatively invert the reality of things; and this is indeed what Marx and Lenin think.[1] But the consequences that follow from this lead to the most blatant and the least perceived contradictions. Chief among the consequences—the others are only its corollaries—is that, if illusion were the result of a genetically faulty consciousness, *it would be forever undetectable*. That is—corollary 1—there was never anyone to reveal it, it being understood that the same causes produce the same effects, and therefore every consciousness is equally genetically faulty; unless we admit that with the millions of human consciousnesses since the beginning only one (that of a Feuerbach or a Marx) has escaped universal genesis by a miracle. And this is to say—corollary 2—that there was no one to whom such a genetic explanation might be addressed and therefore for whom it would have a critical value, for, every consciousness having been deceived, everything would be deceptive, including its own undeceiving, and its blindness would be irreparable; unless we admit that, by a miracle, all human consciousnesses became suddenly upright and lucid at the very moment when Marxist doctrine made its appearance, which confers on present humanity a rather flattering advantage over the congenital imbecility of prior generations.

Although unperceived, these contradictions do not fail however to exercise their constraint on Marxist thought itself: sometimes, when functioning 'Marxistly', it asserts that it shows the genesis of religious ideology's distorting mirror, and sometimes, forgetting its theoretical commitments, it presupposes the existence of this mirror thanks to which alone there is a distorting reflection. Sometimes genetico-criticism leads necessarily to the equations: consciousness = alienation = religion = ideology (this is the case with *The German Ideology*); sometimes, to the contrary, the ideological superstructure is designated, not as a product, but as an *instance*, as the mirror required for every reflection (this is the case with *Capital*, or with the preface to *Contribution to the Critique of Political Economy*). Thus is brought into high relief the incoherence of a doctrine that wants to be at once both explicitly struc-

1. Intellectual knowledge "includes ... the possibility of the transformation (moreover, an unnoticeable transformation, of which man is unaware) of the abstract concept, idea, into a fantasy (in the final analysis = God)"; Lenin, "Philosophical Notebooks," *Collected Works*, vol. 38, 372.

tural (or synchronic) and genetic (or diachronic), and, to do this, rests upon a confusion of cause and effect: consciousness is the cause of alienation, itself the cause of consciousness. This is what is called, with good reason, the *genetico-critical circle.*

But we also see the consequences to which the setting-up of this circle leads, a circle within which Marxists specifically want to shut away religious consciousness, and which in reality closes in around all human reason, sentenced to a definitive illusion. As we now know, and we do not wish to prove anything else, this great shutting-away of humanity was only beginning: madness henceforth abides within reason itself and is only one with it. By blotting out the signs of the sacred, all those lights by which religious faith might be illumined on earth, the great atheists of the modern world have doomed men to the prison darkness of an incurable insanity.

Freud and the Inversion of the Sacred

PSYCHOANALYSIS WANTS TO TAKE RELIGION'S PLACE

If we compare the attitude of Freudian psychoanalysis with that of Marxism regarding the question of the sacred, one difference is immediately apparent: while Marxism seems completely disinterested in religion and treats of it only with reticence, Freudianism to the contrary presents itself as an 'anti-religion'. Intending to give a sequel, fifteen years later, to the lectures on *Introduction to Psychoanalysis*, and casting a retrospective glance over this 'science', Freud writes in *New Introductory Lectures on Psycho-Analysis* (1933): "The theory of dreams has remained what is most characteristic and peculiar about the young science, something to which there is no counterpart in the rest of our knowledge, *a stretch of new country, which has been reclaimed from popular beliefs and mysticism*."[1] This definition seems altogether exact. Psychoanalysis in fact pursues the declared project of substituting itself for religion, of taking its place, of occupying the terrain where the transcendent and sacred held sway until then. And it has actually become the religion of western humanity.

Surely Communism—rather than Marxism, which is just barely an ideological alibi for it—can be considered as a religion by virtue of an undeniable but deceptive formal analogy (a revival of Judeo-Christian messianism, heresies, excommunications, etc.). And yet, as we have

1. *The Complete Psychological Works of Sigmund Freud*, vol. XXII (London: The Hogarth Press and the Institute of Psycho-analysis, 1964), 7. Our italics.

shown, its doctrine is incapable of understanding religious phenomena, and is itself compelled to eliminate it physically.

Freudian psychoanalysis, to the contrary, has set itself the task of laying siege to the religious soul in its depths and very interiority. It will probably be objected that there is nothing to this, that Freud was basically a physician who wanted to alleviate certain mental sufferings resistant to ordinary treatment, and that the 'cultural' repercussions of his discoveries are relatively secondary consequences. And, even if they are important, they do not affect Freud's essential and primary aim, which was exclusively therapeutic in nature. We will not go back over the question of the origins of psychoanalysis here. But these origins are more rooted in religious soil than is commonly mentioned, for this major reason: mental pathology has always been considered, by all human civilizations, as naturally dependent on the sacred.

On the other hand, we cannot forget that the last work published by Freud is devoted to religion and summarizes his entire work.[1] So why not see in this growing importance of religion for Freud the result of a clearer and clearer awareness of the nature of his doctrine, which springs in its own right from the 'Œdipus' discovery?[2] This discovery, as Ricoeur notes, was concerned *from the start* with the individual drama as well as humanity's collective destiny, with the origin of neurosis as well as the origin of culture.[3]

Although it is certain then that Freud's major texts dating from the last part of his life turn to culture and religion (*The Future of an Illusion*, 1927; *Civilization and its Discontents*, 1930; *Moses and Monotheism*, 1937–1939), *Totem and Taboo* dates from 1912. And we can even go back farther. In his book *The Psychopathology of Everyday Life*, published in 1901 and therefore composed around 1899–1900, Freud writes quite clearly: "In point of fact I believe that a large part of the mythological view of the world, which extends a long way into the most modern religions, *is nothing but psychology projected into the external world.* The obscure rec-

1. The writing of *Moses and Monotheism*, which appeared in 1939, was spread out over several years, as Freud explains at the beginning of the second part by saying that he found himself "unable to wipe out the traces of the history of the work's origin, which was in any case unusual" (Ibid., vol. XXIII [1964], 103).

2. After having thought he found, at the origin of many neuroses, an attempt at seduction of the child by a parent, Freud discovered, by self-analysis, that it was actually an attempt at seduction of the parent by the child (of his mother by himself). "The first allusion to the Œdipus complex is to be found in Draft N which accompanies Letter [to Fliess] 64 of May 31, 1897" (P. Ricoeur, *Freud and Philosophy: An Essay on Interpretation*, trans. D. Savage [New Haven and London: Yale University Press, 1970], 189, n. 12).

3. *Freud and Philosophy*, 188.

ognition (the endopsychic perception, as it were) of psychical factors and relations in the unconscious is mirrored—it is difficult to express it in other terms, and here the analogy with paranoia must come to our aid—in the construction of a *supernatural reality*, which is destined to be changed back once more by science into the *psychology of the unconscious*. One could venture to explain in this way the myths of paradise and the fall of man, of God, of good and evil, of immortality, and so on, and to *transform metaphysics into metapsychology*."[1] This programmatic text traces the main axes of what will be the Freudian doctrine of religion and very clearly defines psychoanalysis as a fully conscious interpretation of religious paranoia and metaphysics. The vocabulary is still imprecise, but the guiding themes of his thought are there. They will only have to be clarified and developed.[2]

There should be then no doubt about the 'situation' of Freudian psychoanalysis. By discovering the secret of the unconscious, it thinks it has captured by surprise the secret of the making of religions. That mysterious alchemy by which the whole edifice of the sacred is engendered is finally brought to light: it is an (unconscious) projection of the desires and dramas that our (unconscious) soul has experienced and continues to experience (unconsciously) insofar as psychic analysis has not helped

1. *The Complete Psychological Works*, vol. VI (1960), 258–259.

2. For use as needed, let us recall a few indispensable notions. The psychoanalytic method was at first a therapeutic practice that implemented theoretical concepts elaborated 'on-the-job'. Little by little there was imposed the need to introduce order into the concepts and organize the relationships maintained among them, so as to eliminate confusions and contradictions. Now this therapy was essentially based on the idea that the psychism is the theater of continual conflicts between at-once opposed and inseparable personae that struggle on equal terms. These actors in the drama of the psyche are called 'instances' by Freud. These instances divide up the psyche's entire space, and this is why a schematic representation of their relationships will be called a 'topic' (from the Greek *topos* = place). To elaborate such topics is not the work of psychology but 'metapsychology', since what is psychologically observable is surpassed in order to construct a systematic representation of the psychic apparatus. Freud initially adopted a topic in which he distinguished the instance of the conscious (abbreviated Cs), the preconscious (Pcs), and the unconscious (Ucs). This topical representation for making the psychic dynamic understandable was already proposed in *The Interpretation of Dreams* (1901) and will be subsequently clarified. But Freud recognized that the conscious/unconscious distinction, however indispensable it might be, did not allow the instances of the psychism, some of which present both conscious and unconscious aspects, to be differentiated. Hence, in 1923, we have a second topic that distinguishes id, ego, and superego. We find two figures representing this second topic in *The Ego and the Id* (*The Complete Psychological Works*, vol. XIX, 1961), 24, and *New Introductory Lectures on Psycho-Analysis* (Ibid., vol. XXII, 1964), 78. The id (*das Es*) was originally translated into French as 'le soi' (the self), a quite unfortunate translation since there is nothing very 'personal' about it.

it to become aware of them. With awareness the tension resulting from contradictory desires, of course, does not disappear, since nothing can make them be non-contradictory or not conflict with the impossibility of being satisfied: the desire of the child for its mother cannot be achieved in any case. Eros conflicting with the reality principle can only be wounded. But what disappears, according to Freud, is the false consciousness that we had of this wound. We cease disguising it beneath religious tatters to finally look it straight in the face, such as it is. For Freud seems obsessed with the threat that the 'seething caldron' of the id is always ready to dash at civilization. Morality, science, and the rules of humanity seem extremely fragile, "for these things are threatened by the rebelliousness and destructive mania of the participants in civilization."[1] The aim of psychoanalytic healing is not then—at least in principle—to give free rein to the most immoral instincts, nor even to pacify the battle that various instances of the psychism wage in us, but to teach us, with total clarity, how to live our inevitable drama; for no society can survive without imposing on its members the harsh repression of their instincts. As Freud says in *Psychotherapy of Hysteria*, we must learn how to transform our "hysterical misery into common unhappiness."[2]

RELIGION IS A USELESS DELUSION

We needed to firmly establish this point so to clearly see that *psychoanalysis is situated in exactly the same place as religion*, since it is basically proposing to act upon the signs of the powers that subjugate us, rather than on the powers themselves, or only to the extent that manipulating the signs that express these powers acts on them by stripping away their masks and disguises, thus depriving them of their illusory prestige and apparent transcendence. Now the sacred is precisely the realm of symbolic signs because, as we have shown on many occasions, the sacred cannot exist in the order of sensible realities other than through the mode of symbolic signs. Even when religious faith speaks of a 'physical' presence of the divine, in the case of a miracle, for example, or in that of the Eucharist, miracles like the Sacred Species are still signs.

Such a situation makes of Freudian psychoanalysis the actual completion of Feuerbach's project, and, more generally, of a critique of religious consciousness and a radical subversion of the meaning of symbols. It is the unsurpassable form that this semantic inversion assumes, it presents the perfected model, to such an extent that all the forms preceding it—from which we are by no means saying it derives,

1. *The Future of an Illusion*, 10; also 5–9, 15–17, etc.
2. *The Complete Psychological Works*, vol. II (1953), 305.

although it would not have taken place without them—appear in a new light which emphasizes all that they contain of inverting hermeneutics. Since Freud we can no longer read Marx, Feuerbach, or even Hegel in the same manner. With him we are then at the end of the second stage of the destruction of sacred symbolism. We must now attempt a description of this ultimate reversal.

First let us recall that Freud's hostility with respect to the sacred is total, decisive, and continual. True, it is generally covered by a mask of scientific objectivity, Freud excelling in adopting the respectful and firm tone of someone for whom only the truth matters, and who does not hesitate to give proof of the most modest and appealing courage for its sake. Nevertheless the texts are there: "Where questions of religion are concerned," he writes in *The Future of an Illusion*, "people are guilty of every sort of dishonesty and intellectual misdemeanor,"[1] which does not prevent him from coldly asserting a few lines later that "to assess the truth-value of religious doctrine does not lie within the scope of the present enquiry. It is enough for us that we have recognized them as being, in their psychological nature, illusions. But we do not have to conceal the fact [he confesses with 'great sincerity'] that this discovery also strongly influences our attitude to the question."[2] In another text, dealing with *a certain conception of the universe*, he begins by declaring that "psycho-analysis has a special right to speak for the scientific *Weltanschauung*," a claim for the scientific character of his method that is perhaps the major constant in all his work. But evidently this science has enemies. And, calling together art, philosophy, and religion in a most Hegelian manner, he reveals that "Of the three powers which may dispute the basic position of science, *religion alone is to be taken seriously as an enemy*."[3] What is then the source of this hostility?

The first and most decisive, it seems, keeping to the texts at least, is Freud's scientistic rationalism. To our knowledge this is even the only 'profession of faith' that he ever made. He is an avowed supporter of the 'intellect': "Our best hope for the future is that intellect—scientific spirit, reason—may in process of time establish a dictatorship in the mental life of man.... Whatever, like religion's prohibition against thought, opposes such a development, is a danger for the future of mankind."[4] That this involves a belief—a belief in "Our God, *Logos*"[5]—

1. Page 32.
2. Ibid., 33.
3. *New Introductory Lectures on Psycho-Analysis*, op. cit., vol. XXII (1964), 159, 160.
4. Ibid., 171–172.
5. *The Future of an Illusion*, 54.

Freud willingly recognizes, but this is the only one assured of a future: "The voice of the intellect is a soft one, but it does not rest until it has gained a hearing.... This is one of the few points on which one may be optimistic about the future of mankind."[1] As for the desires in the depths of the psychism, they will remain, just as the constraining order of the real remains: "Our God, *Logos*, will fulfill whichever of these wishes nature outside us allows."[2] Psychoanalysis thus contributes to safeguarding science, for "science has many open enemies, and many more secret ones, among those who cannot forgive her for having weakened religious faith, and for threatening to overthrow it."[3]

True, Freud concurs, religion has a task to fulfill. This task is even triple: "exorcize the terrors of nature... reconcile men to the cruelty of Fate, particularly as it is shown in death, and... compensate them for the sufferings and privations which a civilized life in common has imposed on them."[4] However, in the course of historical evolution, men realize that science fulfills the first task better than religion, and that, as for our mortal destiny, belief in gods does not allow us to escape from that. What is left then is the third function: to help us to practice 'cultural frustration'[5] or renunciation of the instincts through their sublimation, for "it is impossible to overlook the extent to which civilization is built up upon a renunciation of instinct."[6] Now—and this is the second reason for the Freudian condemnation of religion—religion fails at this function. It is a poor way to protect us against ourselves and help us endure 'man's helplessness',[7] poor because misleading: "Its [religion's] consolations deserve no trust. Experience teaches us that the world is no nursery. The ethical demands on which religion seeks to lay stress need, rather, to be given another basis; for they are indispensable to human society and it is dangerous to link obedience to them with religious faith."[8] Not then, as is said nearly everywhere, for the sexual repression that it imposes does Freud reject religion, but for the opposite reason: "Religion has clearly performed great services for human civilization. It has contributed much towards the taming of the asocial instincts. But

1. Ibid., 53.
2. Ibid., 54.
3. Ibid., 55.
4. Ibid., 18. The same doctrine is in *New Introductory Lectures on Psycho-Analysis*, 161, or in *Civilization and its Discontents*, *The Complete Psychological Works*, vol. XXI (1961), 74; more clearly, 83–85.
5. *Civilization and its Discontents*, 97.
6. Ibid., 97.
7. *The Future of an Illusion*, 17–18.
8. *New Introductory Lectures on Psycho-Analysis*, 168.

not enough."[1] But how does it actually go about this? "Its technique consists in depressing the value of life and distorting the picture of the real world in a delusional manner—which presupposes an intimidation of the intelligence. At this price, by forcibly fixing them in a state of psychical infantilism and by drawing them into a mass-delusion, religion succeeds in sparing many people an individual neurosis. But hardly anything more."[2] If civilization rests on such a foundation, it will collapse the day that the foundation itself disappears. And that day has already dawned. This is why it is for psychoanalysis itself to assume this function and save culture from an ever-threatening nature. It is hard to say more clearly that Freudian doctrine aspires quite exactly to take the place left empty by religion and finally succeed at what religion has not known how to lead to a successful conclusion.

We come now to this statement already mentioned in the just-cited text: religion is a delusion. No other term is more frequently repeated by Freud when characterizing this cultural form, perhaps because no one more than himself has had so keen an awareness of religion's delusional nature. Hence the numerous passages in which Freud tells of his surprise, his incomprehension, even his anguish before a religion that has known how to "bring the masses under its spell, as we have seen with astonishment and hitherto [until Freud] without comprehension in the case of religious tradition."[3] It is enough for him to recall for a moment the notions of God, providence, or the future life to feel himself overwhelmed and powerless: "The whole thing is so patently infantile, so foreign to reality, that to anyone with a friendly attitude to humanity it is painful to think that the great majority of mortals will never be able to rise above this view of life."[4] Surely we have already noticed similar judgments with a Spinoza or a Kant, and the Marxists like to speak a great deal about 'fantasm' and 'fantasy'. After all, the anti-religious literature of the eighteenth century from Diderot to Helvetius, from La Mettrie to the 'curé' Meslier, offers a wealth of identical formulas. The tone, however, is no longer the same; whereas then denunciation of religious folly seemed above all the fruit of a polemic, or at the very least inseparable from a polemical intention, now, with Freud, it would have itself be an objective and scientific statement. For the first time, to speak of *religious* delusion truly falls within the bounds of mental pathology.

1. *The Future of an Illusion*, 37. Our italics.
2. *Civilization and its Discontents*, 84–85.
3. *Moses and Monotheism*, *The Complete Psychological Works*, vol. XXIII (1964), 101.
4. *Civilization and its Discontents*, 74.

Polemic gives way to medical diagnostics: the sacred is a sickness. No one had dared go so far, in any case in so serious and systematic a fashion:[1] "[E]ach one of us behaves in some one respect like a paranoiac, corrects some aspect of the world which is unbearable to him by the construction of a wish and introduces this delusion into reality. . . . The religions of mankind must be classed among the mass-delusions of this kind. No one, needless to say, who shares a delusion ever recognizes it as such."[2] Happily we now have Freud and psychoanalysis, which might eventually, and not without difficulty, apply their beneficial method to civilization as a whole: "I would not say that an attempt of this kind to carry psycho-analysis over to the cultural community was absurd or doomed to be fruitless." Difficult, yes, insofar as the reference point for 'normal' is lacking here, given the collective nature of neurosis. "In spite of all these difficulties, we may expect that one day someone will venture to embark upon a pathology of cultural communities."[3] But before applying therapy, a good diagnostic procedure is needed.

Up until this point religion is considered to be in fact a means, worked out by human civilization, to aid it in subjecting the pleasure principle to the reality principle. In this sense such an explanation does not differ greatly from the Marxist theory of the opium of the people. This is moreover what Freud himself affirms: "Life, as we find it, is too hard for us; it brings us too many pains, disappointments and impossible tasks. In order to bear it we cannot dispense with palliative measures."[4] And this is why we find some of his texts very reminiscent of Engels: How does man "defend himself against the superior powers of

1. In many of the circles favorable to psychoanalysis, but also sensitive to spiritual (often eastern) currents, this basically anti-religious dimension of Freudianism is minimized or entirely covered over, under the pretext that this is only a personal problem of Freud, a problem which, for lack of bringing his self-analysis to completion, would have been a life-long search to settle accounts with his father and his religion. As for ourselves, we refuse to psychoanalyze Freud, we can only keep to the texts that are perfectly clear in this respect. Also notice in this connection that, as much as Freud's thinking is expressed clearly and in a very comprehensible manner, just so much is that of the Freudians obscure and complicated. In any case it is contradictory to want to combine the Freudian point of view with the religious, for every contradiction requites itself. Now we might equally be denied the right to speak of psychoanalysis, having never been psychoanalyzed. To which we will respond that, if this is true, then that is false! *Qui potest capere capiat.*

2. *Civilization and its Discontents*, 81. Note the last, utterly terrorist sentence, since it forbids any protest from the deluded subject by wrapping itself up in this hermeneutic.

3. Ibid., 144. In other words: humanity on the couch with Freudian psychoanalysts behind.

4. Ibid., 75.

nature, of Fate, which threaten him as they threaten all the rest?"[1] We also find texts relating how nature is 'humanized' by transforming cosmic forces into divine powers. In this sense religion "comprises a system of wishful illusions together with a disavowal of reality, such as we find in an isolated form nowhere else but in amentia, in a state of blissful hallucinatory confusion."[2]

But, abiding by this schema, there is no *history* of religious illusion. According to the mechanism of hallucinatory psychosis, we understand why there is religious hallucination (illusion), but we still do not understand how this is produced. We have a clear idea of the religious form in general; we do not have a clear idea of its particular contents. For that we must penetrate deeper into the secret of the 'laboratory of delusions'.

THE SECRET OF THE ORIGIN OF RELIGIONS

Freud thinks that he has broken through to a secret, as many declarations bear witness: "We believe that we can guess these events and we propose to show that their symptom-like consequences are the phenomenon of religion."[3] But the problem is difficult to understand. Let us recall how it is posed.

The overthrow of the medieval mythocosm, by depriving the symbol of its referent, made it a sign without significance. However, as absurd as this might be, it was no less culturally present, and this was an incomprehensible scandal for any thinking person who identified himself with Galilean reasoning. Sacred symbolism can be expelled, exiled, marginalized, but it cannot be disowned. Hence the need to account for it, give it a meaning, and therefore, at least to a certain extent, justify it. This is why Hegel begins by making it a figure of consciousness, undoubtedly alienated, but nevertheless a form of *logos*, and, on the whole, almost acceptable to reason. Hegel so clearly exhausted the rational strangeness of religious symbolism that *almost* nothing of it remained. But, as we have seen, this was at the price of the disappearance of the real itself. Feuerbach especially wished to lead its entire reality back to this dialectic of the alienated consciousness, an alienation also full of wisdom provided philosophy knew how to interpret it. But the mechanism for the

1. *The Future of an Illusion*, 16.

2. Ibid., 43.

3. *Moses and Monotheism*, 80; likewise, ibid., 92, where Freud declares that he has brought to light "the unsuspected origin" of religious facts. Even more clearly, he recounts how he was led to write a book in which he would reveal an Egyptian Moses assassinated by the Jews; Freud explains: "I had scarcely arrived in England before I found the temptation irresistible to make the knowledge I had held back accessible to the world" (ibid., 103).

producing of fantasy remained inside consciousness itself and, therefore, basically Hegelian. Marx clearly perceived the dangers of such a hermeneutic: humanity cannot be liberated from religious illusion by situating the source of this illusion inside human consciousness. Indeed the religious effect is produced in the consciousness; however, the cause of this reflection is elsewhere, in the material conditions of economic production, and so reason can, in its essence, believe itself sheltered from religious folly. But then, as we have seen, this hypothesis transforms consciousness into a simple product, an epiphenomenon, the faultiness of which is genetically determined, and the rationality of which is but an improbable miracle. We find ourselves then faced with the same difficulty: how does humanity's cultural consciousness account for these undeniable traces of delusion, when it is ruled out that it might originate either from a non-existent divine region or from human thought that is essentially reasoning? What is left is the Freudian solution: to situate the strangeness in the depths of oneself, to lodge in man himself what is exterior to man, to introduce the enemy into the stronghold and to finally understand this alienated consciousness as truly unconscious, but an unconscious endowed with a formidable 'working' power, capable of that mysterious alchemy that gives life and existence to our imaginings and summons the thousands of interwoven sacred forms out of the indistinctness of the soul's dark hell. For finally there must indeed be this laboratory of delusions out of which spring the lasting phantoms of human history. There must indeed be within us something not us, haunted as we are by this indecipherable director, the author and make-up artist of our inner dramas, the untiring sorcerer whose spells have kept mankind enchanted since ancient times. Now behold: his lair is henceforth discovered, the power of his enchantments is broken; the monster's face is laid bare, it can no longer hide itself beneath a god-mask. Freud, the modern oracle, has revealed to the Œdipus that we are the murder that we have formerly perpetrated and that everyone begins anew. Such is our original sin; with him, with this unbelievable event, human history begins. So true is this that history is inevitably and profoundly tied up with events.

This explanation is no longer of the same order as the previous one. It is no longer in some manner a structural projection, but an act accomplished; as such, it cannot but occur, and therefore, however repressed, because shameful, cannot but 'return' and be repeated obsessively throughout the ages, until the day when humanity will finally reach the adult state and, thanks to the psychoanalytic light, will finally rid itself of it. From this point of view "religion would thus be the universal obsessional neurosis of humanity; like the obsessional neurosis of children, it arose out of the Œdipus complex, out of the relation to the father. If this view is right, it is to be supposed that a turning-away from religion is bound to occur with the fatal inevitability of a process of growth, and that we find ourselves at this very juncture in the middle of that phase of development."[1]

There is no need to explain Freud's discovery in detail, which he already formulates in *Totem and Taboo*, in 1912, and to which he will return throughout his life until his last book, *Moses and Monotheism*, in 1939, this time applying it no longer to the so-called 'primitive' religions, but to the two developed religious forms, Judaism and Christianity. This thesis is well known, so we will merely retrace its major features.

The Freudian theory of the origin of religion is the cultural and collective dimension of the discovery of the Œdipus complex. It is not added to it from without; it is an essential and inseparable component. 'Œdipus' is a 'complex', that is a sum total of psychic elements structured around the central phenomenon of infantile sexuality: the love of the son (and the daughter)[2] for the mother, which leads the child to see in his father an admired and hated rival because he occupies the place that the former would like to take. This corresponds to the family unit, this is the 'primitive horde', the theory of which Freud borrowed from Darwin: involved here is a state that universally characterizes a humanity organized into small groups dominated by a 'robust male'. It is he who possesses all the women; and the situation of the sons is quite difficult: if they awaken his jealousy by becoming his sexual rivals, they are driven out, massacred, or castrated. Then the brothers join together, kill the father and eat him. This murder produces three things. First of all, by joining together the brothers create a social contract, and therefore establish a social institution and its basic laws; they also institute morality, since collective existence is opposed to the satisfaction of their desires; lastly, remorse for the parricide and the resulting sense of guilt,

1. *The Future of an Illusion*, 43. Compare this with the quote from *Civilization and its Discontents*, 74, which says just the opposite. The relationship between obsessional neurosis with its compulsion to repeat and religion with its repetitive ritualism appears quite early with Freud, starting in 1907 with the first text that Freud devotes to the neurosis-religion analogy, *Obsessive Actions and Religious Practices* (*Complete Psychological Works*, vol. IX [1959]), 117–127: "In view of these similarities and analogies one might venture to regard obsessional neurosis as a pathological counterpart of the formation of a religion, and to describe that neurosis as an individual religiosity and religion as a universal obsessional neurosis" (p. 126).

2. Assuming the constitutional bisexuality of the young child, "after the stage of auto-erotism, the first love-object in the case of both sexes is the mother" (*An Autobiographical Study, The Complete Psychological Works*, vol. XX [1959], 36).

by becoming interiorized, engenders religion: "From this point on," writes Ricœur, "we can define religion as a series of attempts to resolve the emotional problem posed by the murder and the guilt to bring about a reconciliation with the offended father."[1] Of these attempts Freud basically studied two: totemism and Judeo-Christian monotheism. Totemism is explained by the substitution of an animal taboo for the father, a taboo with which one can become identified and which is eaten at the time of the ritual meal. Subsequently, after a period of latency,[2] the memory of the parricide becomes more and more clear and conscious; one proceeds from animal god to human god, and then to monotheism, which is "the religion of their primal father to which were attached their hope of reward, of distinction and finally of world-dominion."[3] But the encysted memory demands at times to be actually performed once more to reactivate its repetitive power. "The murder of Moses [that Freud supposes] was a repetition of this kind and, later, the supposed judicial murder of Christ.... It seems as though the genesis of monotheism could not do without these occurrences...."[4] With the death of Christ, however, a new element is introduced. St. Paul, who is according to Freud the inventor of Christianity, in fact substitutes the son's death for the father's; the son taking the place for all the murdered brothers, the crime is expiated, the guilt cast off by redemption. Lastly, religion becomes universal. And so Christianity, which "did not maintain the high level in things of the mind to which Judaism had soared" (because it fell back into a certain magic), marks however "in the history of religion—that is, as regards the return of the repressed... an advance."[5] As we see, the compulsion to repeat that characterizes obsessional neurosis does not exclude certain changes, and this is why there is all the same a history of religions that is due to three causes: a delay in the return of the repressed; its weakening, which requires a reactivation; and lastly the (relatively) more and more explicit awareness that we have of it. From this vantage point Christ's death and the Eucharistic meal bring out the truth of Freudian interpretation.[6] This theme of the cultural 'return of the repressed' is not, in Freud's writings, a minor remark. To the contrary one is struck by its growing importance as Freud advances in the drafting of *Moses and Monotheism*. The second

1. *Freud and Philosophy*, 242.
2. *Moses and Monotheism*, 85. The reason for this is that what is repressed does not immediately return.
3. Ibid., 85.
4. Ibid., 101.
5. Ibid., 88.
6. Ibid., 84.

part, which curiously 'repeats' the first, is almost entirely devoted to it, as if Freud were becoming more and more aware that here was the basic driving force of the history of religions, or of what in many places he calls "religious development,"[1] and the less and less 'disguised' major stages which he strives to identify: first the stage of totemism; next the matriarchal cult, the appearance of the divinized human hero; then the stage of a god, master of other gods (henotheism); and finally, after many hesitations, "the decision was taken of giving all power to a single god and of tolerating no other gods beside him. Only thus was it that the supremacy of the father of the primal horde was re-established and that the emotions relating to him could be repeated. The first effect of meeting the being who had so long been missed and longed for was overwhelming," hence the love that the Jewish people experienced for the God of Sinai.[2] But the memory of the guilty sons must also 'return', under the form of a Son of the Father offering himself as redemptive victim in such a way that the religious phantasm becomes the most legible of all. Surely "men have always known . . . that they once possessed a primal father and killed him,"[3] but only with Christianity did they consent to recognize this, hence the hostility of Christians with respect to the Jews: "They will not accept it as true that they murdered God, whereas we admit it and have been cleansed of that guilt."[4] It should be clear then that the history of religions is a kind of *immense collective psychoanalysis* of religious humanity, but an incomplete one, one that requires completion through Freudian analysis, which alone actually lets us face once more the truth of the unconscious, the murder of the father. From this vantage point psychoanalysis, by taking the place of religion, does not mean to destroy it; it is not merely iconoclastic. But it leads the truth of religion to its end; even more, it *is* the truth of religion, it confers the transparency of reality and reason on what is illusorily expressed in it through so much pain and for such a long time. In this sense Freud can declare: "we too believe that the pious solution contains the truth—but the *historical* truth and not the *material* truth."[5] In other words, for want of the psychoanalytic correction, belief in "a single great god" should be termed "delusion."[6]

However, if we want to appreciate this correction philosophically and

1. Ibid., 132ff.
2. Ibid., 133.
3. Ibid., 101.
4. Ibid., 136.
5. Ibid., 129.
6. Ibid., 130.

understand Freudianism for what it is, a counter-religion, we should first analyze what might be called the 'topological conditions' of the religious structure in general or, if preferred, of sacred symbolism, because it is these that impose their laws on the psychoanalytic correction itself.

THE TOPOLOGICAL CONDITIONING OF THE FREUDIAN CRITIQUE

It is curious to observe what an important role the notion of *place*, and therefore *locale*, plays in Freudian psychoanalysis. Freud not only gives the name 'topic' (*topos* = locale) to his conception of the psychic apparatus, which does not after all seem obvious (we would rather expect something like 'schema'), but he further terms 'displacement' those changes of value that the working of the dream, under the action of the censor, introduces into the relative importance of items in our dream representations: "This accent has passed over from important elements to indifferent ones. Thus something that played only a minor part in the dream-thoughts seems to be pushed into the foreground in the dream as the main thing, while, on the contrary, what was the essence of the dream-thoughts finds only passing and indistinct representation in the dream."[1] We ourselves have from the first understood psychoanalysis as something that would 'take the place' of religion. Everything transpires as if the concept basically at work in psychoanalysis was that of a 'topological inversion', just as the concept of reflection is the one basically at work in the Marxist critique. In all domains psychoanalysis intends to realize a displacement or, more exactly, a symmetrical inversion, in such a way that what is below is above and what is above is below. Certainly this concept is also to be met with inside Marxism and even Hegelianism; moreover, it is implied in every philosophy that presents itself as the righting (in both senses) of an inverted consciousness. But it is at work in the most 'literal' manner in Freudianism, and this is also why Freudianism achieves the most radical inversion of sacred symbolism. Sacred symbolism, as we have observed many times, basically has the closest connection with the vertical scheme of above and below, that is, the scheme of transcendence and immanence. Every true symbol is, by itself, a sign of transcendence through immanent mode. And this is why the true symbol is a sacred symbol, in which this transcendent aim is explicitly posited and by which it is distinguished from the profane symbol, the literary symbol for example. Thus every authentic symbolism can be referred to the first instruction of the *Emerald Tablet*, which declares: "True it is and without lie, certain and most true: what is low-

1. *New Introductory Lectures on Psycho-Analysis*, 21; likewise *Introductory Lectures on Psycho-Analysis* (*The Complete Psychological Works*, vol. XV [1963]), 233–234.

est is like what is Highest, and what is Highest is like what is lowest so to accomplish the wonders of the One Thing."[1] Does this connection between psychoanalysis and alchemy prove the former's 'esoteric' character? This is the claim,[2] but this is untenable. First of all Freud, an atheist and scientist, would have rejected this thesis with disgust. On the other hand, apparent resemblances justify such comparisons only by ignoring (more or less intentionally) the true nature of both psychoanalysis and alchemy.

Returning to the text, it speaks of "What is highest" and "what is lowest." It does not deny then that there is a superior pole and an inferior pole; to the contrary, this is affirmed. Besides, this is what an ancient commentary makes explicit: if the 'Tablet', it explains, speaks of high and low, this is "because the [Philosopher's] Stone is divided into two principal parts by Art: Into the superior part, which ascendeth up, and into the inferior part, which remaineth beneath fixe[d] and cleare."[3] And what is said here of the Philosopher's Stone is obviously valid for the sacred symbol with which the Philosopher's Stone is ultimately identical, since the symbol is, as we have seen, a semantic transmuter.[4] Clearly this is also why the *Emerald Tablet* starts with 'what is below' in order to go to 'what is above'; this is the path of symbolic knowledge

1. We translate the Latin text as found for example in an engraving of M. Merian, below a representation of the alchemical cosmos that is preface to the third part of a treatise by Johann Daniel Mylius: *Tractatus secundi seu basilicae chymicae* (Frankfort, 1620). This text of the *Smaragdina Hermetis Tabula* is translated in *Alchemy: Science of the Cosmos, Science of the Soul*, by Titus Burckhardt (trans. W. Stoddart, rev. ed. [Louisville KY: Fons Vitae, 1997], 196–197). The Latin text is the following: "*Verum, sine mendacio, Certum et verissimum: Quod est inferius, est sicut quod est Superius, et quod est Superius est sicut quod est inferius ad perpetranda miracula Rei Unius.*" This text differs slightly from the Arabic text found in a work of Jabir ibn Hayyan (eighth century) but clearly dates back to the sixth century; Burckhardt (ibid., 196) thinks however that a hermetic, and therefore pre-Islamic origin, is certain. This is also the conclusion of the most recent research that indicates a Greek origin. In the West the Latin text of the *Smaragdina Tabula* has been known since the beginning of the thirteenth century. Les Belles Lettres editions have published, in 1994, the translation of several Latin and Arabic versions of this text under the title: *Hermès Trismégiste—La Table d'émeraude* (preface by Didier Kahn). Françoise Bonardel's book, *Philosophie de l'Alchimie. Grand Œuvre et Modernité* (Paris: PUF, 1993) is today the reference work on alchemy's metaphysical implications.

2. Such is the opinion, for example, of Robert Amadou, "Cahier de l'Homme-Esprit," num. 3, 1973, 99.

3. *A briefe Commentarie of Hortulanus the Philosopher, upon the Smaragdine Table of Hermes of Alchimy*, in Roger Bacon's *The Mirror of Alchimy* (London: 1597), 19.

4. This is the entire meaning and conclusion of our book *Histoire et théorie du symbole*.

that goes from reflection to Model, then descends back from the superior to the inferior pole "so to accomplish the wonders of the One Thing." This designates the process of spiritual realization, for it is by the grace or the 'power of the Most High' that the embryo of the divine is formed in the virginal womb of human nature. In other words, the first part of the precept concerns knowledge, the second concerns being and its realization, the miracle or wonder of the creature's union with the Uncreated, indicated by the absence of a prior comma: "so to accomplish the wonders of the One Thing" (*ad perpetranda miracula Rei Unius*), in such a way that only the second analogy, the one proceeding from superior to inferior, directly establishes the miraculous realization of deifying union. We are therefore in the presence of an inverse analogy.[1] Whatever interpretation we give it, this principle signifies likeness through differences; it implies then a diversity of qualitatively distinct regions. It is simply taking note of this difference and by no means proposing a reversal of respective positions by proclaiming that what is below should be above, and conversely, or still more radically, that below is above and vice-versa. Even when a true reversal is to be achieved, when, for example, Christ says that "the last will be first, and the first last" (Matt. 20:16), it is clear that 'Wisdom Eternal' does not intend to deny the existence of a spiritual hierarchy, to transform those who occupy its various degrees. Moreover, the use of the future tense indicates that this hierarchic reversal is a function of a here-below/hereafter distinction which alone gives it its full meaning.[2] As we see, traditional symbolism rests on the conception of a "qualified space"[3] which is a primary given that is intuitively imposed on the human mind (it cannot be reduced through any genetic explanation) and therefore underlies every symbolic representation.[4] In the qualitative regions of this symbolic space are situated realities whose ontological dignity

1. René Guénon, *Symbols of Sacred Science*, trans. H.D. Fohr (Hillsdale, NY: Sophia Perennis, 2001), 307.

2. This corresponds to the Shankarian distinction between *paramarthika* (viewpoint of absolute reality) and *vyavaharika* (viewpoint of relative reality).

3. René Guénon, *Reign of Quantity and the Signs of the Times*, trans. Lord Northbourne (Hillsdale, NY: Sophia Perennis, 2001), 31–37.

4. An indirect confirmation: when a sacred perspective means to surpass the symbolic order, but still has need of an ultimate symbol to designate this surpassing—which is also one from nature—it utilizes the image of non-qualified space, of aether-space (*akasha*), to represent the Absolute: "By space we must understand here, not the qualified space in which we orient ourselves (*dish*), but a primordial, non-constructed environment (*asamkrita*), at once space and aether, prior to shapes and colors, a kind of primary, quasi-absolute given" (Guy Bugault, *La notion de "prajñâ" ou de sapience selon les perspectives du "Mahâyâna"*, Éd. de Boccard, 138).

either do or do not correspond to the regions they occupy, according to whether they are seen from the vantage point of the here-below or the hereafter, which implies that symbolic topology is distinct from the realities found there and does not change along with them. It is this in fact that symbolizes, or *signals*, the ontological order of things: what is situated above is, in principle, what is truly superior. If these topological distinctions become blurred, the human being will no longer have at hand any mark that enables him to recognize himself.

This is however what Freudian psychoanalysis wants to realize, or at the very least what it would like to do if it did not in this case clash with the nature of things, which can never lose its rights altogether. And that is why this undertaking is contradictory. Psychoanalysis maintains in fact an impossible discourse: first, it denies that there is an 'above' or 'below' as such by showing that 'above' is quite simply reduced to 'below', that is, by making the superego and the 'ego ideal' projections or sublimations of what is lowest in man: desires for incest, murder, and cannibalism[1] that the Œdipus present in all of us attempts to realize and are at the origin of religious neurosis; but, secondly, it postulates that there is an above and below to the very extent that it talks about 'sublimation', superego, ego ideal, and projection, or otherwise one does not at all see how such a process would be possible. Here we have one text among others altogether significant in this respect, where we rediscover all the terms we would like to stress at this time: "Other instincts... are induced to displace the conditions for their satisfaction, to lead them into other paths. In most cases this process coincides with that of the *sublimation* (of instinctual aims) with which we are familiar, but in some it can be differentiated from it. Sublimation of instinct is an especially conspicuous feature of cultural development; it is what makes it possible for higher psychical activities, scientific, artistic or ideological, to play such an important part in civilized life. If one were to yield to a first impression, one would say that sublimation is a vicissitude which has been forced upon the instincts entirely by civilization. *But it would be wiser to reflect upon this a little longer.*"[2] In a certain

1. "Among these instinctual wishes are those of incest, cannibalism and lust for killing"; *The Future of an Illusion*, 10. These are precisely those desires that seek satisfaction in the original murder of the father followed by his manducation by the brothers of the horde. The extraordinary concordance (but also all the distance) there is between this text and Plato's in the *Republic* (IX, 571d) has already been remarked: in its sleep the tyrannical soul "does not shrink from attempting to lie with a mother in fancy or with anyone else, man, god, or brute. It is ready for any foul deed of blood; it abstains from no food" (trans. Shorey, *Plato, The Collected Dialogues*, 798).

2. *Civilization and Its Discontents*, 97; the final italics are ours.

manner all the difficulties of psychoanalysis can be reduced to this basic contradiction: on the one hand it is proposing to reveal the very genesis of this symbolic topology and its regional organization, which presupposes that these regions are generable and that the space they share is strictly neutral before the whole regional organization is deployed; but, on the other hand, the very possibility of this deployment by which the phantasmagoric space of culture and religious neurosis is organized entails that an urge 'recognizes' or knows beforehand that one surmounts, surpasses, or sublimes (translating the Hegelian *aufheben* as *to sublime* has been suggested) only 'from above'. No 'from above' without sublimation, affirms a reductive genesis that aspires to restore divine heaven to its squalid earthly roots; but, right reason answers back, no 'sublimation' without a preexisting 'from above' that is in particular its basis. And so no genesis could ever account for the *meaning* of sublimation, that is for the 'elevated' character, as Freud naïvely states, of the non-sexual goal towards which it causes all or part of the sexual energy to deviate. Surely we are not denying, and this is even altogether obvious, that the energy of *eros* is required for the accomplishment of the most difficult and most noble tasks. Plato says it repeatedly; this is also the significance of *tantra yoga* and the "ascent of the *Kundalini*."[1] But obviously, if this energy can be invested elsewhere than in sexual satisfaction, this is because it is not truly sexual in nature, because the sexual realm is only one possible investment for it, and because, all things considered, it has no set nature: it is energy as such.[2] This is why it requires a hierarchic ontological structure offering its various degrees for its investings and therefore determining them and bestowing on them their meaning: to ascend, a ladder is needed. As for Freud, he clearly wanted to find a theory in which sexual energy would engender the rungs of the ladder to be scaled according to its various investings. But this is altogether impossible. Never will anyone grasp in mere desire the principle of its own limits, the reason for its own surpassing and its 'self-transcendence', for desire is only desire to the very extent that it is unaware of these limits: "There is nothing in the id," Freud himself states, "that could be compared with negation"[3]—unless to see in desire the 'existential consciousness' of our creaturely separateness, that is, to consider desire as the creature's response to its ontological limita-

1. R. Guénon, "Kundalini Yoga," in *Studies in Hinduism*, 2nd rev. ed., trans. S.D. Fohr (Hillsdale, NY: Sophia Perennis, 2001), 15–28.

2. Metaphysically considered, cosmic energy is a trace of divine Infinitude.

3. *New Introductory Lectures on Psycho-Analysis*, 74.

tions.[1] We are therefore once again referred back to the preexistence of a scalar ontology, of a hierarchical order of degrees of being, and, hence, to the primary and foundational given of the distinction between a superior and an inferior pole. A moment ago Freud warned us that, if we reflected on this "a little longer," we would perceive that the constraining power of institutions is perhaps not what we imagined, and is actually the constraining power of diverted *eros*. But, if we reflect *a little longer yet*, we will discover behind the force of *eros* the immutable order of the nature of things.

THE FREUDIAN COUNTER-RELIGION

As a result of the contradictory situation of psychoanalysis, it can by no means succeed in reducing the superior to the inferior in a kind of universal horizontalism. This reductive operation, this horizontalization will be the work of Structuralism, which will bring the entirety of the semiotic apparatus down to the positivity of signifiers. This is what the third and last phase of the critique of the symbolic sign will show. Psychoanalysis being unable to abolish the distinction between superior and inferior, what remained was to invert the meaning of their semantic relationship. Such was the conclusion with which we would like to end. It means this: as signs of the transcendent, sacred symbols are always natural realities of an 'inferior' order designating invisible realities of a 'superior' order. The Freudian inversion consists in revealing that what is symbolic are the so-called invisible, superior realities, and the referent to which they point is the actually invisible 'inferior' realities. In a passage from *The Ego and the Id* Freud observes: "Psycho-analysis has been reproached time after time with ignoring the higher, moral, supra-personal side of human nature.... But now that we have embarked upon the analysis of the ego we can give an answer to all those whose moral sense has been shocked and who have complained that there must surely be a higher nature in man: 'Very true,' we say, 'and here we have that higher nature, in this ego ideal or super-ego, the representative of our relation to our parents. When we were little children we knew these higher natures, we admired them and feared them; and later we took them into ourselves. The ego ideal is therefore the heir of the Œdipus complex, and thus it is also the expression of the most powerful impulses and most important libidinal vicissitudes of the id... the super-ego stands in contrast to it as the representative of the internal

1. "The only infinite thing you possess is the affection and desire of your souls," Christ tells St. Catherine of Siena (*Dialogue* 92, Thorold translation).

world, of the id."[1] Notice how the terms 'representative' and 'expression' clearly designate superior realities, symbols of inferior realities biological in nature (incest, cannibalism, murder): "What has belonged to the lowest part of the mental life of each of us is changed, through the formation of the ideal, into what is highest in the human mind by our scale of values." Truly forming "the higher nature of man," the ego ideal "contains the germ from which religions have evolved."[2]

Here we have come back then to a quasi-Feuerbachian schema, and, besides, we have anticipated this in saying that Feuerbach defined the field of demystifying hermeneutics once and for all. But this is a completed Feuerbach who has substituted the economic work of the unconscious for the illusions of consciousness. Feuerbach's 'brainstorm' is a 'stroke of genius', but in a certain manner somewhat ineffective to the extent that consciousness does not easily admit to being deluded in its very act of consciousness. Hence the necessity, seen by Marx, to contrive this exteriorly. But here, as we said, the exterior is within the interior; without our knowing it, the contriving is within us. Each psychic apparatus carries within itself its heaven and hell: the whole religious phantasmagoria is enclosed within the psychic monad of each individual, and what each of us takes to be the flames of the divine Sun are only deceptive reflections of the id's infernal storms. Still, it would be almost impossible to convince us of such an infernal transcendence if, by a no-less-inspired brainstorm and by a kind of psychic translation of the spiritual universe of Leibniz, Freud had not posited from the outset a strict parallelism and, even more, an actual identity between the ontogenesis of the individual superego and the phylogenesis of the collective superego, the *Kultur-Überich*, in such a way that the Freudian monad participates with its own neurosis in the universal neurosis of the species: "the two processes, that of the cultural development of the group and that of the cultural development of the individual, are, as it were, always interlocked."[3] The result of this is a total enclosing of the individ-

1. Pages 35–36.
2. Ibid., 36–37.
3. *Civilization and Its Discontents*, 142. Freud himself on innumerable occasions borrows the terms 'ontogenesis' and 'phylogenesis' (or 'ontogeny' and 'phylogeny') from the vocabulary of evolutionist theory. Having long been struck by the similarities between the embryonic phases of a particular living being and the anatomical characteristics of a particular species, naturalists have proposed seeing in this the proof that the individual, in his embryogenic development (ontogenesis), retraces the various evolutionary phases of the line (*phylum*) to which he belongs: "As Haeckel has said in a lapidary phrase (1866): *ontogeny is a brief recapitulation of phylogeny*. Or, to say it another way, phylogeny determines ontogeny" (L. Cuénot, *La Genèse des espèces animales*, Félix Alcan, 1921, 376).

ual religious neurosis within the religious neurosis of all human history, and, vice versa, an enclosing of all culture within each psychic monad.

Consequently Freudian psychoanalysis ends up actually reuniting the three elements separated by Kantian rationalism: subject, nature, and culture, if we were to allow that the id is nature within us. True, Hegelianism had proposed such a reunion. However, as we have seen, it situates this at the *end* of the Spirit's history, access to self-knowledge being gained through (Hegelian) philosophy, so that this is clearly rather a theoretical program than a truly attained end (unless everything is reduced to Hegelian discourse)—whereas Freud assigned to the human subject as well as to culture one and the same unique origin, situated in a necessarily immemorial past and, for this very reason, necessarily repetitive and obsessional. Here reconciliation is so well accomplished that it is so for ever more, and the difficulty resides not so much in their reunion as in their distinction. Now, by positing the common source of subject and culture at history's origin, Freud makes of their growth and development, *not a thinking community, but a lived community,* because history constitutes the reality of what we are living. Hence the great superiority of Freud over Hegel as to efficacy: the community of the cultural Superego (of 'Culture-over-ego') and the individual superego is surely a universal reality, but a 'concrete universal' because experienced by each individual, because this community is inherent to the fabric of each soul. Within each of us is the entire form of human history which is continually being remade while we make it. The transcendence of the sacred and its terrifying threat, so oppressive to reason, are finally warded off: God is shut away forever in the cage for neurotic imagination's monsters. Just as Galileo's physical world is without a hereafter, so is Freud's psychic world. But this is also, and by way of consequence, a world without hope. We are in hell because hell is within us, because hell is us.

Here Freudianism reveals itself for what it is, that is to say a counter-religion, that is to say still a religion, but the worst one of all. By ceaselessly denouncing the religious illusion, by even making of this denunciation the essential goal for the therapy of neuroses, and precisely insofar as the Œdipus complex is the major concept of psychoanalysis, Freud indicates at the same time his true God, and this God is the id.

It is in fact our right, in our turn, not so much to interpret the Freudian interpretation, but more simply to *see* it as it is, to put down very objectively what he is actually saying, what 'sticks out', but is not always perceived, because we are generally much more interested in a method's implementation than in its results. Freud claims to reveal the most hidden secret of the immense religious edifice. He wants to disclose what we *really* think, do, and love when we cleave to this edifice. As for ourselves, we have no ambition to reveal anything whatsoever, but only to say what Freud himself really and truly says when he 'undoes' the edifice. Now what does he actually say and do? On the one hand, as we have just seen, far from abolishing religious typology, he positions himself inside this typology, being content to make it function upside-down. On the other, all those characteristics that he denies God in the name of scientistic rationalism—and that oblige him to deny God[1]—he attributes to the id, without the least awareness of a contradiction: eternity, invisibility, omnipotence, and inaccessibility, so many concepts that univocally define God and the id.

First of all it must be observed that the 'hypothesis of the unconscious' (*Ucs* in the first topic) truly goes "beyond the limits of direct experience," but that this hypothesis "is a perfectly legitimate one, inasmuch as in postulating it we are not departing a single step from our customary and generally accepted mode of thinking,"[2] that is, it consists in inferring the existence of a cause from the existence of its effects. But there is nothing in all this that differs from the proof of an invisible God's existence by the visible existence of contingent beings, a proof that utilizes exactly the same 'mode of thinking'. Furthermore, this unconscious is endowed with an actual timelessness: "The processes of the system *Ucs.* are timeless; i.e., they are not ordered temporally, are not altered in the passage of time; they have no reference to time at all. Reference to time is bound up with the system *Cs.*"[3] Here we have then a timeless being, inaccessible in itself, immutable, that knows only its own 'whims', and that is apt to wield its power not only in its own sphere where it knows "no negation, no doubt, no degrees of certainty,"[4] but

1. As already indicated, Freudian atheism is, to a certain extent, extra-psychoanalytic and can only be such: every negation (like every affirmation) of God is a philosophical thesis. But the converse is not true: psychoanalysis is not 'extra-atheist', since, to the contrary, it intends to account for religious dementia. Freud expressed himself clearly on this subject. Never short of irony where God is involved, he declares: "How enviable, to those of us who are poor in faith, do those enquirers seem who are convinced of the existence of a Supreme Being! . . . We can only regret that certain experiences in life and observations in the world make it impossible for us to accept the premise of the existence of such a Supreme Being. As though the world had not riddles enough, we are set the new problem of understanding how these other people have been able to acquire their belief in the Divine Being and whence that belief obtained its immense power, which overwhelms 'reason and science'" (*Moses and Monotheism*, 122–123).

2. *Papers on Metapsychology*, in *The Complete Psychological Works*, vol. XIV (1957), 167, 169.

3. Ibid., 187; that is with the conscious system.

4. Ibid., 186.

even over everything not itself and yet under its control. Confining ourselves to these attributes, there is no way to distinguish it from God. The description of the second topic altogether confirms this conclusion. Certainly the id is not the entire unconscious, since the superego is likewise unconscious and even the ego in large part. But it too is outside space and time: "We perceive with surprise an exception to the philosophical theorem that space and time are necessary forms of our mental acts. There is nothing in the id that corresponds to the idea of time; there is no recognition of the passage of time.... Wishful impulses which have never passed beyond the id, but impressions, too, which have been sunk into the id by repression, are virtually immortal."[1] As such, it constitutes, not only the *fate* of each individual, but also that of humanity as a whole in the entire unfolding of its history.

In this history the role of original sin is performed by the murder of the father, at once the failure and realization of the Œdipus complex: the sons have indeed taken the father's place, as 'Œdipus' urges each of us to do, but at the same time they discover that they cannot perform his role. Just as Adam and Eve, by eating of the forbidden fruit, have indeed succeeded in becoming 'like God',[2] but at the same time their eyes are opened and they perceive that they are incapable however of assuming the function implied by the knowledge of good and evil. As for the Redeemer who comes to bring salvation, this is obviously Freud himself and the psychoanalytic doctrine, the therapeutic application of which doctrine can be brought about only by means of duly ordained 'priests', that is, duly psychoanalyzed, because only an analysis of their *psyche* produces a true '*metanoia*' in them, that is, a conversion which is a change of consciousness, and imprints on them an indelible 'character', rendering them in their turn apt for doing the same for others.

Such are, briefly explained, the chief 'doctrinal' components of Freudian religion. Quite obviously Freud is totally unaware of the topologically religious nature of all psychoanalysis. He is too convinced of the positive and scientific nature of its approach for that. What is left now is for us to inquire about the nature of the Freudian 'God' and its basic relationship to the Œdipus complex, which will enable us to highlight those elements corresponding, in this religion, to a 'spiritual way'.

As for its nature, as already mentioned, it is pure power, blind energy. This 'God' is neither Reality nor Consciousness nor Bliss;[3] it is neither Being nor Wisdom nor Love, but only indeterminate dynamism, abso-

1. *New Introductory Lectures on Psycho-Analysis*, 74.
2. *Genesis* 3:22.
3. The Vedantic *Sat-Chit-Ananda*.

lute violence. As for its basic relationship to the Œdipus complex—to the extent that a whole portion of the id is produced by repression—we now need to grasp its true significance. It is simple and rests upon the 'grain of truth' contained in the Freudian conception.

If, by positing the 'Œdipus' principle and the original murder of the father, Freud succeeds in perverting at its root the relationship of the believer to his God, this is because—and this must be clearly recognized—the religious and the parental relationship are closely linked: the latter is in fact that *living symbol* that tells most directly of the relation uniting the creature to its Creator. Living symbols, or symbols of the human form,[1] borrow their signifying elements from the human condition. Man then, through the various aspects of his existence, is a direct bearer of the sacred: his own life is a signifier of the divine and lived as such. Hence, to affect this signifier in itself is also to affect its significance; this affects the only mode according to which this significance, that is to say this truth, can be vitally present to human experience, the only mode that truly speaks to the soul in its native language. And this is so true in the case of the parental relationship that our very existence is involved. Indeed, the basis for the symbolic rapport that unites sonship and creation is the gift of existence: the temporal relationship of begetting is the image of the ontological relationship of existentiation, which continually sustains our contingent existence in being. Quite remarkably, though, Freud never as it were speaks about such a relationship, whereas it is nevertheless the fundamental and primary datum of all life: we first discover ourselves as *born beings*, that is as dependent beings, beings not having their origin in themselves but exterior to themselves, in an act that precedes their own existence and about which they can do nothing: the union of a man and a woman. That each of us possesses this 'knowledge' within ourselves, even if in an 'unconscious' state, is something that should not be in doubt. Freud himself, moreover, in a way alludes to this when he affirms: "maternity is proved by the evidence of the senses while paternity is a hypothesis, based on inference and a premise."[2] Whatever the pertinence of this remark, we will at least retain the fact that there is, even for him, a 'knowledge of the mother', if not of the entire family, at the root of all our relationships with our father and our mother. Besides, how could it be otherwise? Does not our envying the place of the father with the mother imply first that both are recognized as such in one way or another, that is under-

1. On this expression and the classification of symbols, cf. *Histoire et théorie du symbole*, chap. VII, art. II, sect. 4.

2. *Moses and Monotheism*, 114.

stood as the *authors* of our life? Now, *nothing* of this fundamental and primary fact, itself the true living symbol of man's relationship to God—as the symbolism of all human religions attest, where the Divine Principle (Father or Mother or both) is posited as the Giver of existence—remains in psychoanalytic theory. The father is simply a 'figure', a symbol, but the naturalness of the signifier has completely disappeared.[1] Clearly, as a father, he is admired and envied, but the fact that he is a father, that is to say first someone who begets, a cause of existence, plays strictly no role in the relationships that a son might have with him and that Freud claims to reveal. This determination, which alone makes someone a father, is obscured with Freud by that of sexual rival. What is primary in such a perspective is not, therefore, an 'awareness' of our origin—the fact that we have an *originated being*—but the revolt against our principle of existence. Basically, for Freud, the one who is castrated first is not the son, it is the father; in his theory the father exercises all the functions of paternity: spouse of the mother; master of the children, over whom he has total power; owner of the group's property; head of the family or horde; all except one: that of begetter. All the functions of paternity play a part in the Œdipean drama, give rise to oppositions, and thus contribute to the building up of the psychic apparatus, so that their indelible traces can be rediscovered and identified in the repression of the id, in the ego ideal, and even in the self; all except one, which is as if it were not, and yet this is the one that is the basis for all the others. Freud accepts the begetting father only in an abstract and marginal way; he literally 'does not want to know him'. Now this 'ignorance' is the condition thanks to which it is possible to posit the individual as unbegotten and therefore autonomous, that is, as if he had his origin in himself, as if he were the 'son of himself'. Such is clearly the ultimate significance not only of the doctrine, but also the therapy. In common psychoanalytic doctrine, each one of us is entirely produced by our own individual or collective history, and this explains everything. Therapeutically, this involves completing the Cartesian project within ourselves: *to kill the child*. Not to kill childhood memories, their attributes, their effects which still continue and are felt into adulthood, but to destroy the child-substance, that is, our *state as children of a father*—a state which is the objective subject of these

1. Recall what Ricœur says at the end of his *Freud and Philosophy: An Essay on Interpretation*: "So great is my ignorance of the father that I can say that the father as a cultural theme is created by mythology on the basis of an oneiric fantasy. I did not know what the father was until his image had engendered the whole series of his derivatives" ("Dialectic: A Philosophical Interpretation of Freud," 541–542).

attributes—so to set the adult in childhood's place.[1] We will not deny that this state of being a child is our unconscious: our unconscious is our origin. We also think we should attempt a recovery of everything within us that seems connected, 'from behind', to something other than ourselves. But to make of the adult 'self' as such the subject of relationships experienced by the child can be realized only at the cost of a veritable infanticide, precisely for want of going as far as the relationship that founds all the other ones, the filial relationship. To the original parricide of Freudian pseudo-mythology thus corresponds the terminal infanticide of this spiritual pseudo-way that is psychic analysis. While, in reality, the recovery by the adult 'self' of everything belonging to this alter-ego—our childhood within us—can be accomplished only on the condition that we go right to the end of this childhood, right to its ontological root, that is, our state of being a child, which implies a self-renunciation of the adult 'ego', a conscious dying to its illusory autonomy, and its openness to its own origin, that is, to what is not itself within it: to "become like a child" is to "enter a second time into his mother's womb" and "be born twice." For if our origin is what is not us within us—finiteness and ontological dependence—this is at the same time access to a hither-side of ourselves and to our own infinity. But here we are obviously quite far from anything that Freud was ever able to imagine.

Have we strayed from our initial purpose? Have we lost sight of our 'critique of the symbol'? By no means. The concept of *being originated* dealt with in our remarks constitutes a major theme for a study of sacred symbolism, since a sign is symbolic only by reason of its origin. Similar to the umbilicus, the trace of our being born and proof of our dependence, signifying elements are only symbols by reason of their dependence in being, that is as 'offspring' as Plato says,[2] as visible children of an invisible Father. Hence, by erasing the trace of our origin in our soul, by concealing our psychic umbilicus under the heap of the id's impulses, Freud intends to abolish the subjective foundations of the symbolic order and eliminate from our substance everything that might constitute its beginnings. Perhaps such a conclusion will be surprising: is not Freudianism ordinarily considered as a rediscovery of the world

1. In this respect, there are grounds for taking into account all the remarks made by Descartes in the first three *Metaphysical Meditations* concerning the hypothesis that he should be the cause of himself. In any case these remarks prove that Descartes was altogether aware of his *being originated*.

2. *Republic*, VI, 506e: "But of what seems to be the offspring of the Good and most nearly made in its likeness I am willing to speak if you too wish it."

of symbols? But we think this is a complete misinterpretation. Rather, it is clearly the most radical effort ever attempted to destroy the psychic roots of sacred symbolism.

Conclusion: the Price of Sense

The time has come to draw up our balance sheet and show that the price paid for the definitive reduction of religious symbolism is enormous, even prohibitive; and, to tell the truth, no speculative bank is capable of paying it off. However, before arriving at this major assertion, it will be useful to pass in review, as briefly as possible, some 'minor' doctrinal problems, problems left aside so to let Freud speak for himself and to give the logic proper to his discourse its full force, for it is this logic that interests us; this is what is significant for a critique of the symbol, and we intend to show its absurd and 'suicidal' character. But this does not mean that we are indifferent to the improbabilities of a theoretical construction whose success owes little to rational motives.

The first point to be made concerns the doctrine of symbolism expressly set forth by Freud himself and about which we have said nothing yet, given the slight place it occupies in his work (one chapter from *Introductory Lectures on Psycho-Analysis*). This concept has, moreover, no interest in itself, and clearly Freud never sought to develop it in depth. It amounts to defining the *constant* relation that, according to him, unites specific elements of a dream to their 'translations' (like a 'key to dreams'): "A constant relation of this kind between a dream-element and its translation is described by us as a 'symbolic' one, and the dream-element itself as a *'symbol' of the unconscious dream-thought.*"[1] These symbols or constant substitutes (here we rediscover the theme of topological shift) vary greatly insofar as signifiers, but the realities they signify and stand for are few in number: this basically involves the body and the various aspects of the sexual life.

Two remarks are imperative. First, the Freudian symbol is just a camouflage, a disguise, which only makes its appearance because it hides something, and so there is merely a negative interest in it.[2] Second, since

1. *Introductory Lectures on Psycho-Analysis, The Complete Psychological Works*, vol. XV (1966), 150.

2. For example, in *The Future of an Illusion*: "We have become convinced that it is better to avoid such symbolic disguisings of the truth" (44). This is also the thesis of Ernest Jones, that is, the disciple of Freud who has reflected most on symbolism. In *Papers on Psycho-Analysis* (Boston: Beacon Press, 1967) he writes: "Only what is repressed is symbolized; only what is repressed needs to be symbolized. This conclusion is the touchstone of the psycho-analytic theory of symbolism" (116).

symbolism is the language of the unconscious, and since its lexicon is constant and universal, the question of knowing where knowledge of this language comes from must be posed, a question that leads us to the second point: that of the circularity of the psychoanalytic explanation.

Here is, in fact, how Freud responds to the question posed: "we learn [this knowledge]," he states, "from very different sources—from fairy tales and myths, from buffoonery and jokes, from folklore (that is, from knowledge about popular manners and customs, sayings and songs) and from poetic and colloquial linguistic usage."[1] We are therefore sent back from individual dream symbolism to the collective symbolism of culture in general. And Freud shows that the same symbolic relations are at work in this culture. Here, culture interprets the dream: the proof that the dreamed-of 'house' symbolizes the body is that "we greet an acquaintance familiarly as an '*altes Hans*' ['old house']."[2] But a few pages later we learn that the proof of culture's unconscious sexual symbolism is supplied by the psychoanalysis of oneiric symbols: "The mental life of human individuals, when subjected to psycho-analytic investigation, offers us the explanations with the help of which we are able to solve a number of riddles in the life of human communities or at least to set them in a true light."[3] In a more general way, circularity characterizes the relations of ontogenesis and phylogenesis, individual neurosis and collective neurosis, personal Œdipus and social Œdipus. The one referring to the other, each mutually conditioning the other, one can neither leave nor be admitted to this vicious circle.

We have already pointed out that this principle of onto-phylogenetic identity evades any historical verification and raises major difficulties. From the vantage point of positive science this amounts to what is called an 'anthropology-fiction'. Freud does not waver before any unlikelihood and provides hypotheses to order. As he does not hesitate to declare, "Audacity cannot be avoided,"[4] so much the more unavoidably as there is no proof to what he claims to reveal. To this must be added Freud's great ignorance in religious matters. He knew a little about Judaism and Egyptian religion, at least from the historical point

1. *Introductory Lectures on Psycho-Analysis*, 158–159.

2. Ibid., 159. As far as we know this German 'greeting' does not exist elsewhere.

3. Ibid., 168.

4. *Moses and Monotheism*, 100. He is insensible to ridicule, and even to what might be seen as an explaining mania. For example, he informs us that, if man has known how to domesticate fire, this is because he has renounced extinguishing it by dousing it with his urine in the course of a kind of 'homosexual joust', and that woman was appointed its guardian because her anatomy prohibits her from doing likewise (*Civilization and its Discontents*, 90, n. 1).

of view; he read some works on totemism.[1] About Christianity he knew almost nothing. To prove the survival of totemism, he appealed to the tradition that associates each Evangelist with "his own favorite animal,"[2] which has no significance, since, first, one of these animals is a man (or an angel); we are not dealing with a favorite animal, but an emblem whose origin dates back to St. Irenaeus (*Adversus Haereses*, III, 11, 8) interpreting a vision of St. John in the *Apocalypse* (4:6–9), itself recapitulating the vision of Ezechiel (1:5–14), which does not associate an animal with a man, but describes the four mysterious 'living creatures'. All that Freud knew of Greek and Latin patristics is reduced to the overly-famous "*Credo quia absurdum*"[3] that was never actually voiced by Tertullian. As for the eastern religions, it seems that distance authorizes the most aberrant approximations: "The apparently rationalistic religions of the East are in their core ancestor-worship and so come to a halt, too, at an early stage of the reconstruction of the past."[4] This assertion, to some extent applicable to Chinese 'religion', is hardly appropriate for Hinduism or the various forms of Buddhism.

But the chief problem with this hypothesis resides in the contradiction that it introduces at the root of analytic theory, examples of which abound in Freud; we could summarize it this way: no collective neurosis without individual neurosis, but no individual neurosis without collective neurosis. And, as needed, one serves as guarantee for the other: "the analytical study of the mental life of children has provided an unexpected wealth of material for filling the gaps in our knowledge of the earliest times."[5] In a general way "this remarkable feature [of religions that repeat the unconscious past after a time of latency] can only be understood on the pattern of the delusions of psychotics."[6] But, conversely, the reactions of a neurotic child with respect to its parents "seem unjustified in the individual case and only become intelligible phyloge-

1. Freud's theses on totemism as religion's primal form, on the original parricide, and on ritual cannibalism have raised from the start very specific refutations on the part of ethnologists and historians. Between 1920 and 1939 (the date of *Moses*) authors like W. H. Rivers, F. Boas, A.L. Kroeber, B. Malinowski, and W. Schmidt have shown that neither totemism nor cannibalism were originally practiced, and that the theory of the primitive horde subject to a 'super-male' with ownership of the women should be abandoned. But "Freud did not take these objections into consideration" (M. Eliade, *The Quest: History and Meaning in Religion* [Chicago: University of Chicago Press, 1984], 20–21, which summarizes all these objections).

2. *Moses and Monotheism*, 85.

3. Ibid., 118; *The Future of an Illusion*, 29.

4. *Moses and Monotheism*, 93.

5. Ibid., 84.

6. Ibid., 85.

netically—by their connection with the experience of earlier generations," experiences that subsist under the form of "memory traces"[1] and are transmitted by 'heredity' and not by 'oral tradition'. It must be admitted, however, "that at the time we have no stronger evidence for the presence of memory-traces in the archaic heritage than the residual phenomena of the work of analysis."[2]

This hermeneutic or methodological circularity obviously implies a real or objective circularity. Is an individual's neurosis produced by the collective neurosis that is religion, or else is religious neurosis only the result of an initial individual neurosis? Where did it begin? Freud's answers proceed perhaps in divergent directions. Sometimes he even seems to confess his ignorance. Can renouncing the satisfaction of sexual instincts be explained by the authority of the father, when "the father ... is only elevated into being an authority by the advance" that this renunciation constitutes?[3] We think there is with Freud a psychobiological, positivist inclination that leads him toward an etiology of an individual type: it is the anatomo-physiological nature of infantile sexuality that accounts for the Œdipus complex; and a hermeneutic and increasingly marked antireligious bent leads him towards an etiology of a culturalist type. In the closing pages of *Moses* he rejects the Jungian solution of a "collective unconscious"[4] which would account for certain features of neuroses. The reason he gives for this is curious. First, he recognizes that "it is not easy for us to carry over the concepts of individual psychology into group psychology." But above all he sets aside the notion of a collective unconscious being added to an individual unconscious, because *the unconscious, even the individual unconscious, is always collective*: "The content of the unconscious, indeed, is in any case

1. Ibid., 99.
2. Ibid., 100.
3. Ibid., 118.
4. Perhaps we will be reproached for having unreasonably bypassed an examination of Jungian doctrine, the considerable work of which in many ways sidesteps our criticisms addressed to Freudian psychoanalysis. True, on the one hand, Jung's knowledge of the religious domain is incomparably greater than Freud's, and, on the other, his ideas as a whole approximate 'traditional doctrines' in a certain manner. However, we think that there is with Jung a confusion of the psychic and the spiritual, specifically borne out by his theory of archetypes. (We have explained the confusion of these two orders of reality in *Amour et vérité*.) To dispute this theory would have added to the already quite extensive dimensions of our work. But, from the point of view of the general development of twentieth-century ideas, it is incontestable that Freud has played a more important role historically than Jung. Only this general development and its logic interests us, not the thought of a particular person considered in itself.

a collective, universal property of mankind."[1] The conclusion is obvious, and confirms the guiding theme of the present section on every point: Freudian psychoanalysis is basically a counter-religion and intends to substitute itself for true religion, it is a global and antireligious hermeneutic of all human culture through the complete neutralization of sacred symbolism, and only by chance does it come forward to care for some suffering individual.

We come then to our third and final point, by which we will only restate and summarize everything said since the beginning of our study, and therefore presented as the balance sheet of three centuries of cultural and philosophical development. The Galilean revolution, inaugurating the reign of scientistic rationalism, brings out in a flagrant manner the irrational, that is to say insane, appearance of religious belief and its symbols. Inherited by European culture, this 'situation' inevitably condemns philosophy to rationally account for the irrational, just as physics physically accounts for the lie of cosmic forms: *mundus est fabula*. There is no other means than to introduce the same lie into the human soul and show that *anima est fabula* so to account for sacred appearances: to at last make of man this 'fabulous opera' that Rimbaud became when he sojourned for a season in hell. Now to account for something is always to make something intelligible. To account for religious nonsense is to discover the sense of nonsense, and this is clearly in one manner or another what Hegel, Feuerbach, Marx, and Freud wanted to do. But the solution is dual: either to show that apparent nonsense really makes sense, and this is Hegelian pseudo-gnosis and his panlogism; or to show that apparent sense is real nonsense, and this is in a certain manner the Freudian solution. For ultimately, all things considered, does not psychoanalytic truth simply reveal man's universal insanity? *Such is the price of sense at any price.* That *one* man could calmly declare that the entire world since the beginning of time and over the whole surface of the earth was clinically insane, that he was not only listened to but even believed, and that his doctrine was acknowledged, recognized, admired, celebrated, and spread through the entirety of the western world is surely a great mystery, but it is a fact. It is even one of the major facts of contemporary civilization, and by this it also appears to be a 'sign of the times'. By it our civilization and our culture express *their deepest thoughts* about themselves. And what they say is that they prefer to side with the universal insanity of the human species, rather than recognize for an instant the truth of the signs of God. And let no one object about Freud's profession of intellectualist faith, this

1. Ibid., 132.

religion of the *logos* some evidence for which we mentioned at the beginning; for this profession of faith has a reality only insofar as it is directed against religious neurosis: the rational, for Freud, is *all about* the irrational. He believes in the intellect only to the extent that it can be applied to the destruction of the sacred. In itself, in its own nature, the intellect is the plaything of the neurotic imagination. "It has not been possible," he declares, "to demonstrate in other connections that the human intellect has a particularly fine flair for the truth or that the human mind shows any special inclination for recognizing the truth. We have rather found, on the contrary, that our intellect very easily goes astray without any warning, and that nothing is more easily believed by us than what, without reference to the truth, comes to meet our wishful illusions."[1] The second proposition is a rather trite remark which, although arguable, raises no very serious difficulty. But the first has quite another import. Philosophically contradictory, it leads us to the limits of the absurd: by wishing to expel the sacred that it bears within itself, with renunciation after renunciation the human intellect comes to proclaim its own annihilation. The destruction of the *mythos* is also the killing of the *logos*.

1. Ibid., 129.

5

A Supposedly Rediscovered Symbolism

Introduction

Our conclusions seem severe. They clash with the generally held notion that a rediscovery of symbolism was due to the efforts of psychoanalysis. And in fact, by conferring a meaning on what seemed devoid of it in the eyes of scientistic dogmatism, it gave symbolic speech the freedom to be heard, and at least made it possible for this 'something' to be named, something banished from western cultural space as unworthy of inclusion in the program of intelligent discourse. On the other hand, the continual progress of knowledge in ethnology and the history of religions made the immense importance of sacred symbolism throughout the world more and more obvious. Mircea Eliade is right to see in this massive arrival of information a major event for western culture itself, a fact henceforth inscribed in its own history and irreversibly changing the course of this history.[1] What a wonderful conjunction! Psychoanalysis subjectively prepared minds to welcome the objective knowledge brought by the science of religions.

Christians were obviously especially interested in the question. As stressed at the beginning of our work, the scientific revolution and the resulting disappearance of the mythocosm put them in a very awkward position, as Bultmann recognized in his first masterpiece. More than anyone else in the West, they were the ones who made use of symbolism, and in such a manner that it is actually impossible to separate it from faith itself. This is why the profound changes brought by the spread of psychoanalysis and the history of religions should be welcomed by them

1. *Mephistopheles and the Androgyne: Studies in Religious Myth and Symbol*, trans. J.M. Cohen (New York: Sheed and Ward, 1965), 9–15.

to the very extent that they seemed to offer a solution. To tell the truth, at the very beginning the attitude of Christians with respect to the history of religions differed markedly from the one adopted with respect to psychoanalysis. While we see, starting with the second half of the nineteenth century, eminent Christians becoming eminent historians and practicing a positive method on a par with the most antireligious university teachers,[1] analytic psychology was at first violently rejected as an even more dangerous enemy than Feuerbachian demystification. This was because it destroys religion in its objective roots and scandalizes sensibility by its reduction of the instincts to sexuality. A treatise on theology in 1950 summarizes in this way, in order to deplore them(!), Catholic reactions to Freudianism: "The positions taken by psychoanalysis, above all at its beginnings, have doomed it to the most unfortunate vicissitudes: a scandal to the cultivated man, a hog-wallow for literature, etc."[2] The most enlightened Christians, Maritain for example, all while recognizing that the method was 'brilliant' and its psychological observation 'admirable', thought that the whole was spoiled by an aberrant metaphysics: "What is to be said of a theory of the soul developed in common by Caliban become a scientist and Monsieur Homais become an irrationalist?"[3] But the very violence of Christian reaction gave a forecast of its brevity: accusations of mortal sin would soon change to blessings. "For some years now," we can read in the Preface to a collective work inspired by Catholicism and devoted to symbolism, "our curiosity has been aroused about the mysterious world of symbols. We had remained impervious to this mode of expression making use of laws other than those of analytical thought. But, thanks in particular to the historians of religion, to the thorough research of the psychologists, and to works about the poetic imagination, the world of symbols is today open to people."[4] Echoing this is Jean Chevalier's declaration in his introduction to the *Dictionary of Symbols*: "Symbols today are experi-

1. J. Bellamy, *La théologie catholique au XIXe siècle*, G. Beauchesne et Cie, 1904, 100–141 and 182–209.

2. *Initiation théologique*, tome III, *Théologie morale* (Paris: éditions du Cerf, 1961), 216.

3. *Quatre essais sur l'esprit dans sa condition charnelle*, Paris, Alsatia, 1956: "Freudisme et psychanalyse," 46. [Trans. — The passage is omitted from the English translation of this essay in *Scholasticism and Politics* (Garden City, NY: Image Books, 1960), 139–161. Monsieur Homais, an arch-rationalist, is a character in Flaubert's *Madame Bovary*.] Likewise in *The Degrees of Knowledge* (Notre Dame, IN: University of Notre Dame Press, 1995), 159. His judgment is less severe in *Moral Philosophy* (New York: Charles Scribner's Sons, 1964), 450–451.

4. *Polarité du symbole*, "Études carmélitaines," D.D.B., 1969, 9.

encing a new vogue. Imagination is no longer disparaged as a *flight of fancy.* Twin sister of reason, it is rehabilitated as the inspirer of discovery and progress."[1] But, without speaking of the clergy who take up psychoanalysis through a therapeutic bent, the supreme sanction would come from Vatican Council II, which, in the pleasantly superficial style that the Church has thought it should adopt, respectfully saluted depth psychology: "[M]odern culture is characterized as follows: the 'exact' sciences foster to the highest degree a critical way of judging; recent psychological advances furnish deeper insights into human behavior; historical studies tend to make us view things under the aspects of changeability and evolution." Thus "new ways are opened up for the development and diffusion of culture."[2]

All these assertions, chosen from among so many others, undeniably reflect an aspect of the truth. In all this there is something positive for the symbolist perspective it would be wrong to neglect. However, neither can we underestimate the part played, in these opinions, by the illusions of a desire for the sacred repressed for so long and so severely that this desire takes delight in the most deceiving satisfactions. And above all we too easily forget that this return of the symbolic is only possible through its radical neutralization and a complete subversion of its meaning. Far from having rediscovered the authentic sense of the symbol, modern culture has to the contrary so definitely strayed from it that it can specifically mention it and linger over it with total impunity. Symbolism has become inoffensive: it can no longer encroach on the real. Culturological reduction has impoverished it, the Freudian hermeneutic has denatured it. One no longer risks anything in giving oneself up to the delights of the imaginary; having done this, the western soul can even assuage with little cost the remorse that pierces it for having slain God and denied the sacred. But, in reality, access to the symbolic is by no means, for the Christian, this broad and easy way providentially opened up by psychoanalysis and the human sciences. Surely things are so presented nearly everywhere: the Christian thinker, unchanged in his doctrine, would only have to add the happily rediscovered symbolic dimension. This is obviously not so. This slack optimism, to be encountered (alas!) even in conciliar prose, plays fast and loose with all intellec-

1. *Dictionnaire des symboles*, Seghers, 1973, tome I, Introduction, xi.

2. *Vatican Council II: the Conciliar and Post-Conciliar Documents*, ed. A. Flannery, O.P. (Northport NY: Costello Publishing, 1975), Pastoral Constitution on the Church in the Modern World (*Gaudium et Spes*), Part Two, Chap. II, Sec. 1, §54, 959. The Latin text declares: "*recentiora psychologia studia humanam activitatem profundius explicant*," that is to say: the most recent psychological studies explain human activity *in a more profound way.*

tual rigor. And just think of the extreme gravity of Bultmann's undertaking to gauge the difficulty of the task imposed on us. Actually, for anyone who today intends to engage in the way of symbols, with the formation and intellectual requirements of a *modern* man ... we would have to say this is rather about an abandoning of religious doctrine. And this is what an Antoine Vergote lucidly recognizes, for example, when he proposes to study religious language: "our contributions," he declares, "suspend the mass of theological knowledge in order to delve into the human sources and cultural complicities of the Christian signifiers. They are a research into the archaeology of Christian data.... A theo-logy that would reject the insights of the archaeological sciences would condemn itself ... to cultural solitary confinement.... Cut off from living culture, religious thought would be nothing but a cultural neurosis...."[1]

Such is in fact the choice offered to modern man: either the age-old clinging to symbolic forms which, they tell us, truly indicates a 'cultural neurosis'; or their archaeological analysis, which implies the suspending of "the mass of theological knowledge," or, to say it more clearly, the abandoning of dogmatic formulations. This is why it has seemed of interest to try to appreciate *in concreto* the value of this supposed return to symbolism. Two domains have especially held our attention for various reasons: the liturgy and modern art. In our opinion, both domains undeniably bear witness to the profound inability of post-Freudian humanity to enter into the symbol's living reality.

A Recent Restoration of Religious Symbolism

Here we would like to briefly consider the reform that the Roman authorities have imposed on the celebration of Catholic liturgical rites, chiefly on the celebration of the Mass. It seems that this important event, which has thrown the religious life of millions of human beings into turmoil, could not be ignored when reflecting on the symbol, even if we are unaccustomed to seeing such considerations appear in a work of philosophy. Is there not some paradox in ethnology, sociology, the phenomenology of religions, anthropology, and history being interested in the beliefs and symbols of an Australian or South American culture that at times involves only a few hundred individuals, and refuse to study the rites and religious behaviors involving a sixth of humanity?

But this is only the first of the reasons that make an examination of liturgical reform so necessary. Two other reasons help to make this

1. *Interprétation du langage religieux* (Paris: Seuil, 1974), 9–10.

reform the preferred laboratory for a direct observation of symbolic consciousness in the modern world.

The first of these other reasons consists in the fact that the reformers are precisely men formed (in principle) by the discipline of the human sciences. They have remembered the lessons of Freudian and Jungian psychoanalysis, have read the sociologists and the historians, are open to the symbolic systems of the entire world, have listened to van der Leeuw, Rudolf Otto, Mircea Eliade, Paul Ricœur, and all the thinkers seeking to rediscover the symbol's meaning. And not only are they open to all this through speculative interest, but religious authority itself commits them to it under one of its major forms, an ecumenical council: "In pastoral care, not only theological principles, but also the findings of secular sciences, especially psychology and sociology, should be sufficiently acknowledged and put into practice."[1] If reform is necessary, this is because humanity has undergone such changes that it can no longer religiously live with the same rites and the same behavior. On this subject the council speaks of "the spiritual uneasiness of today" and "the changing structure of life" connected to "a broader upheaval" that favors the predominance of the scientific mentality: conversely "the scientific mentality has wrought a change in the cultural sphere and on habits of thought."[2] And so the religious reformers find themselves in an ideal situation to accomplish their task. Not only do the social sciences give them a profound understanding about the nature of liturgical rites and symbols, and therefore the object of the reform, but even more they teach in a most precise manner about the psycho-social requirements and needs of modern man, and therefore about the subjects to which the reform should be applied. Never has an undertaking been presented under such favorable auspices.

The second reason consists in the fact that the reformers, who should rather be called revolutionaries, intended a complete overhaul of the more than thousand-year-old rite of the Catholic Mass in such a way that, virtually unnoticed, "huge chunks of the old order would be refashioned beneath their eyes," as Jules Romains expresses it; nor was the perfection of what they wanted to construct impeded by the imperfection of the older structures. True, all continuity with the previous liturgy was not utterly abolished, at least nominally. However, and here is a fact that must be considered with the greatest attention, by their importance and their extent the changes produced in the rite of the

1. *Gaudium et spes*: "The Church in the Modern World," § 62, *Vatican Council II: the Conciliar and Post-Conciliar Documents*, 966–967 (translation partly modified).

2. Ibid., 906.

Mass are a unique event in the history of Catholicism and, perhaps, in the history of religions in general. Not that history never exhibits striking ruptures; far from it. But these ruptures either appear after very gradual and often imperceptible changes, and are then 'ruptures' only to eyes of the historian who compares two quite distant states after a slow development, or they are the act of a 'prophet' who precisely wants to change the tradition, to break with some of its forms, either to return to an 'original purity', or to found a new religion. This is the case of Luther relative to Roman Catholicism, or of Christ relative to Palestinian Judaism. But what we have never seen is a break by the religious authority itself, decisively and of set purpose, with a ritual tradition of an impressive antiquity, and the imposition of a new rite in a thoroughly obligatory way, all the while asserting that the same faith, the same religion, and even the same liturgy were still involved.[1]

Such are the premises for which the conclusion could only be an extraordinary renewal of religious symbolism, a liturgical 'Pentecost' finally in accord with the needs of modern man, whose "circumstances of life have undergone such profound changes on the social and cultural

1. This is what the great German liturgist Klaus Gamber confirms in *The Reform of the Roman Liturgy: Its Problems and Background* (trans. K.D. Grimm [San Juan Capistrano, CA: Una Voce Press, 1993], chap. 3). A considerable literature on this question already exists. The clearest work is Louis Salleron's *La nouvelle messe* (Nouvelles Éditions Latines, 1976). Some of our assertions will be contested, either with respect to the antiquity of the Roman Mass, or to the immutability of this rite, or to the importance of the changes that the new rite caused it to undergo: "the old Mass is quite recent," "the old Mass has changed continually," "the new Mass is similar to the old one." As for its antiquity, it must be admitted that "at least the core of the Roman canon must have existed by the end of the fourth century" (J.A. Jungmann, *The Mass of the Roman Rite: Its Origins and Development* [New York: Benziger Brothers, Inc., 1951], 51). For the Mass as a whole, its "framework . . . must therefore have been essentially determined by the turn of the fifth century" (ibid., 58). As for its immutability, it is true that in the course of a millennium and a half changes in detail have occurred, such as the addition or suppression of a prayer, and even a word or a rubric. But, apart from a few nuances, the Mass heard by John the Scot, St. Bernard, St. Thomas Aquinas, or St. Theresa of the Child Jesus was exactly the same. As for the essential identity of the two rites, we will only recall the conclusion of Cardinals Ottaviani and Bacci in their *Short Critical Study of the New Order of Mass* (trans. A. Cekada [Rockford, IL: TAN Books, 1992], 27): "the *Novus Ordo Missae* . . . represents, both as a whole and in its details, a striking departure from the Catholic theology of the Mass." To become convinced of this it is enough to refer to the definition of the 'Mass' given by the *Institutio generalis missalis romani* which, in article 7 of the *editio typica* of 1969, deliberately set aside the notion of sacrifice, which constituted a complete break with prior tradition (*Les nouveaux rites de la messe. Références*, Éditions du Centurion, 1969, 23–24). Before the astonishment and the scandal prompted by such an omission, the Liturgical Commission was compelled to publish, in 1970,

level that one is entitled to speak of a new age of human history."[1] Now, quite the contrary, an extreme impoverishment of liturgical symbolism is undeniable. Any ethnologist who took 'Masses for the Dead', for example, as an object to study ritual forms and their development in the liturgical celebrations of the Catholic Church between 1950 and 1980 would observe the almost total disappearance of gestures, actions, words, and chants proper to the liturgy of the deceased. The reformers and those charged with applying and implementing the reform had in fact extremely precise ideas on what the sacramental forms should be: they may be summarized as a general aversion to the sacred (which they deemed pagan), along with an adherence to a more-or-less well-understood Bultmannism. Churches in particular must disappear, then, with the most beautiful turned into museums: "We reject in fact," writes Pierre Antoine, S.J., "every intrinsic or ontological valuing of any place whatsoever as sacred in itself, which would amount to localizing the divine. Desacralization has a not-to-be-ignored spiritual and mystical dimension."[2] Neither sacred place, sacred time, sacred objects, sacred language, nor sacred vestments are to remain. "The liturgy should seek out forms of a bare and primitive simplicity [as if the primitive were not complex].... The general hallmark of the office should not be sacral. As for the priest's clothing, let it show that he is the spokesman, but also that he is connected to the community (he can, for example, wear a long grey cloak) ... [etc.]."[3]

Notice how the enthusiastic words of Christians and the Council in view of the "marvelous discoveries of the social sciences," discoveries that reopen the door to symbolism, do not seem to agree with the requests of those liturgists most concerned about modernity. At times the same people are involved. Christian discourse thus presents a curious ambiguity: sometimes it is lost in wonder before the rediscovery of the folklore of foreign cultures, which is its external practice; sometimes, to the contrary, in its internal practice it proscribes all sacred

another *editio typica* in which this omission was rectified. On the sacrificial reality of the liturgical act that is at the root of the quarrel over rites, the essential work, from the historical point of view, remains *L'idée du sacrifice de la messe d'après les théologiens depuis les origines jusqu'à nos jours*, by M. Lépin, (Éd. Beauchesne, 1926). To this should be added the remarkable study of Father Joseph de Sainte Marie, *L'eucharistie—Salut du monde* (Éd. du Cèdre, 1981).

1. *Gaudium et spes*, §54; op. cit., 959.

2. Pierre Antoine, "L'église est-elle un lieu sacré," in the Jesuit review *Études*, March 1967, 437.

3. Rev. H. Oosterhuis, *La liturgie*, in the joint anthology *Les catholiques hollandais*, Desclée de Brouwer, 1969, 96–98.

forms, and the sacred in general, as a 'pagan' adulteration of the pure faith and a fall back into mythology. The cohabitation of these antinomic spheres in the Christian mindset cannot but pose rather considerable problems. In any case, it provides material of the highest interest for philosophical reflection on Catholic behavior with respect to religious symbolism.

Clearly, such principles can only lead to a setback. This is what one of the reform's authors, the illustrious French liturgist Father Joseph Gélineau, sees: "The change in the liturgy was so sudden and so radical, that it could truly be called a crisis. . . . Whole walls began to crumble. . . . In fact it is a different liturgy of the mass. We must say it plainly: the Roman rite as we knew it exists no more. It has gone."[1] After this Father Gélineau engages in a most remarkable analysis of the process: what is altogether new, he says, in this reform is the "relationship between law and life." The old liturgy had codified and legalized immemorial practices. "Vatican II reversed the situation. First the books were composed. . . . prefabricated."[2] An interesting idea which might to some extent account for the ambiguity pointed out above. *The social sciences open to us, perhaps abstractly, the doors of the symbolic world, but, practically, this theoretical mediation only ends up with an increasing rationalization of ritual symbolism and ultimately its destruction.* It is necessary, says the author, to talk about a "general impoverishment of the forms of the cult itself."[3] Surely he is right in attributing to modern western society the loss of "symbols necessary to any festival," but he also observes that "the post-conciliar liturgical reform is based on a very fine theology of Sunday . . . but paradoxically it seems to have produced a leveling down of festivals."[4]

Father Gélineau finally concludes that there has been a total loss of religious symbolism. The reformers have eliminated symbols that are too concrete, and have replaced them with signs as declarative and precise as possible. "Many symbols were removed from the liturgy because they were thought to be inexplicable, incomprehensible to people today, when they clearly still had living force."[5] "The reaction against the hieratic or grandiloquent ceremonialism of a certain type of 'solemn' liturgy, the desire to have only 'authentic' (?) signs, the Council's wish to

1. *The Liturgy Today and Tomorrow*, trans. D. Livingstone (New York: Paulist Press, 1978), 9–11.
2. Ibid., 15–16.
3. Ibid., 20.
4. Ibid., 61.
5. Ibid., 98.

make the rites understandable, the resistance by a certain contemporary attitude to the haziness of symbols and poetry ('It's not serious,' 'it's not logical'), and other reasons brought about a fury of spring cleaning, throwing out 'rubbish,' and even 'deritualization'. People now prefer the kitchen table for the eucharist, the common glass, ordinary red wine and bread."[1]

To summarize: phenomenally rich, ancient sacred forms have been destroyed, and in their place an abstract, verbose, and barren symbolism has been concocted. Not only in Africa, America, and Asia have cultural murders been carried out, with millennia of beauty and splendor, treasures of imagination and sensibility vanishing, but also under our own eyes, in the West, a similar devastation has been ushered into the sanctuary.

And yet, thirty years later, it seems that the iconoclast fever of a certain brand of clergy has abated somewhat. Has the venture of liturgical desacralization exhausted itself, just as a fire is extinguished when everything is burned? Undoubtedly. But the phenomenon is more complex than a simple tipping of history's scales, and the restoration taking shape calls other actors into play. There is certainly the research of the academics, ethnologists, anthropologists, historians, and musicologists who have studied the nature and development of symbols and liturgical rites, as well as sacred music. There is above all, especially in the Germanic and Anglo-Saxon worlds, worlds only minimally involved in the French quarrels between 'progressives' and 'integrists', the work of theologians and liturgists who have dispelled the hasty certainties of the iconoclasts; in the first rank of these works is *The Spirit of the Liturgy* by Joseph Ratzinger, a work of astonishing power and depth.[2]

1. Ibid., 101.

2. Translated by John Saward (San Francisco: Ignatius Press, 2000). We will also mention some of the principal recently-published titles on this subject: Aidan Nichols, O.P., *Looking at the Liturgy. A Critical View of Its Contemporary Form* (San Francisco: Ignatius Press, 1996); *idem*, *Redeeming Beauty: Soundings in Sacral Aesthetics*, Ashgate Studies in Theology, Imagination and the Arts (Aldershot, UK: Ashgate, 2007); *Autour de la question liturgique avec le Cardinal Ratzinger*, Actes des Journées liturgiques de Fontgombault, 22–24 July 2001, Abbaye Notre Dame, Fontgombault, 2001; Abbé Claude Barthe, *Le Ciel sur la terre: essai sur l'essence de la liturgie*, F.X. de Guibert, 2003; Jean Duchesne, *Retrouver le mystère: Plaidoyer pour les rites et la liturgie*, Desclée de Brouwer, 2004; Alcuin Reid, O.S.B., *The Organic Development of the Liturgy*, with a Foreword by Joseph Cardinal Ratzinger, 2nd ed. (San Francisco: Ignatius Press, 2005); Uwe Michael Lang, *Conversi ad Dominum: Zu Geschichte und Theologie der christlichen Gebetsrichtung*, mit einem Geleitwort von Joseph Cardinal Ratzinger (Einsiedeln-Freiburg: Johannes

Modern Art and Symbolism

Actually the liturgical reform of the Catholic Mass is not the only domain in which the failure of a supposed return to the symbol is definitely in evidence. The development of modern western art testifies to it in a more indirect but just as undeniable manner—indirect because sacred symbolism is rightly no longer involved in this case; undeniable because, from the moment a work is produced, one can no longer lie: every work 'betrays' its author, because it is only what it is, and because its creator can no longer hide behind his intentions. Surely it will seem the height of pretension to wish to pass judgment on five hundred years of modern art in a few lines. Fortunately, such is not our purpose. Firstly, and this is where the liturgy is distinct from modern art, we must make allowances for genius: the liturgical reform was decided and processed in a deliberate manner and with an intention's objective abstraction. The modern history of painting, music, poetry, and sculpture has been left to the randomness of the creative genius whose sudden appearance no one can foresee or hinder. And this is why beauty and success are not absent from our civilization. But they are precisely the improbable fruit of especially gifted individuals, and no longer the continuous and general expression of a culture. When we think about it we realize that there have existed and there still exist human societies *that create nothing ugly*, as extraordinary as that might seem—which in no way signifies that all their creations attain the beauty of a work of Bach or Gauguin. Secondly, we only want to attempt to characterize, philosophically, in a few words the underlying tendency that seems to orient the movement of the arts in Europe. This tendency is already the one manifested in Hegel's esthetic, and so we rediscover, at the close of our analysis, the truth we glimpsed at the beginning.

Here is how we think we can, very synthetically, describe things. Classical European art, until the end of the eighteenth century, around 1770 in any case, is wholly an expressive art, that is eminently an art for society. This basically means that the primary rule is to please, and therefore

Verlag, 2003): *Ever Directed Towards the Lord*, edited by Uwe Michael Lang (London/New York: T&T Clark, 2007); *Beyond the Prosaic: Renewing the Liturgical Movement*, edited by Stratford Caldecott (Edinburgh: T&T Clark Ltd., 1998); Jonathan Robinson, of the Oratory, *The Mass and Modernity: Walking to Heaven Backward* (San Francisco: Ignatius Press, 2005); László Dobszay, *The Bugnini Liturgy and the Reform of the Reform* (Front Royal, VA: Church Music Association of America—Musica Sacra, 2003); Laurence Paul Hemming, *Worship as a Revelation: The Past, Present and Future of Catholic Liturgy* (London/New York: Burn & Oates, 2008). Lastly, we cannot help but refer to the *Motu proprio 'Summorum Pontificum'* of Pope Benedict XVI, promulgated July 7, 2007.

that, in a certain manner, the esthetic act (the making of a work of art) is spent in a handling of the form. There is art each time a 'happy inspiration' imposes a well-ordered form suitable to the expression of an idea, a feeling, a thing, or an event. Only the handling of the form enables the work of art to be unquestionably recognized, identified; and the primary aim of a work of art is precisely to be recognized *as such* and not be confused with what is not art. This is why Monsieur Jourdain's philosophy teacher can declare: "Everything not prose is verse, and everything not verse is prose."[1] For that it is enough to observe whether discourse follows the dictates of meter and rhyme or not, which obviously excludes the existence of a poetic prose or a prose poetry. In music, painting, and theater, this principle is everywhere verified; everywhere we see the artist occupied with arranging his material in the most agreeable and most acceptable fashion. This is an 'engineer' whose talent is composed of skills and who triumphs in the execution. To appreciate the *art* with which a work is executed is to taste mastery, subtlety, variety, the spell of its formal arrangement. The great artist is basically thought of as someone doing things well, not as an eccentric individual obsessed with some disregarded secret of beauty. Neither La Fontaine, the greatest French poet, nor Racine and Molière, the greatest French dramatists, had any notion of belonging to an essence superior to the common run of mortals: for them it simply involved, as Pascal says, a "better placing of the ball." At times we also see them—something almost incomprehensible to us—show a kind of indifference to the position of their work: they leave it without having the feeling of betraying a mission or deserting a priesthood. Basically, the word *art* still retained its primary significance of 'work with a material'. And not until 1542 was this term applied to painting and sculpture.[2] This usage will take long to be established and, little by little, will lead not only to distinguishing the liberal arts from the mechanical arts, as has been done for a long time,[3] but also the 'fine arts', mingled at first with the mechanical arts. The term 'fine arts' prevailed more and more in the seventeenth century, although it entered into the *Dictionnaire de l'Académie* only in 1798.[4]

Such a 'conception' of art, that of a whole society rather than particu-

1. Molière, *Le Bourgeois gentilhomme*, II, 3.

2. F. Brunot, *Histoire de la langue française*, tome VI, 1st partie, 2nd fascicule, "Le XVIIIe siècle," A. Colin, 1930, 680.

3. H.I. Marrou, *Saint Augustin et la fin de la culture antique*, E. de Boccard, 203–204.

4. F. Brunot, op. cit., 681. The word *artist* develops later. In the seventeenth century it is indistinguishable from the word *artisan*. Only in the nineteenth century does the distinction become standard (ibid., 682). And this is clearly not without significance.

lar individuals, cannot but reveal one day or other its insufficiency, its unilateral character, and therefore in a certain manner its contradiction, to the very extent that it reduces all art to just one of its dimensions. This contradiction is the following. Art in every domain (liberal, mechanical, and fine) is defined as a form imposed on nature, since there is art only wherever there is a transformation of nature.[1] The form in question is necessarily a secondary form of such kind that, for art, the natural or primary forms themselves are always 'material'. Art therefore comprises, in its essence, a relationship of differentiation with respect to nature. This is why this act requires of necessity so much more a reason, an end, that justifies its estrangement from the fact of nature as a function of a 'superior' nature; for, in the end, we never leave nature, but we can return to it through more or less lengthy detours. The liberal arts (the sciences) and the mechanical arts (technology) have their justification in the need to know nature in order to use it: "one governs nature by obeying it," and "art without science is nothing."[2] It is the same with the fine arts, the raison d'être for which is, *naturally*, not so apparent, which especially brings out the artificial or gratuitous character of their activity: what does painting add to the beauty of a rose? We are unsure if it makes us see it better, and it can also make us forget it. The raison d'être for the transformation of nature into art is therefore a searching beyond 'ordinary' nature, and is necessarily situated at the level of the divine and sacred, that is, at the level of primordial, perfect, and changeless Nature. The imposition of a secondary form on nature has for an end to reveal in this very nature the divine essences found hidden there, and of which man alone can be conscious. The fine arts thus connect the accidental and peripheral multiplicity of cosmic forms to their eternal prototypes and save them from their own evanescence. Art is basically metaphysical and sacred. This perspective was abolished in Europe at the end of the Middle Ages: the sacred content of high art was completely lost from sight. This is moreover why, art no longer having

1. This is the title of Ananda K. Coomaraswamy's work: *The Transformation of Nature in Art* (New York: Dover Publications, 1956).

2. True, the technical detour can assume such proportions in some civilizations—this is the case with ours—that one has the impression of living in a totally artificial world. This impression is at once true and false. In reality there are always human needs that govern technical activity, that is to say nature, and no amount of advertising can succeed in making a product of lasting demand for which there is absolutely no need. But the manner in which they are satisfied can be artificially produced indefinitely: the cybernetics of needs is not so rigorous that it excludes choices and therefore, ultimately, *moral options*. No human activity operates in a closed circuit in a purely homeostatic manner, and this 'by definition'.

any raison d'être, it appeared in its own right as fine art, whereas in traditional cultures it is hidden and absorbed by its very function. Hence the birth of the 'fine arts' concept and that of the 'artist'. But, on the other hand, with art reduced to its act, that is, to an operation informing material, it was perceived more and more as gratuitous and artificial. Thus the artist, by his activity, is placed more and more in an exceptional situation within the sum total of the society's activities; he is more and more 'marginalized', but, at the same time, finds in his work of art fewer and fewer reasons that justify this situation. An individualization of the artistic function is the result. The artist is turned back upon himself. Genius ceases being engineer; it becomes priest and prophet of a new religion, of a new divinity, the sole bearer of which it is, and its mission is to reveal it to the world.

This change begins to occur in the second half of the eighteenth century. The most striking example of the former conception is provided by Voltaire—for us it is absolutely incredible that he was considered the greatest European poet and owed the greater part of his legendary glory to a dramatic work that no one can read any longer. Poetry was for him "reason set to harmonious verse."[1] But with Diderot, for whom the conception remains as yet confused, another exigency is brought to light. Poetry is not only a form, it also requires a proper matter. "What does the poet need?" Diderot asks. And he replies: "Poetry must have something of the barbaric, vast, and wild."[2] Poetic beauty is no longer a shaping, it is a secret that resides in things themselves and which the order of the world conceals from all eyes except those whom a fatal destiny has marked with the sign of genius. Clearly then form itself, which was the chief concern of the artist, also becomes his chief enemy. And, by form, we must understand not only the order of language, but also the order of nature, the first being moreover only an expression of the last. In every domain form is the intelligible structure, the coherence of design; to understand it is to see its mathematical and rational character at every turn. The Galilean reduction of nature to the visible, to the geometrically demonstrable, knew no slackening in the course of this period: there was no romantic revolution in science; the nineteenth century prolongs and accentuates the eighteenth, itself only a working-out

1. Replying to Helvetius, who was quite wary with regard to Boileau, Voltaire, in defending the poet, also expressed his conception of poetry. He writes: "He has put reason into harmonious verse" (Letter of June 20, 1741, in: Chassang and Senninger, *Recueil de textes littéraires français. XVIIIe siècle*, Hachette, 117).

2. *De la poésie dramatique*, chap. XVIII: *Des mœurs*, in Chassang and Senninger, ibid., 171.

of Galileo and Newton. If the beautiful, the poetic, is truly something other than an arrangement of parts and formal elements, such a science leaves no chance for the artist to discover it unless it is elsewhere, "just so it is out of the world."[1] And rhetoric leaves no chance for the poet to sing it except by "dislocating this great simpleton of the Alexandrine," by "twisting the neck with eloquence," by "giving a purer sense to the words of the tribe," in short by destroying syntax and overthrowing vocabulary. Beginning then was that lengthy revolt of modern art against the form of things and expressions (linguistic, pictorial, musical, sculptural), which would take it from the brief audacities of Romanticism up to the systematic destructions of the surrealists and moderns. Most telling in this respect is the famous definition of poetry given by Hugo in his *Odes et Poésies diverses* in 1821[2]: "Beneath the real world there is an ideal world which reveals itself in all its brilliance to the eye of those accustomed by serious meditation to see in things more than just things. The beautiful works of poetry of every kind, in verse or prose, that have brought glory to our century, have disclosed this heretofore hardly known truth: poetry lies not in the form of the ideas, but in the ideas themselves. Poetry is the intimate essence of all that is."[3] Surely, by itself, such a definition entails no particular consequences; even more, it only expresses the pure and simple truth. Moreover, once we become cognizant of the debates that it has stirred, we realize that the symbol is clearly involved here.[4] Finally, it is all too obvious that genius, by its inmost acts, can only ignore theories. But it cannot really change the objective conditions of its exercise either. Certainly Hugo sees 'more than things' *in things*, and no French writer has loved and known how to play with words as he did. And yet the general movement of western culture inexorably leads modern art (with reactions like Parnassus) to seek out the depths of the real beyond the real itself, behind the world and the rational series of its components, to the very extent that the syntax of the world, like that of language, was posited by classical art as the boundary and model: not only can we not breach it, but we should obey and follow it on every point. Moreover, are these depths,

1. Baudelaire, *Paris Spleen*, XLVIII (New York: New Directions, 1970), 100.

2. *Œuvres complètes*, Éditions chronologiques Jean Massin, tome II, 5. This collection will receive its definitive title *Odes et Ballades* in 1828.

3. J. Massin rightly compares this definition to those of Coleridge, for whom poetry is "the faculty of evoking the mystery of things," and Novalis : "this is the mysterious path towards the interior," etc.

4. We are referring to the letters that V. Hugo and a certain M.Z. exchanged in the columns of *Journal des Débats*, in 1824, (V. Hugo, *Œuvres complètes*, tome II, 533–548).

this 'ideal' world, nothing but the alibi or support for a revolt? As we have shown, classical art quite truly places the artist in a contradictory situation, by depriving him of the spiritual function—and therefore of the supernatural end—conferred on him by sacred cultures, marginalizing him all the while with respect to the useful functions of society. He can no longer, then, truly have access to the transcendent realities that religion maintains present here below and by which it assures further communication. On the whole art can be no more than a substitute for religion, and Hugo's ideal world only the pale copy of intelligible splendors. The only proof that it remains so will be precisely the denial of the natural world and the common language: in the very destruction of things, the order of the world and language, and in the subversion of forms it will be able to vouch for the rightness of its intentions to itself.

The developments just briefly retraced lead us to characterize modern art as an angelism, as much in the context of a disincarnated portrayal of the spiritual and divine as in that of its destructive revolt against the formal order of things and words. Every angelism also has two faces as inseparable as the *recto* and *verso* of the same medal, a 'good' and 'evil' angelism: for the former words weigh as heavily as the body, the divine floats in ectoplasm and utters the inexpressible; everywhere, in painting as well as in literature, the material becomes the enemy that must be combated and against which nobility of soul and the 'ideal' must be expressed. Hence the constant drabness of religious imagery and the depictions of sanctity. The vaporous, the celestial, the hazy, the fluidic, this whole phantasmagoria that is more 'spiritist' than spiritual, to which the Balzacian mysticism of *Seraphitus-Seraphita* bears witness, betrays the deep-seated angelism that inspires it.[1] As we see, we are at the antipodes of true symbolism for which the 'metaphysical transparency' of the sensible world is so evident that it is only one with its most carnal presence. But we are not deceived by this; the full acceptance of corporeal reality is only possible precisely as a function of

1. Balzac's novel dates from 1835. Spiritism was born in the United States in December, 1847 and arrived in France only a little later (R. Guénon, *The Spiritist Fallacy* [Hillsdale, NY: Sophia Perennis, 2004], 16). Balzac was probably then unaware of it. But analogies and affinities are not lacking, if not as to the doctrine of communication with the dead, unknown to Balzac, at least as to the general climate, which is that of a 'confusion of the psychic and the spiritual' and a kind of 'fluidic materialism' of the extracorporeal, the inevitable counterpart of the disincarnation of the spirit. As is known, the major source of this confusion stems from Swedenborg, who at least in part inspired Balzac. We do not deny that the Swedenborgian doctrine of symbolism, essentially based on the (active) correspondence of all the degrees of reality among themselves, may be true. This correspondence is moreover an expression of the proceeding

its metaphysical transparency. If one conclusion is to be clearly drawn from all our studies it is this: *materialism and angelism are mutually related*. Far from corresponding to a victory over the hazy and the imaginary and a conquest by the real and concrete, the materialist reduction of the world to its sheer corporeality is actually a 'disincarnationism', a veritable loss of the sense of the real and the triumph for abstraction and the conceptual. This is nothing paradoxical, for sheer corporeality is only a limit, that of the here-and-now; this is the last degree, beyond which there is simply nothingness. The question may be debated in as many directions as one would like, but there is no escaping this consequence: the purely spatio-temporal is only a 'being of reason', a mental 'term', the limit (actually inaccessible) that sensible reality ideally attains when reduced to its conditions of existence.[1] And yet for three hundred years western civilization has by and large lived with the contrary certainty. Materialist terrorism has achieved such a degree of power over all minds that the identification of the real with the material has passed into a state of speculative reflex, and this is experienced by every thinker who dares challenge it as a feeling of guilt: he faces the main interdict of scientific culture, he sets himself against the thousands of scholars and philosophers, and he is not far from admitting, deep down, that his struggle is only a concession to the human being's irrational assumptions, while it should be obvious that materialism, whether scientific or philosophical, is necessarily the last stage of intellectual decline, the mind-destroying limit of an exhausted intelligence. And bearing witness

of inferior realities from superior ones: "Whatsoever appears anywhere in the universe is representative of the Lord's kingdom, inasmuch that nothing exists anywhere in the ethereal and starry universe, or in the earth and its three kingdoms, but what in its manner and measure is representative ... for from the Divine proceed the celestial things which are of good, and from these celestial things the spiritual things which are of truth, and from both the former and the latter proceed natural things." And so everything that exists "cannot be otherwise than representative of those things whereby they came into existence" (*Arcana Cœlestia* [London: The Swedenborg Society, 1904], vol. IV, §3483, 348). But this classical doctrine is coupled with a scorn for the corporeal on the one hand and a materialization of the spiritual on the other. With regards to man, he declares that "his body is only a superadded external" to the soul, which itself contains internal and external (*The New Jerusalem and its Heavenly Doctrine*, trans. J. Whitehead [London: Swedenborg Society, 1938], §46, 29); whereas "unless spirits [had organs] and unless angels were organized substances, they could neither speak, nor see, nor think" (*Arcana Cœlestia*, vol. II, § 1533, 167–168). This is why, for example "all the angels have their own dwellings in the places where they are" (*Arcana Cœlestia*, vol. II, § 1628, 211).

1. As Leibniz has shown, and as the *philosophia perennis* of the entire world teaches, there are no two beings that only differ *solo numero*, that is in a purely quantitative manner.

to this epistemic guilt are, first and foremost, artistic productions characterized by a deep-seated angelism.

As for the 'dark face' of angelism, evil angelism, it emerges from the very mystery of satanic being. For Satan is also an angel, even an angel reduced to its simple 'angelness', ontological angelism. We propose the following description. Satan, who is at first Lucifer, the 'Light-Bearer', was dedicated, like every angel, to the cosmic reverberation of the Divine Glory: he is himself only to the extent that he turns towards the principial Light in order to reflect It. But, having turned towards himself, he then discovers his own splendor, which dazzles and blinds him. Forgetting that he is only a reflection of the divine radiance, he wishes to seize it and appropriate it for himself. Through a veritable *angelic cogito*, he identifies his being with the possessive awareness that he has of it. As a consequence, eclipsing the radiant source that illuminates him, he 'actualizes' the 'dark side' of the mirror along with his misunderstanding of the Absolute. For every other creature appears unworthy of himself and the Creator, no one having more care for *God's honor* than himself. How admit that the Most High also creates this 'lower world' and "an essence of mud, filth, and hair"? He therefore wants to stop with his own mirror the radiance of the Divine Beauty. Having done this, everything below him is found covered with his shadow and plunged into night: evil is thus as if the dark reverberation and shadow of Satan upon the world. Not having understood that to love the Absolute is to consent to the relative, he hopes to be able to blot out the lower creation that his jealous love of God would not support. Such is the angelic essence of evil. This is why the revolt against all forms and their esthetic destruction clearly derives from angelism and its obscure face. This satanism is explicit in Baudelaire, who, however, still did not implement it esthetically, but rather tried to actualize the monstrous presence in himself, in his sorrowful and corrupted soul. It appeared much more clearly with Rimbaud, who not only behaved, in his life, in a 'satanic' manner, making himself Verlaine's 'demon',[1] but also sought, in and through his work, to realize a veritable destruction of the world. With him the work of art, the poem, becomes the instrument and place of revolt. The concern of 'Dada' and Surrealism was to draw out the

1. Even as 'restrained' an author as Antoine Adam was obliged to recognize that "Verlaine placed himself in the school of his diabolic companion.... Rimbaud... [who] very coolly urged him to reject all discipline, flout moral law, and detest servitude. He taught him to be ashamed of his remorse.... He found it amusing to set free, in his friend, those demonic forces that he had had the joy of discovering in the author of *la Bonne Chanson*. And he later specifies that there was (but not solely) in Rimbaud "a satanic master" (*Verlaine*, Hatier, 30–31).

consequences of this development, to attempt to synthesize it so to bring it to its completion by realizing a total psychological and social as well as literary and artistic revolution. We know what happened—what could not fail to happen—to Surrealism, and how it collapsed beneath the honors lavished on it by a secretly delighted society, a society that they had gone to so much trouble to shock.

But the Surrealist endeavor is more interesting for its objective significance than for its actual accomplishments. And this significance is, it seems, perfectly in keeping with the lineage we have traced. The term 'surrealist' (which for Apollinaire first meant supernaturalism) clearly indicates a will to discover and reach that reality hidden in reality itself, that unimaginably interior dimension of a world which is yet wholly exteriority (for Breton is—and wants to be—a materialist), and which ultimately sends us back to that mechanist cheating with which, since Descartes, scientific explanation is identified.[1] But be not deceived. This universal revolution, this war declared (on paper) against all the forces of order, is not in pursuit of a transcendent reality. The exclusion that struck down the sacred symbol's supersensible referent is entirely upheld. It is the divine order as a whole that is rejected with violence. Feuerbach, Marx, Engels, Lenin, and Freud are constantly invoked by Breton: "Not only must the exploitation of man by man cease, but also the exploitation of man by the so-called 'God' of absurd and exasperating memory."[2] Our poet is also rather heavy-handed: "Everything that is doddering, squint-eyed, vile, polluted and grotesque is summoned up for me in that one word: God!"[3] The signs of the sacred are no more than superstitions: "The absolute limit has been reached in the domination by the symbol of the thing signified," he states in connection with

1. "We have long proclaimed" declares Breton, "our adherence to dialectical materialism, all the theses of which we have made our own: the primacy of matter over thought, the adoption of Hegelian dialectic, etc." (Cited by G. Durozoi and B.Lecherbonnier, *Le surréalisme* [Paris: Larousse, 1972], 226). Likewise, A. Breton, *Manifestoes of Surrealism*, trans. R. Seaver and H.R. Lane (Ann Arbor, MI: University of Michigan Press, 1969), 141–142.

2. *Manifestoes of Surrealism*, 285. There are obviously several ways to co-opt this atheism, the worst of which is to not mention Meister Eckhart's apophatic theology. Does not Breton himself approvingly cite René Guénon and *Multiple States of the Being* (*Manifestoes of Surrealism*, 304)? But, as Durozoi and Lecherbonnier write: "Although he reminds us of the importance of the Kabbalah, of the gnostic thinking of the Zohar, of Eliphas Lévi, etc., Breton always distinguishes their spiritualizing *a priori*, which he rejects outright, from their ambition" (*Le surréalisme*, 157).

3. *Surrealism and Painting*, trans. S.W. Taylor (London: Macdonald and Company, 1972), 10.

"worldly beliefs."[1] It should be perfectly clear then that surreality does not border on 'hinter-worlds'. There is a definite similarity between mystical and poetic analogy, both of which transgress the laws of logic. But poetic analogy "is fundamentally different from mystical analogy in that it in no way presupposes the existence of an invisible universe that, from beyond the veil of the visible world, is trying to reveal itself."[2] The surreal is therefore not the transcendent. It is even its complete negation and lays claim to nothing less than being its substitute by realizing, on the level of sensible reality itself, what the supernatural claims to succeed in doing in the hereafter. The surreal "actually unites in itself all forms of the real."[3]

So then, ultimately and to the extent that all the currents of specifically modern art meet in Surrealism, it should be now altogether obvious that, far from rediscovering an understanding of traditional symbolism through it, we are definitely alienated from this symbolism. Surrealism is in fact first and essentially based on a rejection as total as possible of the naturalness of signifiers, which it means to destroy and make distasteful from top to bottom; whereas, as we have shown, sacred symbolism always rests on a recognition of the world as willed by God, on the nature of things, which is of itself a divine message. To introduce disorder into this theophanic nature, to overthrow the nature of things, is to spoil forever any possibility of a sacred symbolism. On the other hand, Surrealism[4] had in view precisely what Freudianism itself did not dare undertake: *the surrealizing of objective reality.* With Freud the hinter-world, which claims to restore to human life the *depth* that scientific materialism had deprived it of, is situated in the unconscious soul, in the 'no matter what' of the *Id.* Surrealists, to the contrary, want to situate this depth on the very surface of things. With Freud a kind of transcendence as yet remains, but inverted. The sign's referent is 'below'. Now this transcendence has disappeared completely and is spread out in simple immanence. Basically, the surrealists' ideal would be that, according to Taine's expression, the perception of the real world is our hallucination. This is why, in such a system, the sign's referent is only another sign. Either symbols are empty, or everything is symbol, but in a purely horizontal relationship where everything refers to everything else, through a kind of entirely indeterminate correspondence. Already,

1. Ibid., 156.

2. "Ascendant Sign," in *Free Rein*, trans. M. Parmentier and J. d'Amboise (Lincoln & London: University of Nebraska Press, 1984), 105.

3. Durozoi and Lecherbonnier, op. cit., 87.

4. Ibid., 26.

for Baudelaire, the theory of analogical correspondences undergoes a remarkable deviation, since it merges "perfumes, colors, and sounds," that is, realities of the same order. With Surrealism this horizontalism becomes systematic and anarchic. This is what was exhibited in the 'One in the Other' game, developed by the surrealists in 1953, which meant to show that "no matter what object . . . is 'contained' in no matter what other object."[1] It is possible to show in this way that there is an analogical relationship of every thing to every other thing which, Breton claims, enables us "to go back to the source of rites,"[2] provided however that we abandon ourselves to a thinking freed from logical constraint, and from that of the principle of identity in particular. We recover, then, the Hegelian dialectic that is said to ignore non-contradiction and bypass antinomies. Here is manifested, with all needed clarity, the inseparable relationship that joins the destruction of the *logos* to the destruction of the *cosmos*. "It will be asserted that dialectic legitimizes symbolism, and that symbolism (analogy) completes dialectic. . . . Analogy 'particularizes' the connections of simple 'resemblance' then generalizes them (through its inductive effect) as far as a total metamorphosis where differences break down into what is inert in them, just as dialectic 'analyzes' the connections of the old logic and, dissipating the antimonies that impede it, assumes the amplifying destiny of 'Reason.'"[3]

Conclusion: From *Symbolon* to *Diabolos*

We have now arrived at the end of the second phase critical to sacred symbolism. The subversion of meaning is accomplished and religious consciousness definitely neutralized. The sacred symbol as such has survived. But it has disappeared from western culture in how it is thought of after the decisive shock imparted to it by the Hegelian explosion and its series of aftershocks. Each of the strategies developed to reduce the scandal of religious insanity by western reason ends up more and more clearly dissolving the structures of this reason and making insanity the mind's most fundamental norm. Surrealism, not content to declare like Freud that religion is a collective neurosis and take note of it so to limit its effects, gives the best proof of this when it proposes to activate this

1. Ibid., 125. The game consists in asking one person, who has secretly chosen to be identified with a set object, to describe himself by identifying himself with another set object indicated by someone else, in such a way that the object thought of in secret can be guessed.

2. Ibid., 127.

3. G. Legrand, "Analogie et dialectique," in the review *La brèche*, n° 7, 28–29; cited in *Le surréalisme*, 127–128.

insanity in everyone and establish it as a daily derangement: "the definition of Surrealism given in the first Manifesto merely 'retouches' a great traditional saying concerning the necessity of 'breaking through the drumhead of reasoning reason and looking at the hole,' a procedure that will lead to the clarification of symbols that were once mysterious."[1] In this dissolution of the rational and cosmic order, the bond that unites the thing signified to its sign is obviously broken. Breton even specifies that it is impossible to 'go back' from the first to the second. What is left is no more than "to bring language back to true life . . . to go back in one leap to the birth of that which signifies,"[2] the spontaneous fruit of uninhibited desire. Here we have the final phase of a critique of the symbol's meaning. Under the traditional system, the raison d'être of the signifier is to be found, in the last analysis, in the nature of the intelligible and transcendent referent the manifestation of which is the signifier. Under the system of the Galilean *episteme*, the raison d'être of the signifier is situated in a neurosis of the human mind, in such a manner, however, as we see with Feuerbach and Freud, that the bond uniting signifier to its psychic root is recognizable by virtue of some analogy. Now, with Surrealism, everything transpires as if this identifiable analogy were repudiated. The cause of the signifier is always situated 'within' the soul, in the 'region of desire'. But, being pure freedom, this desire would not give birth to any *foreseeable* signifying form. No longer is there between 'thing signified' and signifier any intelligible link; a path from one to the other is impossible. But what is the bond that unites referent to signifier if not precisely meaning? Once meaning has disappeared we have reached, then, the outer limit of the critique of the symbolic sign's second pole.

In a certain fashion we have also reached the outer limit, if not of the symbolic sign, at least of its function. After the neutralization of meaning, nothing is left of the symbol except its corpse, with which Structuralism in particular will be occupied. For the function of the symbol being to unite, to gather, as indicated by its etymology (*sumballo*), it can assume this only *if it at first exercises this function within itself*, if it is by itself the union, the gathering together of a referent and a signifier, of a 'word' and a 'flesh'. Here, to the contrary, the symbol becomes what separates, what divides, what disunites: instead of being 'the divine knot that ties things together', it makes itself the place of their untying, their dispersal, their indeterminate 'liberation', their explosive dissolution, their indefinite drifting through a shattered universe. In other words,

1. A. Breton, *Manifestoes of Surrealism*, 300.
2. Ibid., 299.

symbolon is changed to *dia-bolos*, that is into a 'divider' (from *diaballo*,[1] which signifies: to divide, and hence to accuse, calumniate, precisely because calumnies and lies consist in cutting the linguistic signifier from its referent by perverting its meaning). Just so does the above-mentioned *angelist cogito* take on its most precise significance. The satanism of modern art (in its general trend and not necessarily in each one of its works) is not only a feature adjacent to the neutralization of meaning; to the contrary, it constitutes its essential act. The reversal and inversion of the *symbolon* is actually realized in the *dia-bolic* sign.

Lastly, we should not pass over in silence the manner in which Sartre's existentialism (that of *Nausea* and *Being and Nothingness*) illustrates philosophically, as an unimpeachable witness, the conclusions of our critique. The dividing of flesh and word attains here the rigor and necessity of a proof. It is seized upon as the expression of the very structure of the real and acquires by this fact universality and objectivity. The whole famous analysis developed by Antoine Roquentin while meditating on a chestnut tree root, in a public garden, amounts to asserting that the meaning of things conceals their existence from us, because each thing finds its intelligibility only insofar as it refers to something else: as with the root that feeds the trunk which is itself supporting the branches, and so forth. In this syntax of finality, each thing finding its exact place and being joined together with every other thing, it disappears somewhat like the piece of a puzzle in the design of the whole, and becomes invisible by the very fact of its exact arrangement. To the contrary, if beings are stripped of the semantic varnish that covers them, they then rise up in their monstrous absurdity, their overwhelming existence at last revealed to consciousness: "I couldn't remember it was a root anymore," Roquentin recounts. "The words had vanished and with them the significance of things, their methods of use...."[2] No philosophy, we think, has pressed so far with the separation of existence from essence; none has so completely divided word from flesh, that is: none has so rigorously and scrupulously deduced its consequences: "the world of explanations and reasons is not the world of existence,"[3] Roquentin concludes. The world of existence is one of cold absurdity. Not the artificial absurdity of a universe carefully disordered by its creator to make madness intelligible, but the peaceful and indifferent

1. The inverse relationship between *symbolon* and *diabolos* has been especially pointed out by Evdokimov in *Woman and the Salvation of the World*, trans. A.P. Gythiel (Crestwood, NY: St. Vladimir's Seminary Press, 1994), 144.

2. *Nausea*, trans. L. Alexander (New York: New Directions, 1964), 126–127.

3. Ibid., 129.

absurdity of raw existence, one of "soft monstrous masses, all in disorder—naked, in a frightful, obscene nakedness."[1]

Obviously one can stress the literary character of these descriptions: neither posturing nor making an impression are absent from them. Even the seriousness and 'sincerity' of their author can be doubted. And we would be the first to acknowledge in Sartrean existentialism a kind of 'philosophy-fiction', that is a philosophy that plays fast and loose with the sense of the real for the pleasure of scandalizing or provoking, as well as by an 'adolescent' fascination with the games and paradoxes of an extreme logic. Too often we have the impression that Sartre lets himself be led by his thoughts rather than leading them (and also at times he forces his thinking when he has nothing to say). He is often visibly 'possessed' by a dialectical theme and, not without a joyous ferocity, he will expand it to its most incredible and destructive consequences. And, most likely, the use of stimulants and drugs contributed to enhancing this tendency. But that takes away nothing from the testimony it represents from our point of view. To the contrary, his lack of 'good sense' and his taste for extreme logic guarantee its significant value. A philosophy more attentive to reality would be less of an example. And this is why we speak of an unimpeachable testimony. Sartrism speculatively leads the negation of signs of the Transcendent to its unavoidable conclusion: he denounces the incompatibility of being and *logos*; he finally puts atheism face to face with its truth. This time meaning has truly disappeared, collapsing upon itself by implosion, like those 'black holes' out of which light no longer shines.

1. Ibid., 127.

PART III

The Reign of the Signifier or the Elimination of the Symbol

6

Structuralism as 'Anti-Metaphysics'

Problematic of the Third Critique

We have now arrived at the last phase of the critique of the symbolic sign. Last in fact since, of the three poles that govern its structure, only one remains, that of the signifier. The critique of sacred symbolism having proceeded successively with the elimination of the ontological referent, then with the reversal of the religious sense of sacred signs, can now only attack the very form of the symbolizer, its sensible reality. However, we should not be misled by this linear development: this third phase of the symbolic sign's critique, although directed by the very need for a triadic structure of the symbol, is not altogether situated within the continuity of the previous phases, in the sense that it can no longer be a critique of sacred symbolism. Such a critique was in fact achieved with the neutralization of religious consciousness. Nothing essential can be added. As shown, the general framework for this was provided by Feuerbach's Hegelianism and only the factors at work inside the general scheme of the alienation of consciousness can vary. This obviously does not mean that western humanity has definitely stopped believing in the truth of religious symbols, but just that the process of such a belief is definitely learned, except for a few details, and therefore in the eyes of the western *episteme* there is, quite simply, no longer any sacred symbolism, unless in the context of an ethnological category: sacred symbolism subsists as a fact, in the West or elsewhere, exactly wherever the critique has not been accepted, but it no longer has any right to exist in a rational consciousness. This is not the case with Feuerbach's or Freud's demystification, which has meaning only on condition of still recognizing some *philosophical* importance in what is religious, of somehow taking it seriously, because one demystifies only what retains some chance of being believed: it would not occur to anyone to deny the existence of

'Santa Claus', this myth obviously no longer having any connection with objective reality.[1]

By this we see that the second moment of our critique is its most essential moment, at least from the point of view of sacred symbolism as such. In fact, the ontological neutralization of the symbol is only the consequence of a cosmological revolution that is not at first concerned with the problematic of symbolism, but with the general conception of the universe. True, as we have shown, there is no cosmology that is not a mytho-cosmology, and therefore the Copernico-Galilean revolution was actually in league with religious symbolism. But this relationship only becomes apparent with reflection, so that the problematic of sacred symbolism was experienced by the participants in this history as something marginal, at least insofar as the Catholic Church did not intervene in the debate. Only after this tremendous event did a questioning of the religious as such and its sensible manifestations arise in European consciousness as a major problem that, from Spinoza to Kant, from Kant to Hegel and Feuerbach, and from Feuerbach to Marx and Freud, will become more and more acute. Only at that time will the sacred be critiqued in itself, and the claims of symbolism, in itself, be dismissed. In other words, the *cosmological revolution has necessitated the critique of symbolism*. But it is Enlightenment philosophy that, through the notion of alienated consciousness, has made this possible and actually realized it.

However, this second critique was carried out, as we have said, *from the point of view of sacred symbolism*, and therefore, in a certain manner, by ignoring the symbolic sign as such. Besides, this was inevitable to the extent that, once the rejection of the supernatural (and its immanence) necessarily expressed by the scientific revolution is admitted, the unacceptable claims of religion still remain. The second critique is nothing but a reaction elicited by the sacred itself in every mind anxious about scientific rationality. It is therefore determined by what it challenges, not by no matter what symbolism (mathematical or poetic, for example), that is, by symbolism in general, but only by religious symbolism. This is why moreover, as we have seen, none of the demystifying hermeneutics proposed exemplify an in-depth analysis or a rigorous elaboration of the concept of symbol. The viewpoint of the sign's structure was entirely overshadowed by that of its significance, that is, its function and religious use. And yet a critique of symbolism is clearly involved, to the extent that the divine and the transcendent rightly presentify themselves only by means of symbols. Philosophical atheism might indeed think a

1. This does not prevent a Freudian or Jungian significance from being conferred on it, or even being contested in the name of a psycho-pedagogical prophylaxis.

speculative refutation of God satisfactory, relegating all else (sacred scriptures, pious imagery, customs, festivals, etc.) to the superstition of the ignorant masses. The very idea of God, as Descartes says, is by itself already a symbol. And the more atheism becomes rigorous and exacting, the more it perceives that, step by step, all of human life, all culture, all society should be rid of religious impurities: a vast, exhaustive cleansing that especially demands that symbolism as such and as a whole be taken into account so to achieve an ultimately liberating hermeneutic. Basically, this critique's second phase, by itself, verifies the *inseparable nature of the sacred and symbolic*, not only because, as we have stated, the transcendent divine can be presentified in no other manner, but also because every symbol, its function being to make known something other than itself, realizes its function best in the case of the invisible and the sacred, that is, in the case of an essential wholly *other*. Otherwise the symbolic sign is only a substitute which, in a strict sense, one could do without. Now everyone will agree that the best definition of anything whatsoever is to be found 'at the limit', at the moment when, going from distinction to distinction, one at last arrives at a 'remainder' impossible to subtract from without lapsing into nothingness.[1] In other words, the pure essence is what remains when nothing more can be removed from a thing. It is the same for the essence of a function, its truth only becoming apparent the moment it is absolutely indispensable. This is why pure symbolism by itself implies the divine transcendent or what is essential and invisible.[2] Yet what remains is that this implication is seen in reference to the sacred and that, by very virtue of the two terms' ('sacred' and 'symbol') indissociability, the critique of the first is founded on specifically neglecting the critique of the second.

But this neglect (or forgetfulness) would not last long. The neutralization of religious consciousness through a demystifying hermeneutic, by depriving the symbol of its function, only allows for the existence of its corpse, and this will be called the *structure of the symbolic sign*. As mentioned earlier, this sudden appearance of the symbol in the nakedness of the signifier was already noticed by surrealists. However, it is all too obvious that, by itself, the second critique would not be expressly interested in it. Absorbed in its demystifying task, it cast no glance beyond. Besides, a critique of the signifier would make *no sense* to it. The critique is, in fact, basically an act of rational consciousness and is

1. From this viewpoint, the definition of a thing is what ultimately separates it from nothingness.

2. The result of these considerations is that the sign is religious in essence, or again that the primary sign is the sacred symbol, the linguistic sign being only its 'mental trace'.

concerned with the contents of such a consciousness. A symbolic signifier is one thing, an image, a relationship, in short a 'sensible' factor that is only able to be critiqued insofar as an intention of the human mind is expressed in it. Outside of that the materiality of the signifier is what it is, but in any case 'non-signifying', without interest for a religious critique. And there is not in fact any critique of the signifier as there was a critique of referent and meaning, that is, on the one hand a rejection and on the other a reversal. To reject the materiality of the signifier would be equivalent to suppressing the critique by suppressing its object. Under these conditions a critique of the signifier can be realized in only one way: change the meaning of the signifier itself with respect to the function that it fills insofar as symbolic sign. In other words, this involves showing that the demystifying critique of sacred symbolism rests on an implicit conception of the symbolic sign's structure, which is itself never called into question, while it is however perhaps in contradiction with this critique's basic intention. Such is in fact the nature of the problematic with which the third critique should be confronted. On the one hand, it can only, as we have said, be concerned with consciousness; on the other, insofar as 'third' it can only deal with the signifier. This will then necessarily involve a critique of the consciousness of a signifier. But at the same time—and this is what proves that we have reached a limit—this will also be a critique of consciousness *by* the signifier, a critique so radical that it will explicitly conclude with the *dismissal of consciousness itself*. And so the 'deconstructing' critique of the symbolic sign as followed throughout our study is enough to provide the essential themes for an analysis of the development of European philosophy, showing by this the central although too-often unperceived role of sacred symbolism in this evolution. Having done this, the critique of and by the signifier will have the intention of leading the critique of religious consciousness to its ultimate end and completing what this latter critique could not entirely carry out.

It is quite true then, as we said at the start, that this third critique is not exactly situated along the line of the previous phase: it would be a simple continuation and truly accomplish this only to the extent that it breaks away from this line. In other words, it is not enough to only want to repeat, while radicalizing it, the antireligious project of Feuerbach and Freud. Not from the will of atheism alone, even the most unmitigated atheism—as proven by Sartre's example—can there arise a final solution to the problematic of sacred symbolism. This will come from an *epistemological revolution*, foreign to the religious-symbol system, which is concerned with the *linguistic sign as such*. This revolution, carried over next into the realm of the social sciences in general, will then allow for the overthrow of the sign's implicit conception proper to the second critique, and will lastly be crowned by the philosophic systemization that this necessarily entails. Thus the critique of sacred symbolism is both inaugurated and concluded by an epistemological revolution: Galileo in the seventeenth century and Saussure in the twentieth, the very ones taken as examples[1] to illustrate the epistemic closure of the concept—that is, as we will see better in the last part, the disappearance of the concept as symbol.

Structural Surfacialism

We have just shown, as a function of the necessary succession of various phases in a critique of the symbol, what conditions are imposed on the third critique and determine its problematic. And so we must characterize the nature of the answers that it claims to provide. Now this problematic is organized around two kinds of difficulties raised by this demystifying hermeneutic without being posed explicitly. In the course of the previous chapters two series of objections were formulated: first, we have called into question the possibility of passing causatively from alienated consciousness to the symbolic form such as it is, and have shown the genetico-critical method incapable of engendering sacred signs, the genesis of which it nevertheless claims to reveal—a difficulty which, from Feuerbach to Marx and Freud, has only become more pronounced; next, we have brought to light the diallel implied by the thesis of alienated consciousness, namely that this alienation is at once the cause of alienated consciousness and its effect, or again that alienation presupposes itself (the genetico-critical circle).

As for the first difficulty, recall what was said about the extreme wealth of symbolic signifiers (in a traditional liturgy for example), the complexity of their arrangement, contrary to the poverty and monotony of the causes thought to produce them. This genesis, to the extent that the symbols utilized by the sacred are taken seriously, explains nothing. This is why, when ethnologists and philosophers see that linguistic structuralism by itself offers them a model for organizing the semiotic field, without the intervention of an impotent genesis, they hasten to apply this model to the field of myths and symbols, and show

1. *Histoire et théorie du symbole*, chap. IV, art. 1. Recall that structuralism consists in explaining the form of the signifiers (linguistic or otherwise) not diachronically by their history, as with classical philology, but synchronically by their structure, that is, the differential relationships the constituent elements maintain among themselves: the raison d'être for the 'M' in 'Mater' is not to be the 'P' in 'Pater', and vice-versa. In language, says Saussure, there are only differences, since a sign is only for what is distinguished.

how one can account for the nature of signifying entities out of the structures that they form and that govern them through differential relationships. True, in order to be treated structurally, symbolism has been deprived of one of its essential characteristics: it has quite simply ceased being sacred, and is found to be entirely reduced to the linguistic sign. This is what might be called the semiotic reduction of the symbol, so that we actually fall short then of sacred symbolism.[1]

But, in the eyes of the structuralists, a genesis is not only impotent or useless. It is likewise illusory, and, even more radically, consciousness itself is an illusion insofar as it is constituted as the center and source from which the signifier proceeds. As 'fantastic' as such a conclusion might seem, it is already in a certain manner contained in the theory of alienated consciousness. This theory is in fact, as indicated on several occasions, like a subjective counterpart of the 'faked' perspective of objective reality for Galileo's followers. We repeat: just as the mechanistic outlook wants to 'lodge' in a world without ontological density a depth to account for deceptive appearances, so every demystifying hermeneutic resorts to lodging a secret depth in the transparency of a consciousness, by the magic of which this consciousness becomes its own victim. In other words consciousness is the center-source where lies the hidden mainspring of religious appearances, and, at the same time, this subjectivity is itself put off-center and 'engineered' by something within it that is inexplicably unseen. Once more we reiterate—for this is a subtle point and begs to be considered at length—this does not involve an accidental, contingent, and episodic (or an *unconscious*) ignorance, but an essential ignorance basic to consciousness itself, a *structural*... or, better said, ontological ignorance, since it 'creates' its own being. Consciousness is at once posited as the true locus of every explanation, and dispossessed of itself at its very core. We therefore must go further, or rather let the truth being articulated here unawares speak for itself. This truth is that the illusion by which consciousness is victimized is consciousness itself as pseudo-center-source: *consciousness is the illusion of consciousness.* Self-presence, and the identity that characterizes it, is only the structural effect of being other than the other, the differential of an alterity, pure *difference* without any *interiority.*

And so the victory of the signifier is complete. Not only is sacred symbolism not the product of consciousness, not only are its organization and structure not the result of a sick mind projecting itself in a veiled way, so that the history of this mind could be known by a simple deciphering of the semiotic field, but even more it is the field itself which,

1. 'Semantics' designates what relates to meaning; 'semiotics' what relates to sign.

because structured and governed according to the rule of differential entities,[1] necessarily structures and governs consciousness and human thought. Hence the sacred is revealed for what it is: the illusion of a consciousness that thinks of itself as transcendent, the center-source of all things, and by the same stroke thinks that Transcendence and Center are within itself. Feuerbach is at once rediscovered and inverted: rediscovered since the divine is only consciousness itself, but inverted because the illusion of consciousness is not inexplicably lodged within consciousness, the illusion is consciousness itself insofar as locus of interiority, and the plot, of which it is in some way the victim, is hatched from 'without', from the functioning of a structure of which subjectivity is only an effect.

One could say, after a fashion, that we are witnessing the definitive triumph of Galileo. The myth of depth—this 'spatial' dimension not embedded in space—is completely exorcised. The subject, its transcendence secured by the Cartesian mechanistic system, confronted as it was by indefinite exteriority, finally falls back into line; it rejoins a surface without thickness to which everything is reduced; it sunders and disperses its interiority to the four corners of structural relationship. This is why it seems possible to characterize structuralism as a 'surfacialism'. True, the real nature of this surfacialism, and the consequences that spring from it, are actually only perceived starting with the semantic principle, with surfacialism being used, *a contrario*, to demonstrate its proof. But for now we can be sufficiently assured of its reality by understanding the need for it from second-critique contradictions, and also by its telling us something about the structuralists themselves whose declarations, as we shall see, are altogether explicit in this respect.

Culture Against Metaphysics

Structural surfacialism surely deems itself the last word in the anti-metaphysical struggle as the agent of metaphysics' utter destruction and its finally successful expulsion from speculative discourse. And, just as we have recently observed how the second critique proved the necessary correlation to symbolism and the sacred, so now this expulsion, a result of semiotic reduction, proves the inseparable character of the relationship that joins metaphysics to symbolism. Even so, it is not always recognized from either side. Famous examples are not lacking of philosophers, such as Aristotle and Descartes, who reject symbolism or make

1. See what we have said about linguistic structuralism in *Histoire et théorie du symbole*, chap. IV, art. II.

it cosmologically difficult, and who nevertheless cannot help but construct what they themselves term a metaphysic.[1] Conversely, an entire anti-rationalist current sees in symbolism the model for a thinking freed from logic's 'narrow' categories, from the clumsiness and blindness of concepts. Nevertheless, the more western thought becomes cognizant of its own requirements, derived from scientific rationality, the clearer the need to combat the metaphysical illusion in terms of sacred symbolism. With each new philosophical age, at least starting with Kant, the champions present themselves as those who will save the thinking city from the metaphysical hydra. Unfortunately, like all hydra, no sooner is one of its heads cut off than another takes its place, and so the task is repeated. But there is no dearth of heroes. Heirs of a new tradition as well as the mantel of philosophy, their first mission is to fight this monster. Now, by a truly astounding marvel unknown to legend, if we look closely we see that in reality each newly grown head is none other than that of the hero who cut off the previous head!—as if the fabled animal mainly grew from all those who would cut it down. Has there ever been a stranger metamorphosis? This odd tradition obliges its inheritors to kill those who teach it. Thus Heidegger regards "Nietzsche, with as much lucidity and rigor as bad faith and misconstruction, as the last metaphysician, the last 'Platonist'. One could do the same for Heidegger himself, or for a number of others."[2] In any case, structural surfacialism clearly intends to rid Freudianism and Marxism of their metaphysical impurities, of their quasi-religious beliefs in the centrality of consciousness as alienation's privileged place and in the hinter-worlds of the unconscious or 'superstructure', by reinterpreting their doctrines with the help of the linguistic model provided by Saussure. And besides, is not Saussure himself actually the last of the metaphysicians? This is what one specialist declares: "By his systematics Saussure is an eminent metaphysician, and perhaps nowhere better than with him is there realized the desire for purity that nourishes the philosophic myth."[3] As indicated by Derrida, this game can in fact go on for quite a while. But the time comes for the Foucaults, Lacans, Althussers and Deleuzes to make their appearance as frightful metaphysicians, travestied Platonists irremediably compromised by the objectivist illusions of an ill-justified first philosophy.

1. Thus Aristotle condemns Plato as guilty of using metaphors: "To say that the Forms are patterns, and that other things participate in them, is to use empty phrases and poetical metaphors" (*Metaphysics* A, 9, 991a 19–22). Notice however that the possibility of Cartesian metaphysics is tied to the existence within us of the idea-symbol of the Infinite.

2. Jacques Derrida, *Writing and Difference*, trans. A. Bass (Chicago: University of Chicago Press, 1978), 281.

3. R. Strick, article "Structure linguistique," in *Encyclopædia Universalis*, vol. 15, 432.

Surfacialism is anti-metaphysical to the extent that, as we have just seen, it considers language, symbolism, and lastly all culture as a system of purely differential and therefore structure-forming entities. The system is self-structuring, obeying only the rule of differentiality. The illusion denounced by the first critique consists in believing that symbolism reflects the nature of things, and is therefore determined by it, whereas it is only a reflection of human nature, as shown by the second critique, for which it is generic (Feuerbach) or social (Marx) or unconscious (Freud) man that determines the symbolic and ultimately the social order. But, with the third critique, it is symbolism and the structural system that govern and determine man and the world, subject and nature. And so we rediscover the 'critical triangle' identified at the beginning of the fourth part of chapter 3: nature-man-culture. It will be useful now to present a synthetic description of this, which was previously impossible because it had still not reached its full development, just as it was still not possible, at the time of the first critique, to simply formulate it, since only beginning here, that is, beginning with the Galilean revolution, does it arise as a problematic subjacent to the philosophical field. Each pole of this triangular problematic can be seen as exercising in turn dominance over the philosophic field.

In the first critique we see the *nature pole* dominant. It is this that obliges the subject to be distinguished from itself, as the interior and the extra-mundane are distinguished from an entirely spatialized cosmos. The subject is then more or less merged with culture insofar as it is itself more or less identified with reason; but the *Discourse on Method* already outlines the program for a cultural revolution. In this critique nature also forces a rejection of an ontological supernatural referent: this is the struggle of physical reality against metaphysical unreality. In the second critique dominance pertains to the *subject*. This is what, for Kant, governs nature and posits it within itself, while at the same time, for Rousseau, it is defined against culture, of which it becomes sole creator[1] in Hegelian and post-Hegelian philosophies. However the thesis of culture's dominance was already outlined by Wilhelm von Humboldt. In this critique it is also the subject that declares itself producer of all significance and by this rejects the so-called sacred meaning inherent to symbols: this is the struggle of *rational consciousness* against *religious consciousness*. According to the third critique, dominance is conferred on the *cultural pole*, considered to be a self-structuring system. It is this that governs and constructs the subject as well as nature and causes them to be distinguished from each other: *insofar as he differs from the world, man is an*

1. But a creator only revealed to itself by its own creation.

effect of cultural structure. In this critique, it is also the structure of differential unities that accounts for the symbolic order as a whole and compels a denial not only of its referent but also of its meaning, even under the form of the sacred as an illusion and alienation of consciousness, and therefore the symbol is rejected in its entirety. Henceforth the symbol is completely erased; there is no longer anything to distinguish it from any other sign. In the second critique, the sacred still has an existence, that of an illusion in the reality of consciousness. In the third, the sacred has entirely disappeared; hence the critique of a particular symbolism clearly no longer has a raison d'être. In the last analysis, this shows that structuralism has strictly nothing to say about what it has excluded in so total a fashion. Certainly one can still speak about myth, religion, and symbol, but these are purely extrinsic terms in no way concerned with the nature of objects so designated. What is religion? What is myth? What is a symbol? Structuralism never answers these questions, although it claims to carry out a systematic analysis of them. It could almost assert, in the manner of doctors Binet and Simon ("intelligence? this is what our text measures"): "myth? this is what we decompose into mythemes."

However, this promotion of cultural structures to the rank of final explanatory principle is not without significance from our own philosophical point of view, if only to the extent that it leads to acceptable scientific results, which is precisely the case with Lévi-Straussian analysis, at least in part. We recognize in fact that the decomposition of a myth into coordinated structural unities (*mythemes*) enables incontestably positive data from the reality of mythic discourse to be brought to light, and from this can be inferred the structuring and organizing power of every cultural system in general. This is precisely why we must show what that means from the vantage point of sacred symbolism, and how its importance can be appreciated philosophically, which is only possible provided that a fourth term is reintroduced into the critical triangle, the very one spoken about by symbolism, its *ultimate referent*: *the divine transcendent*. It is then that the critical triangle, by the activity of the symbolic triangle, is transformed into the metaphysical pyramid, and so resolves the antinomies that form its structure. But for that it must accept what it has always excluded: the *truth* to which the sacred symbol bears witness. In other words, the critical triangle, after having *deconstructed* the symbolic triangle, finds itself in its turn 'worked' from within by the symbol that thus saves it from its own contradiction.

To clarify these considerations, we will represent them with the help of two successive diagrams. The first situates the symbolic triangle with respect to the critical triangle, and therefore according to the phase of deconstruction, while the second expresses the transformation effected by the symbol's internal semantic activity and the uplifting that it achieves so to unite what was separated. As to the first diagram, we will recall that the primary effect of the critique of the symbol was to completely amputate it from its semantic transcendence, and therefore reduce it to a simple sign. This amputation or decapitation is basic to critical activity as such. This is why the critique of the symbol deconstructs of the sign only what can be reached, that is what is projected from the symbol onto the horizontal plane defined by the three critical instances. The rest, the semantic 'volume', eludes it completely. Such is also the reason why, after having given a tetradic definition of the sign in *Histoire et théorie du symbole*, we have however developed a critique of a triadic symbolic sign. This is because such a triad forms in fact the critique's whole *horizon*, all that it can perceive of it. Now, as for the positioning of the three apices of the symbolic triangle inside the critical triangle, the solution is simple and follows from all our analyses: the signifier is situated on the segment between nature and culture since it stems from a withholding of culture from nature; meaning is situated on the segment between culture and man since it is the activity of a consciousness informed by culture; lastly, the referent is situated on the segment between man and nature since it stems from nature as seen by man (it being assumed that the supernatural referent is not 'visible' from the critical point of view). We therefore end up with the following diagram:

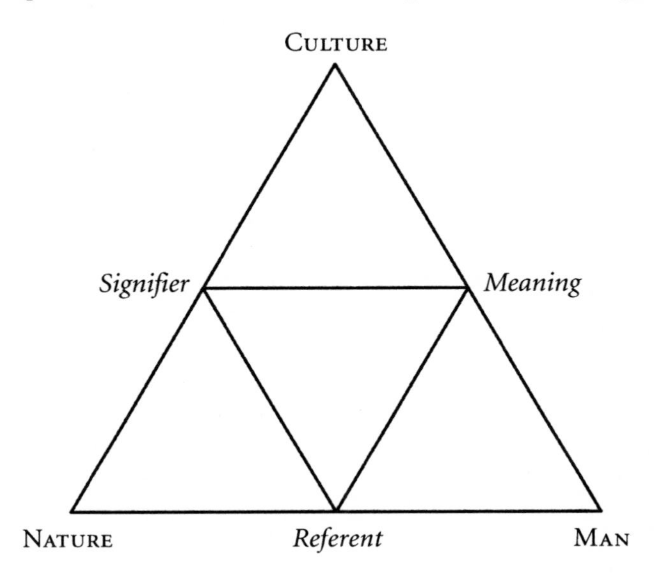

The symbolic triangle embedded in the critical triangle

With the metaphysical pyramid things are somewhat modified. There the symbol will be represented by a point of convergence rather than a triangle, because it has a unifying function. It is itself the mediator through which the unity of the critical triangle is achieved: first a 'horizontal' and centripetal unification, then a vertical and 'exalting' one that the symbol effects by uplifting all things, through its semantic dynamism, towards the Divine Principle. The triangle corresponds to an analytical view of the symbol, whereas, according to the second diagram, the symbol is seen according to its 'synthetic nature', since, as we have established, it is itself 'union' or 'synthesis'. On the other hand, criticism produces not only an expressly negative fruit, that of the symbol's deconstruction (and therefore destruction); it also produces a paradoxically positive fruit, which becomes apparent precisely from the vantage point of the metaphysical pyramid, and which brings to light a triple correlation: one that unites symbol to cosmos, the symbolic to the sacred, and symbolism to metaphysics. Actually, by reducing the world to mere nature, criticism establishes, *ab absurdo*, that there should be no cosmos that is at the same time a symbol, which we have called a mythocosm. By claiming to reduce man to a rational subject, it establishes at the same time that there should be no consciousness of symbolism which is at the same time consciousness of the divine and the sacred. Lastly, by reducing all symbolic and religious consciousness to an effect of purely structural arrangements of a cultural system, it establishes—and this will be the chief object of our fourth part—that there should not be in the end any possibility for symbolism outside the knowledge of metaphysical and transcendent Reality. Thus it quite precisely provides us with the prolegomena to a *metaphysics of sacred symbolism*. Such is the coherence of our approach summarized by the figure of the metaphysical pyramid. By introducing this now, we are anticipating what will only be explained by the semantic principle. We might have left it to close our critique, but, quite rightly, the critique that structuralism has carried out, by its very radicalism, does not close it: it opens directly onto proof of the semantic principle. Besides, in the course of a long hike it is good to offer the traveler an overall view of the countryside, in the form of a *metaphysical pyramid* [see following page], so that he can more easily grasp its unity when he is tramping through the undergrowth of the trails.

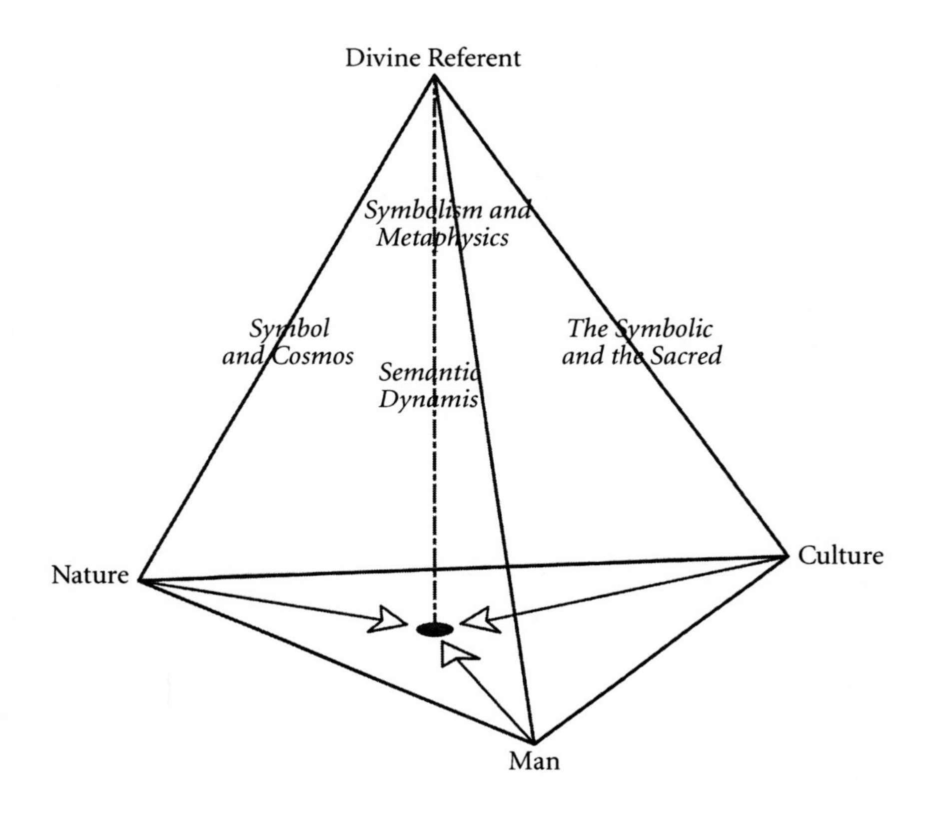
Divine Referent
Symbolism and Metaphysics
Symbol and Cosmos
Semantic Dynamis
The Symbolic and the Sacred
Nature
Culture
Man

7

Structuralism and Surfacialism

The Anthropology of Claude Lévi-Strauss

INTRODUCTION

The examination of structuralist theses will not detain us as long as the one devoted to the problems posed by the first or second critiques. The Galilean crisis actually demanded in-depth historical research to shed light on the reality of its not-always-perceived philosophical substructure. To the contrary, structuralism has arisen almost immediately from the theoretical shaping—and critiques—on the part of Michel Foucault and Jacques Derrida. On the other hand, the second critique has assumed, almost from the outset, through Hegelian gnosticism, a hyper-philosophic character which was therefore directly concerned with the philosophy of the symbol and required a sustained speculative endeavor, while structuralism wants to be exclusively scientific and hence sidesteps our own discourse. From this point of view we have strictly nothing to say about it, except to possibly subject structuralism to an equally scientific examination.[1] Besides, this is obvious and recognized by Claude Lévi-Strauss, since he does not stay on a strictly scientific level and since a certain philosophy is 'implied' by his own research.[2] It is above all in this direction that we focus our attention.

1. Among others, read the studies gathered by Philippe Richard and Robert Jaulin in *Anthropologie et calcul* (Paris: UGE, collection '10–18', 1971). These studies especially show the very approximate character of Lévi-Strauss's use of mathematics in the social sciences. But we would obviously not subscribe to all the theses expressed there.

2. The review *Esprit*, Nov. 1963, 'Réponses', 652.

EPISTEMIC CLOSURE AND ETHNIC CLOSURE

We think that the primary and essential intention animating the works of Lévi-Strauss as a whole is scientific in nature. Basically this involves finding a method that enables the vast data provided by ethnology to be treated scientifically—data which until now, on account of its complexity and even its contradictory character, had eluded any satisfying explanation. Or rather let us say that the explanations proposed bring in causative factors that, in any case, would only consist in hypotheses forever unprovable because eluding all observation and therefore every experimental verification. In other words, the social sciences, and chiefly ethnology, found themselves in the situation of linguistics before Saussure: the recorded phenomena (prohibition of incest and marriage-rules in 'primitive' societies, for example) are taken up by many disparate sciences (biology, psychology, history, etc.), the discordant explanations of which led to an indefinite dispersal of causality.

It was necessary then, as it is anytime someone complies with scientism (that is, with the general form of 'science'), to carry out the epistemic closure of the ethnological concept, to consider a given society as an irreducible whole, *sui generis*, closed upon itself and referring to no extra-ethnic etiological type: human nature, biological needs, laws of history, geographic or meteorological determinisms, in the same way that Saussure decided to consider language in and for itself. Once this closing-off of the social-grouping concept is carried out, we are necessarily led to see that the phenomena observed in the group and collected by ethnography are entirely explained if made the means by which *the group itself assures its own closure*. Here as well, it must be clearly recognized, there is a kind of vicious circle. For Saussure, it is by one and the same act that language is defined as an autonomous whole and that the entities composing it are entirely explained solely by the lateral relationships differentiating them, as if the task of the laws of language was to side with the linguist who had enclosed them in his own concept. Once the relationship of language with the world and with mankind has been severed, there actually remains nothing but differing lateral relationships between entities. Likewise for Lévi-Strauss. Once the ethnic group is considered in itself and reduced to its concept, the rules that determine the relationships of group members among themselves no longer have any need for external explanation: their raison d'être is in the fact that they formally allow for the group's existence and subsistence. In other words, these rules have as their sole end to form the individuals to whom they are applied into members of a group.

But, before one group is distinguished from another, it should be distinguished first from the absence of a group. Now men are actually 'non-grouped' in distinct unities, insofar as they all share the same biological functions, that is insofar as they are subject to universal nature. To be formed into a group is therefore to be distinguished from nature. On the other hand, among the biological functions, there is one which, by itself, depends in some manner on the 'social' realm; this is sexual union, which requires the meeting of two partners. This is why it provides the 'site' where, preeminently, a *relevant* difference between group and nature might be instituted. As a result, every society is first formed by regulating marriage alliances, by forbidding (and therefore authorizing) a particular category of wife, which is expressed by the prohibition of incest, it being given that the primary 'forbidden' women are at first only the closest. Thus marriage rules, like every rule, institute a culture's particularity (since that is its essential function)[1] by which the group defines itself and is able to exist as such. But, on the other hand, it is at the same time as universal as the nature it must govern. Such is the solution Lévi-Strauss brings to the problem of incest prohibition, at once a particular (cultural) rule and yet a universal fact which all explanations until now had failed to justify. The rule has no other raison d'être than itself, it is justified by itself, its necessity follows from its factual existence, it is creator of its own legitimacy. Incest-prohibition is not incest-prohibition: its essence is not constituted by the immorality or dangers of consanguineous unions, but by the rule itself for which prohibition is only an accidental support. For there to be culture and group, a rule that differentiates culture from nature and 'encloses' the group is needed. The marriage alliance "is the hinge, or more exactly the notch where the hinge may be fixed. Nature imposes alliance without determining it, and culture no sooner receives it than it defines its modalities." The 'Rule' thus appears to be at once "the permanent and general substance of culture ... the prime role [of which] ... is to ensure the group's existence as a group, and consequently, in this domain as in all others, to substitute, to replace chance by organization."[2]

The analogy with the methods of linguistic structuralism, for that matter, is not only with respect to the differential function of the rule, but is also concerned with the members of a group and the relationships that the rule enables them to establish among themselves. For the forbidding of a certain category of women entails the need for alliances

1. "Everything subject to a norm is cultural and is both relative and particular" (*The Elementary Structures of Kinship*, rev. ed., trans. J. H. Bell and J. R. von Strumer [Boston: Beacon Press, 1969], 8).

2. Ibid., 31–32.

with women of other categories, and therefore for matrimonial exchanges altogether similar to linguistic exchanges: "That the mediating factor, in this case, should be the *women of the group*, who are *circulated* between clans, lineages, or families, in place of the *words of the group*, which are *circulated* between individuals, does not at all change the fact that the essential aspect of the phenomenon is identical in both cases."[1] This 'identity' obviously exists only on the condition that formal structures alone are considered, that we totally disregard the content, and that an absolute independence of the former with respect to the latter is posited, which is bordering on untenable. However that may be, this identity enables us, by degrees, to view society as a whole, and especially culture and religion, as a structural system of differential unities, the language of which offers the primary model and fundamental locus: "Language, from this point of view, may appear as laying a kind of foundation for the more complex structures which correspond to the different aspects of culture."[2]

However, the relationship to language is still much more obvious if we focus on this characteristic manifestation of traditional cultures that is myth, for "myth is language,"[3] but a language that, much better than ordinary languages, verifies the principle of structuralism. Myth is, let us say, hyperstructural, in the sense that, as opposed to poetry, the words employed have no importance in themselves, and only the structural oppositions between the grand unities that compose it matter, oppositions that seem all the better for being able to mark out a continuity across a greater variety of versions. This is moreover why the model Lévi-Strauss utilizes is much more a phonological model, developed by Trubetzkoy, than one of Saussurian linguistics.[4] Phonemes, precisely because devoid of signification, are more obviously closer to simple structural relationships than signifying unities (morphemes). It is therefore necessary to define these greater unities and the manner in which they are combined. In fact, "if there is a meaning to be found in mythology, it cannot reside in the isolated elements which enter into the composition of a myth, but only in the way those elements are combined."[5] Now we can already distinguish among language-forming uni-

1. *Structural Anthropology*, trans. C. Jacobson and B.G. Schoepf (Garden City, NY: Anchor Books, 1967), 60.

2. Ibid., 67.

3. Ibid., 206.

4. Ibid., 32.

5. Ibid., 206.

ties: (1) phonemes, (2) morphemes, and (3) semantemes.[1] The unities of myth will come then in fourth position; these are the 'large unities' that Lévi-Strauss calls mythemes, and are themselves composed of *relationship bundles*. However, these mythemes, or 'words' of mythic discourse, are not used by this discourse for their own meaning, but are treated as mere signifiers. This is why Lévi-Strauss can assert that "myths and tales use language hyper-structurally.... While remaining terms of the narrative, the words of myth function in it as sheaves of differential elements. From the point of view of classification, these mythemes are not located at the level of the vocabulary but at the level of the phonemes."[2]

STRUCTURAL REJECTION OF SACRED SYMBOLISM

We will not, even briefly, enter into the structural analysis of the myths developed by Lévi-Strauss in the four thick volumes of *Mythologies*, or broach the numerous problems that these continues to raise. Our point of view spares us from this, since what is of interest here is a study of the philosophical consequences of a critique of sacred symbolism. Now, as announced and as we will attempt to show, ethnological structuralism is directly founded on the rejection of such a symbolism, that is of true symbolism. Besides, to express our thinking more simply, we will readily admit that in our eyes—as surprising as this might seem—the structural analysis of myths, as such, holds no interest. The only positive result to be retained would be its having drawn attention to the possibility of isolating segments in the mythic account, and therefore encouraging research into such segments (mythemes). This method may possibly contribute to interpreting myth by shedding light on unperceived analogies, and therefore helping to understand tenuous correspondences, but always based on their proper symbolic significance, and provided that the segmentary unities be true unities, and

1. By *phonemes* (non-signifying unities) is meant, not the sounds, but the formal elements of relevant articulation proper to a language and identified as such by speakers of this language. Thus the phoneme 'r' corresponds [in French] to several different *sounds* (guttural r, rolled r, normal r, etc.), but is identified as the same linguistic entity (a score in French). As for signifying unities, they include: morphemes or monemes, unities marking the point of view under which the referent is envisioned, and the *semantemes* which designate the referent itself. In 'departing', */de/* and */ing/* are morphemes, */part/* is a semanteme. Some linguists do not distinguish between morpheme and semanteme.

2. *Structural Anthropology, Volume II*, trans. M. Layton (Chicago: University of Chicago Press, 1976), 141–42.

not more or less arbitrarily divided up by the ethnologist.[1] When reading Lévi-Strauss or others,[2] it is hard to ward off the impression that they have been caught at their own game: the pleasure in orderliness. And this indefinite pleasure is an end in itself, since it satisfies one of the two basic demands of human intelligence, the combining of relationships, but to the total detriment of the other, spiritual vision. One can thus spend life cutting apart and putting back in order what seemed fantasy and even disorder, and congratulating oneself on the work accomplished, like someone who would interminably set a library in order, not only without ever opening a book, but even by deciding that not one book had any meaning in itself. Basically myths exist for the structuralist ethnologist to analyze and recombine. Also, and still more radically, they only exist as such through him, which is truly the height of ethno-centrism.[3] But is not the constructing of its object also a characteristic of every science?

According to E. R. Leach,[4] theories on myth are divided into two categories: symbolist theories, inspired by the German school, and functionalist theories (Malinowski). With the first—which "have always captivated mythology's amateurs"—the focus is on the narrative in which "familiar symbols" are recognized and thus "deciphered"; but the sociological context is ignored. The second strives to account for myth through the function that it fills at society's core. As for structuralism, it "considers myths, all myths, as answerable to a symbolic *system*. Like the symbolists, it is primarily interested in the text.... But the symbolists, who would work upon the object proper, do not seek a system in it.

1. This is moreover a method by no means ignored by traditional hermeneutics: just recall the typological analogy between Baptism and the crossing of the Red Sea; the Burning Bush (in which the word of God was manifested) and the Virgin (in whom the Divine Word was incarnate); the old Eve born from the sleeping Adam's side and the Church, new Mother of the living, born from the opened side of Christ crucified, etc.

2. For example, the articles contained in num. 22 of the review *Langages* (June 1971), entitled "Sémiotique narrative: récits bibliques," under the direction of C. Chabrol and L. Marin.

3. True, Lévi-Strauss justifies this paradox by appealing to the philosophy of his approach: "For if the final aim of anthropology is to contribute to a better knowledge of objectified thought and its mechanisms, it is in the last resort immaterial whether in this book the thought process of the South American Indians take shape through the medium of my thought, or whether mine takes place through the medium of theirs" (*The Raw and the Cooked: Introduction to a Science of Mythology:* 1, trans. J. and D. Weightman [New York and Evanston: Harper and Row, 1969], 13). But, as we will see, this justification is contradictory.

4. According to Dan Sperber, *Le structuralisme en anthropologie*, Seuil, 49–50; we are following his account, and the passages in quotes are his.

Conversely, functionalists seek a system where there should be none: in the varied conditions of the uses of myth." This definition is surely exact, provided that one specifies that, by reducing symbolism to a system, one is purely and simply suppressing it.[1] This method, from the side of the subject, practices what might be called the *epoche* of the symbolist intention, that is, not recognizing what exactly someone making use of a symbol wants to do, and on the other hand, from the side of the object, ignores the symbolic significance of the signs utilized, explaining them only by the structural relationships they hold in the mythic system. But is it possible to simply isolate a mytheme without taking its significance into account?

As an example we will take the analysis of a Tereno myth concerning the origin of tobacco.[2] This myth tells the story of a man warned by his son that his wife was a witch and an ogress who was poisoning him and giving him food contaminated with menstrual blood to eat. He fled to escape her and drew her toward a hole he had dug to catch game. She fell into it and was killed. He filled in the hole and perceived that an unknown plant grew there, the dried leaves of which he smoked "at night, with great secrecy." The ethnologist analyzes this myth and breaks it down into a series of oppositions, some of which are obvious (of the high-low, night-day, man-woman type), while others are often amazingly shrewd and subtle. The myth is then entirely broken down, its structural oppositions entirely displayed: oppositions "between nature and culture, between what in culture remains subject to nature on the one hand, and what escapes it on the other; there is the woman, too natural in culture, the cooking too feminine, both too bound up with man for the latter to be altogether culture." It is therefore necessary to separate nature from culture, which the myth accomplishes by sending the woman back to nature, to animality. "Then human society is rediscovered, a culture impassive to seasonal ordeals. And so is explained, not the origin of tobacco, but what makes it so good to smoke, 'good to think about.'"[3]

However, this explanation has left out a vital element—and not just some accessory factors recognized as not being taken into account. This

1. This is moreover what Sperber himself recognizes in his most important work: *Rethinking Symbolism*, trans. A. L. Morton (Cambridge and New York: Cambridge University Press, 1975), where he criticizes the Lévi-Straussian conception (for example, 52–63, 78–84, etc.).

2. This myth is cited several times in the *Introductions to a Science of Mythology* (*The Raw and the Cooked*, 100, *From Honey to Ashes*, 360). We are following Dan Sperber's presentation of it in *Le structuralisme en anthropologie*, 52–60.

3. Ibid., 57–58.

vital element is that tobacco is shown to be a *woman transformed*. This mythic fact is blindingly obvious; it forms the conclusion, and therefore the major key, to the whole narrative. *This* is what it wants to tell us. And it is rightly this that is entirely unnoticed and appears in none of the analyzed mythemes. By no means do we deny that here and there mythematic analysis comes across interesting elements, or that it brings to light pertinent and significant oppositions. But, for want of taking traditional symbolism into consideration, which alone could guide it in this search for significant elements, this kind of analysis is led to ignore the most important. And certainly ours is not an isolated example. If ethnologists had recognized then that the myth specifically taught about the origin of tobacco—is there not some inconsistency in concluding that a myth on the origin of tobacco teaches us nothing about this origin?—they would have understood that the woman is not at all "thrown back upon the side of animality," but is alchemically transmuted into a ritual plant in this athanor that is the filled-in hole. Tobacco actually plays a primal role in American traditions: the sacred pipe represents the sacrificial altar upon which the plant, that is to say the earth, is burned and transformed into spirit under the form of the breath inhaled and projected to the four cardinal points.[1] This is a matter then of tapping, through the man's 'priestly' ministry, the virtue of the lower forces, and putting them in the service of spiritual ascension. For these forces either serve or dominate us. At first the man is poisoned by what comes from the wife-sorceress (menstrual blood mixed with food); he limps, that is, he is wounded in his verticality, the mark of his humanity. He is warned by 'his son', who signifies the intellect endowed with discernment. Then the wife-sorceress (the telluric energy in each of us) seeks to devour him. But he flees and gives her impure honey,[2] snake embryos, and nestlings to eat, that is three 'products' of three animal categories symbolizing the 'virtues' of all life forms. By devouring them (the devil's share), not only does the ogress destroy them, but she

1. We are referring in particular to Black Elk (*Hehaka Sapa*); see *The Sacred Pipe* (Baltimore: Penguin Books, 1971). In this work a great Sioux chief, in the evening of his life, explains for the first time the true significance of some of the most important rites of his religion. Black Elk died at Pine Ridge, South Dakota in 1950. This teaching had been orally transmitted to him by Elk Head (*Hehaka Pa*), the keeper of the sacred pipe. Frithjof Schuon's introduction to the French translation, reprinted in *The Feathered Sun* (Bloomington, IN: World Wisdom Books, 1990), 44–70, is to our knowledge the best synthesis of American Indian metaphysics.

2. He keeps the pure honey (solar and luminous food) for his son: the intellect is nourished by the intelligible core of cosmic realities.

also concentrates them within herself.[1] Dead, through the invisible fermentation of the hole, she gives birth to the tobacco plant (whereas at first she poisoned a plant) that issues from her grave; this plant is the quintessence of telluric energy, concentrating within itself all the 'virtues' of the life-forces. This plant-earth is then dried in the Sun, that is *earth* is separated from the *water* that is still mingled with it, then it is burned by the *fire* and transformed into *air* (water corresponding to the snake, fire to honey, and air to the birds). As we see, the essential theme of this myth is that of transformation, of the passage from birth to death, from death to life, from light to darkness, etc., that is, basically, that of a spiritual process.

We have not explained everything in this myth. And our interpretation in no way exhausts its significance, which can be applied to different levels more or less mixed together here. But at least it can be observed that it is in no way inferior to the structural method as for attention to text, and is even much more scrupulous.[2] In other respects, it shows what benefit there is in listening for a myth's basic symbolic intention, which, alone, can yield up its key. For this intention exists. Whether one likes it or not, this is integral to the very existence of myth and symbolism in general. We previously mentioned in a note the case of Black Elk among the prairie Indians. There is also the case of the most illustrious sage Ogotemmêli, as recounted by Marcel Griaule,[3] who has revealed the extraordinary ways of Dogon metaphysics. Lastly, we could mention Muchona the Hornet, an "elderly gnome," whom Victor Turner met among the Ndembu of Zambia, and who one day agreed to explain the significance of all the rites and symbols. This led Turner to conclude that "in an Ndembu ritual context, almost every article used, every gesture employed, every song or prayer, every unit of space and time, by convention stands for something other than itself. . . . The Ndembu are aware of the expressive or symbolic function of ritual elements." They

1. We encounter a similar symbolism in a great number of legends, in "Tom Thumb," for example, where the ogress devours her own daughters (the telluric power must be turned against itself); after which her stolen strength is tapped by Tom Thumb (= intellect or spirit, with his brothers: the mental faculties). This tale could be the object of a quite precise commentary.

2. This demonstration is even more convincing if we turn to the exegesis of a written text like the *Torah*, as found in the Kabbalah. Contrary to what is sometimes claimed, there is no hermeneutic more attentive to the text, down to its minutest details. The subtlest structural analyses seem plainly infantile by comparison.

3. *Conversations with Ogotemmêli: An Introduction to Dogon Religious Ideas*, trans. R. Butler (London: Published for the International African Institute by the Oxford University Press, 1965).

even have at their command a term to characterize whatever is endowed with a symbolic function: the word *chijikijilu*, which signifies a landmark, a blaze, that is to say basically a symbol; "a *chijikijilu* has . . . a name (*ijina*) and it has an appearance (*chimwekeshu*), and both of these are utilized as the starting points of exegesis (*chakulumbwishu*)."[1]

Obviously, to read so many analyzed, decomposed, and recombined myths in so many silent works without a hint of living or lived experience, we might indeed end up believing that a myth is actually nothing but mythemes intersecting on paper, that everything is there, and that everything proves—indefinitely—that we are right. But this is untrue. Symbolism is basically spiritual life, a serious and religious activity, a sacred approach for anyone intending to enter into relationship with their Principle and making it present insofar as this is granted to us. Symbolism's truth is in Ogotemmêli's noble countenance, in Black Elk's attitude full of dignity and grandeur, in the profound simplicity of their speech wherein resounds the echo of every human voice lifted up with prayer and thanksgiving to God since the beginning of time. Let us lend an ear to this mighty, universal song, uninterrupted for ever-so-many millennia, over the entire surface of the globe, and the babbling of western ethnologists will appear for what it is, a mumbling of little interest.

Nevertheless, if we needed proof for the importance of the symbolist intention, it would be enough to turn to ourselves and our direct experience of decultured Europeans: when the symbolist intention has disappeared or has been altered, as is the case with contemporary Catholicism, which has lost *all sense of the sacred*, symbolic structures by themselves do not subsist for long; they crumble with a growing rapidity, and soon nothing remains. And that is valid for all civilizations, for we are not naïve enough to think that everything non-western preserves its traditions intact. Throughout the world, and especially wherever western counter-civilization has polluted cultures, there are symbolic systems degenerating or dying for lack of support by a truly spiritual intent, the disappearance of which is precisely the modern world's aim. In *Rethinking Symbolism*, Sperber poses the following question: if the structural message is fruitful, it should be possible to formulate *a priori* forecasts of symbolic practices that will be verified then in the field. Now the results are positive. Studying the ritual use of butter among the Dorze, an Ethiopian ethnic group, he establishes an initial opposition between dietary butter that is eaten and sacred butter that is not eaten,

1. *The Ritual Process: Structure and Anti-structure* (Aldine Transaction, 1995), 15. Cf. also Turner, *The Forest of Symbols*, Cornell, 1967, as well as *Muchona the Hornet, interpreter of religion*, in Casagrande, ed., *In the Company of Man*, New York, 1959.

in other words between the consumption and non-consumption of a preeminent food. Is this 'hole in the pattern' foreseen by the structural method actually fulfilled? Yes, the Dorze sacrificer eats a portion of the sacrificed ox's intestines.[1] True, Sperber declares, it was quite hard to find this element. Besides, both the banality of the oppositional system resulting from the method (high-low, hot-cold, nature-culture, etc.) and its irrelevance as far as symbolism is concerned need to be acknowledged: "since we do not know what symbolism would have to be like for it not to work, it does not tell us anything about what symbolism is not, nor consequently, about what it is."[2] In other words, the principles of structuralism not being falsifiable, they truly do not constitute a theory. However, it is made clear,[3] this irrelevance concerns the principles of structuralism, not the analyses of detail which can be, themselves, invalidated by the discovery of a new element to be integrated, which proves that there is indeed a systematic symbolism.

This is, however, just what we find unconvincing; for the concepts of structure and function are inseparable, which is equivalent to saying that, outside of function, structure in unintelligible, and therefore that pure structuralism has no significance: quite precisely, it speaks in order to say nothing. And this is the insurmountable impression to be drawn from its analyses: they are ingenious, but they could be true only if they had a meaning, which is exactly what they deny. Only the symbolic function, then, can account for the structure and its possible systematicity. And so the example chosen by Sperber seems extraordinarily artificial. What need is there for the eating of sacrificed ox-intestines to support a structural relationship with the non-consumption of ritual butter? With much research one can always find how to fill a 'hole in the pattern'. But what is the significance of all that? How does one not see that these two factors are so remote from each other that their opposition proves irrelevant and their instructional value strictly nil? It is a fact that ritual everywhere uses consumable substances sacrificed to the divine. It is a fact that sometimes these substances are consumed, or a portion thereof (the rest being burned), or not consumed. Equally it is a fact that there is—in sub-Saharan Africa above all it seems[4]—sometimes ritual consumption of a non-food, which in a certain way corresponds to a 'sacrifice from below', that is an act by which, on the one

1. *Rethinking Symbolism*, 59–63.

2. Ibid., 64.

3. Ibid., 67–68.

4. We have seen a movie filmed in the area of Dakar, in the bush, where participants in an extremely violent 'rite' ate the intestines of a living dog.

hand, the sacrificer sacrifices himself in his turn, and, on the other, integrates into himself creation's most inferior elements so to lead them back to the Principle: this is a kind of sacrificial 'descent into hell'. In a more general manner, symbolism makes use of elements that the corporeal world offers it according to their correspondences with its signifying intention. All possible uses are more or less realized (we say 'possible', for there are innumerable substances much more inedible than ox intestines), and one can, if one wishes, arrange them in a systematic table, but still these practices are obviously determined by the nature of the natural elements and the symbolic meanings to be expressed.[1] The structural method thinks of itself as scientific and positive because it only takes factual data into account, and because it implies no thesis on the origin and nature of symbols. But this is at the expense of a genuine theoretical neutralization, if by that it precisely means renouncing saying anything about its subject. What does a human 'mean' when he symbolizes? What is the meaning of symbolic behavior? The structural method begins by putting this question in parentheses (the *epoche* of the symbolist intention) under pretext of objectivity, and claims to identify an unconscious structure that functions all by itself, without the knowledge of its users. But in that case, about *what*, then, is it speaking? And just who has proven that an *epoche* of the symbolist intention is possible and constitutes a condition for objectivity? We ourselves have shown how, for lack of taking this into consideration, one might cross to the essential side of a symbolic narrative. And we have in mind here not only the fact that the mythographer claims to understand the functioning of myth better than someone who utilizes it. For Lévi-Strauss admits quite forcefully that "many 'primitive' cultures have built models of their marriage regulations which are much more to the point than models built by professional anthropologists."[2] But we are speaking of the symbolist intention as such, which is from the outset disqualified as a principle of intelligibility, because its acceptance would be tantamount to a negation of structuralism as such. The respect and love for 'primitive' peoples to which Lévi-Strauss has always

1. A multitude of structural analogies are noticeable in various domains. We can, for example, first observe a structural opposition between using one's feet to move and not using one's feet to devote oneself to prayer (the stylites or certain *yogis*). Next another type of opposition is to be defined: walking with one's non-feet. And in fact certain pilgrimages are made on the knees; or indeed, like Our Lady's juggler, one can walk on one's hands. And so forth. But all this obviously does not present the least interest, other than (involuntary?) amusement.

2. *Structural Anthropology*, 274.

referred—and about which there is no doubt—does not go so far, however, as to honor in them what, in their own eyes, is the most important. To the contrary, this respect and love ignores, scorns, and denies this 'why' for which these peoples live and die.

Responding to some critics who reproached him for "elaborating the syntax" into a "discourse that says nothing," Lévi-Strauss declared: "I too, of course, look upon the religious field as a stupendous storehouse of images that is far from having been exhausted by objective research; but these images are like any others, and the spirit in which I approach the study of religious data supposes that such data *are not credited at the outset with any specific character*."[1] At the same time he thinks that, better than naturalism and empiricism, the structural method "feels better able . . . to explain and validate the place that religious feeling has held, and still holds, in the history of humanity." He even reproaches critical 'believers' for having rejected it "in the name of the sacred values of the human person," whereas they should have understood that they could draw 'theological' consequences from a thesis that bestows the objectivity of the unconscious on cultural structures. "However," he adds, "it should not be assumed that I am trailing my coat, since this would be inconceivable on the part of someone who has never felt the slightest twinge of religious anxiety."[2] If this is so, what is to be understood then by what he calls 'religious feeling'? Undoubtedly he is supposing that this involves the sentiments? But what if the intellect were also involved? However this may be, what remains is that sacred symbolism is only accessible to him by means of its complete semiotic reduction, which is indeed the height of misunderstanding.

THE ERADICATION OF METAPHYSICS OR LOGIC WITHOUT *LOGOS*

We see how far we are from the two previous critiques, which, under various headings, take into account the symbolist intention, either to denounce it as a superstition, or to see it as an illusion of consciousness, the aim of which is to be mistaken about the object. Henceforth symbolism, in the proper sense of the term, has totally disappeared. What remains are *structural effects* by which a *cultural field* is organized.

At this point we ought to shed some light on the concept of the

1. *The Naked Man: Introduction to a Science of Mythology:* 4, trans. J. and D. Weightman (London: Jonathan Cape, 1981), 639. Our italics. True, he does not hesitate to declare the opposite a few pages later: "This is not to say that I underestimate the specific nature of ritual" (670). As for religion's 'stupendous storehouse', he writes in *The Savage Mind* (95) that "The poverty of religious thought can never be over-estimated."

2. Ibid., 687–688.

human mind implied by this method and the philosophical consequences that flow from it, which we will do first by comparing it with Lévy-Bruhl's theses about primitive mentality. Now, as surprising as it may seem, the distinction between a prelogical mentality governed by the notion of participation and a logical mentality governed by the principle of non-contradiction is closer to a metaphysical theory of sacred symbolism than the thesis of their *structural* indistinctness, because it poses the true problem. For Lévy-Bruhl, that men believe what they hold to be totally impossible poses a real philosophic problem, that is to say has a bearing on reality. Is that true or not? No, it is not true. In that case how can men, necessarily endowed with reason, believe it since, precisely, reason constrains them to reject it? Are they insane? No. We must conclude then that they are not reasoning like ourselves since, faced with a contradiction, they do not feel "the necessity of affirming one of the terms if the other be denied, or vice versa."[1] Lévy-Bruhl's error is clear (but with a clarity seldom noticed *in actuality*): it consists in identifying logic with modern western reason, or more precisely with what reason holds to be possible, that is, with the philosophic concept that it has of reality. The debate, and this is its importance, is situated at the level of the confrontation of two metaphysics, even if the formulation is erroneous; or rather the designation of prelogical mentality expresses the fear, a thousand times hidden away and a thousand times springing back into sight, that a positivist and scientistic mentality experiences before the sheer existence of the symbolist mentality: the logic of participation is perceived as a direct objection to its totalitarian materialism and rationalism.

However, there are at least two different ways of responding to the difficulties obviously raised by the prelogical mentality thesis; in other words, there are two ways to show logic's universality. On the one hand, while recognizing its formal character, and therefore independent of its contents, its necessary relationship with being will be maintained. Surely logic also receives the material for its operations (from sensible or mental experience), yet it is not a simple functioning: if it feels the obligation to be obedient to the principle of non-contradiction, this is because it is still oriented by an 'ontological concern'. Hence the hypothesis of a pre-logic becomes useless. One can—and we think: one should—affirm the universality of logic and reason, along with its necessary relationship with reality. But what is called into question is the

1. *How Natives Think*, trans. L. Clare (Princeton, NJ: Princeton University Press, 1985), 77.

very metaphysics of being, identity and the real.[1] There is only one logic used by all men, but it is not everywhere dependent on the same *ontology of reference.* This is because the evolution of western mentality has led to an extraordinary limiting of a being's identity, reduced solely to the bodily or physical individuality, and so the identifications of religious symbologies appear scandalous. It is not the primitive who does not understand western logic; it is the westerner who does not understand (no longer understands) the primitive's logic. No matter which African or American Indian of average intelligence is capable of using a computer, even the most intelligent and cultured westerners are most often incapable of entering, *truly,* into the universe of a Sioux or a Dogon, that is to say grasping the logic and metaphysics upon which it is based. Such is mainly the path we have chosen, and one sees with what clarity the very existence of sacred symbolism poses the most essential metaphysical questions.

But on the other hand the universality of logic can also be affirmed by considering it as simply a structural functioning. This is obviously the solution of Lévi-Strauss. First, how can we disagree with him in his attempt to rehabilitate the so-called primitive mentality, which is nothing but symbolist thinking? When in the first chapter of *The Savage Mind*[2] he shows, in a rather irrefutable manner, that 'primitives' have nothing to envy us for when it comes to a capacity for abstractions, to a knowledge of nature, its observation and use, we are delighted to see the plain and simple truth recognized in this way by such an illustrious writer. But it must be noted that this recognition is surely still more dangerous than the Lévy-Bruhlian dichotomy, because it is done in the name of a purely 'mechanist' conception of logic, which is reduced to the unconscious workings of a combining device; in other words, because *there is quite simply no longer any logic.* What structural anthropology focuses on, beyond the immediate ethnological interest, is the total eradication of any metaphysical possibility on the part of the human mind. And this is why it so *violently* rejects the very notion of participation, which therefore makes Lévi-Strauss, in reality, a super-

1. Specifically, if logic had no relation to being, to the sense of reality, neither could it be a function of a particular philosophy. It would work all alone, blindly. But the existence of sacred symbolism proves the contrary, since the symbolist is logical. After all, we do not think that cases of indifference to the principle of non-contradiction—except for a possible degeneration—have been truly established. There is identity (of two distinct beings) from a certain point of view, under a certain rapport. Here again we will refer to Black Elk's teachings.

2. "The Science of the Concrete," trans. J. and D. Weightman (London: Weidenfeld and Nicholson, 1966), 1–33.

Lévy-Bruhlian: "The 'savage' has certainly never borne any resemblance," he rightly states, "either to that creature barely emerged from an animal condition and still a prey to his needs and instincts who has so often been imagined nor to that consciousness governed by emotions and lost in a maze of confusion and participation." And he relates this thinking, "experienced in all the exercises of speculation," to that of the "alchemists of antiquity and the middle ages: Galen, Pliny, Hermes Trismegistus, Albertus Magnus . . . ,"[1] which is altogether exact. But why would participatory thinking be confused, if not because such is Lévy-Strauss's idea of it? Has he subjected it to a critical analysis? Not at all. A scornful rejection is enough: "Nothing here calls for the intervention of a so-called 'principle of participation' or even for a mysticism embedded in metaphysics which we now perceive only through the distorting lens of the established religions."[2]

The structuralist position is clear then, and it confirms our basic thesis in a remarkable manner. *For three centuries the major problem for European thought is clearly that of the total elimination of sacred symbolism.* Having said that, we by no means want to reduce the whole history of modern philosophy to the central theme of our work so to increase its importance, but we are simply conscious of stating a fact. Behind the numerous debates human reflection pursues with science, history, law and politics, morality and society, etc., we encounter a more fundamental conflict: man coming to grips with his religious soul. How do we rid ourselves of this? The Galilean revolution has divided man from himself and cast the sacred toward superstition and insanity; the second critique has sought the causes of this insanity in consciousness itself, so that the line of separation between the normal and the pathological is at once established and denied, at the risk of drawing the entire man into a demented state. As for structuralism, it intends to completely erase the demarcation by discovering, for both sides, the identity of a classifying function. But the price to be paid is obvious: that of the very meaning of reason, that is, its relationship to being, since to be reasonable always consists in striving to take into account what is, and therefore to distinguish the real from the illusory. Let the awareness that reason has of its ordering towards being perish, rather than admit the authentically religious significance of some human behavior: such is the guiding principle followed by the whole history of western thought and which has led it as far as this veritable *speculative suicide*.

For structuralist theory by no means hides the consequences that

1. *The Savage Mind*, 42.
2. Ibid., 38.

spring from its method. On nearly every page of *The Savage Mind* Lévi-Strauss explains that the correct meaning of terms placed in relation play no role in their combination and a purely classificatory form is involved: "Arbitrary as it seems when only its individual terms are considered, the system becomes coherent when it is seen as a whole set.... Nevertheless when one takes account of the wealth and diversity of the raw material, only a few of the innumerable possible elements of which are made use of in the system, there can be no doubt that a considerable number of other systems of the same type would have been equally coherent and that no one of them is predestined to be chosen by all societies and all civilizations. The terms never have any intrinsic significance. Their meaning is one of 'position'—a function of the history and cultural context on the one hand and of the structural system in which they are called upon to appear on the other."[1] The basic principle that intervenes in classificatory operations is one of opposition: "The logical principle is always to be *able to oppose* terms which previous impoverishment of the empirical totality, provided it has been impoverished, allows one to conceive as distinct." What matters is "evidence differentiating features" and not their content: once in evidence "they form a system which can be employed as a grid is used to decipher a text," an apparently unintelligible text. "The grid makes it possible to introduce divisions and contrasts, in other words the formal conditions necessary for a significant message to be conveyed." In other words, these systems "are codes suitable for conveying messages which can be transposed into other codes, and for expressing messages received by means of different codes in terms of their own system."[2] Thus the correspondences that Lévy-Bruhl would explain by the principle of participation are described here as "a homology between *two systems of differences*, one of which occurs in nature and the other in culture."[3] Natural phenomena are not then what myths explain, but "they are rather the *medium through which* myths try to explain facts which are themselves not of a natural but a logical order,"[4] which would be equivalent to saying that *the logic of myth is self-signifying.*

A formal thought being independent of its contents should not be characterized by what it thinks. Whether it thinks about myth or does mathematics, the same combinatory form is still involved. The savage mind "is not attributable of itself to something, be it a portion or a type

1. Ibid., 54–55.
2. Ibid., 75–76.
3. Ibid., 115.
4. Ibid., 95.

of civilization.... By the name savage mind I am designating the system of postulates and axioms required to set up a code, enabling 'the other' to be translated into 'ours' with the least inappropriate rendering possible, and vice versa."[1] The logico-mathematical apparatus[2] used by structuralism to extract the formal array of combinations at work in the composition of myths, to the very extent that it succeeds, proves that this system of combinations is logico-mathematical in nature, for otherwise it would not lend itself so perfectly to such a treatment. And so the frontiers between the different 'ages of intelligence' are effaced: "Nothing could be more false than to postulate opposite types of knowledge, mutually irreducible throughout the ages, with a sudden and unexplained switch from one to the other."[3] This is, in the opinion of Lévi-Strauss himself, one of the major advantages of his method: "The false antinomy between logical and prelogical mentality was surmounted at the same time. The savage mind is logical in the same sense and the same fashion as ours, though as our own is only when it is applied to knowledge of a universe in which it recognizes physical and semantic properties simultaneously."[4] It differs from 'tamed' thinking then only by reason of its unconscious, that is to say spontaneous (non-critical), functioning, in which logical structures are never extracted for themselves, but are always present in the concrete forms with which they are invested.

That having been established, the result is what Ricœur has called "a Kantianism without transcendental subject,"[5] in other words the concept of a human mind functioning according to quasi-autonomous structural categories that universally determine it. For we must finally come to the philosophical consequences of this approach, whatever the depth of Lévi-Strauss's scorn for philosophy. As a matter of fact, this Kantianism without subject, without *cogito*, is not always presented

1. 'Réponses', the review *Esprit*, Nov. 1963, 634.

2. The relevance of the mathematical apparatus used by Lévi-Strauss is considered highly controversial by certain specialists. Relying on one declaration by Lévi-Strauss recognizing "the very loose senses in which [he has] employed" mathematical concepts (*The Raw and the Cooked*, 31), André Régnier concludes the chapter "Mathematizing the social sciences" in *Anthropologie et calcul* thus: "In other words, this prefabricated 'awesome conceptual apparatus' is only a mirage" (32); cf. likewise same author, same work: "From group theory to savage mind," 271–298, which is very informative and has lists of numerous misinterpretations in logical and mathematical vocabulary by Lévi-Strauss.

3. *The Naked Man*, 637.

4. *The Savage Mind*, 268.

5. Lévi-Strauss has moreover accepted this definition: 'Réponses', *Esprit*, Nov. 1963, 633.

with perfect coherence. It is even denied by Lévi-Strauss in *The Savage Mind* when, after having shown that symbolic representations determine group distinctions, his 'Marxist conscience' calls him back into line: "It is of course only for purposes of exposition and because they form the subject of this book that I am apparently giving a sort of priority to ideology and superstructures.... Only the reverse is in fact true. Men's conception of the relations between nature and culture is a function of modifications of their own social relations."[1] Later, it seems that Lévi-Strauss tends towards a less historical materialism. The mind's immanent structures seem to be henceforth the homologous product of those found in our cells under the form of the famous genetic code of the DNA macromolecules. For the philosophers who criticize structuralism do not even perceive—dunces thrice-over—that structural linguistics "has been given a natural and objective status through the discovery and the cracking of the genetic code: the universal language used by all forms of life, from microorganisms to the higher mammals, as well as by plants, and which can be seen as the absolute prototype, the model of which is echoed, on a different level, by articulate language: the model itself consisting, at the outset, of a finite group of discrete units, chemical bases of phonemes, themselves devoid of meaning, but which, when variously combined into more complex units—the words of language or triplets of nucleotides—specify a definite meaning or a definite chemical substance."[2] That this conception is devoid of the least scientific or philosophic intelligibility does not disturb our author.[3] These

1. Page 117. Next he justifies himself by saying that he is "merely studying the shadows on the wall of the Cave without forgetting that it is only the attention we give them which lends them a semblance of reality" (ibid.), which cannot help but pose some quite curious epistemological problems.... The image of the shadows on the cave is taken up again in the "Finale" of *The Naked Man* 643. Here Lévi-Strauss is referring to what is called, in Marxism, 'historical materialism' (to be distinguished from 'dialectical materialism'), a principle formulated by Marx in 1859 and consisting of explaining history by material, that is to say economic, causes.

2. *The Naked Man*, 684–685.

3. Conceptualizing DNA or RNA activity as a code is only a metaphor, a 'model' for understanding what occurs, or rather predicting it, for much is still unknown (cf. Joël de Rosnay, *Les origines de la vie: De l'atome à la cellule* [Paris: Seuil, 1966], 67–68). On the other hand, even to suppose that this model is relevant, that is corresponds to some reality, there still remains to be shown what intelligible relation there might be between a genetic structure and a particular life form, a hand or eye for example. Now such a relation is quite simply inconceivable, the extreme mechanicism of a Monod devoid of any speculative consistency: "the chains of DNA in the genes are not, as is imagined, coded messages, instructions for the making of organs. Genes are tools for the first indispensable step to begin fabricating" (R. Ruyer, "La quasi-information," published in *Revue*

are only, he explains, "free-ranging intellectual musings,"[1] which is rather convenient for excusing oneself from thinking rigorously as some out-moded philosophers still strive to do. Meanwhile, we have to be content with what has been disclosed: there is but a "pre-existent rationality which has two forms: one is immanent in the world and, were it not there, thought could never apprehend phenomena and science would be impossible; and, also included in the world, is objective thought, which operates in an autonomous and rational way, even before subjectivizing the surrounding rationality, and taming it into usefulness."[2]

As we see, in such a perspective and despite the protests of Lévi-Strauss, we must indeed recognize that meaning has completely disappeared. For it is not enough to speak, for thousands of pages, about symbols, significations, and semantic correctness. It is also necessary to speak in a manner that does not make their existence impossible, for then meaning becomes the most improbable of miracles. Besides, he himself affirms this: in phonology "meaning is always the result of a combination of elements not themselves signifiers," just as in anthropology "meaning is always reducible . . . behind every meaning there is a non-meaning, and the converse is not true."[3] But how could non-meaning ever produce meaning? This is apparently a question to which he has never thought of responding. Basically preoccupied with eliminating the thinking subject, "that unbearably spoilt child who has occupied the philosophical scene for too long now, and prevented serious research through demanding exclusive attention,"[4] incapable of imagining the existence of a divine *Logos*, culture's true agent intellect, he does not see that, with one stroke, he has made every logic, semantic, and symbolism strictly absurd. The immense mythic discourse signifies nothing, or rather, *it is self-signifying*: "Each matrix of meanings refers to another matrix, each myth to other myths. And if it is now asked to what final meaning these mutually significative meanings are referring—since in the last resort and in their totality they must refer to something—the only reply . . . is that myths signify the mind that evolves them by making use of the world of which it is itself a part."[5] This is moreover why "there is no real end to mythological analysis, no hidden unity to be

philosophique, CLV, 1965, 293). The entire article should be cited. If, in addition, it is necessary to pass at once from genetic structures to linguistic and cultural ones, then the absurdity of the thesis nearly reaches the 'n'th degree.

1. *The Naked Man*, 692.
2. Ibid., 687.
3. 'Réponses', 637.
4. *The Naked Man*, 687.
5. *The Raw and the Cooked*, 340–341.

grasped once the breaking-down process has been completed."[1] In the end, "if it is now asked where the real center of [the myth] is to be found, the answer is that this is impossible to determine."[2]

We see that, despite every precaution that he has not ceased taking with respect to philosophy, Lévi-Strauss has been led to produce a certain conception of man and reality. This involves, as someone has said, overturning the tradition of philosophical humanism, the last refuge of metaphysics. For this humanism human consciousness is, as Descartes has shown, the transcendent locus where the sense of reality is achieved and revealed. True, mechanicism at first dismissed the *Logos* immanent to things and through which they escape from their own contingency. In its place it set up the *cogito* of a reason universal by right and transcendent to the world. Structural anthropology in its turn dislodges the subject from its rational observation post by revealing that 'man' is himself a contingent structure inserted into a world that socio-cultural forms only replicate indefinitely. And, since the perfect type of an actualized structure is the machine, then the *anthropos* must be seen to be a machine: "The world began without the human race and it will end without it. The institutions, manners, and customs . . . are an ephemeral efflorescence of a creative progress in relation to which *they are meaningless*, unless it be that they allow humanity to play its destined role. That role does not, however, assign to our race a position of independence. Nor . . . are his vain efforts directed towards the arresting of a universal process of decline. Far from it; his role is itself a machine brought perhaps to a greater point of perfection than any other. . . ."[3]

The disappearance of the thinking subject, and of the meaning of the world and being; the reduction of the anthropological to the structural: such are some inferences of Lévi-Strauss's work that have deeply shaken the philosophic problematic of our time. The entire universe is reduced to an *immense 'surface' of structural arrangements*. All substantial content has disappeared. One is accustomed to defining form "by opposition to material other than itself. But structure has no distinct content; it is content itself, apprehended in a logical organization conceived as a property of the real."[4] Symbolism and metaphysics, united in life as in death, sink into the impersonal insignificance of indefinite structures. The Sun of the *Logos* has finally set; its last reflections have been extinguished: the scandal of divine immanence has been abolished.

1. Ibid., 5.
2. Ibid., 17.
3. *Tristes Tropiques*, trans. J. Russell (New York: Atheneum, 1968), 397.
4. *Structural Anthropology*, vol. 2, 115.

Foucault and Derrida or the Infra-Parmenidian Exploding of the *Logos*

The philosophies of Michel Foucault and Jacques Derrida have nothing essential to add to the critique of the symbolic sign as developed in the preceding works.[1] Besides, this is not their object, which is rather to draw out the philosophical consequences implied by structuralist surfacialism. But also, to the extent that they have formulated axioms—untenable in our opinion—with special vigor, axioms which lay the groundwork for thinking about structure when it claims to be nothing but the structure of thought, they cannot be ignored. We will leave aside, then, those objections raised by the many assertions about the history of ideas and civilizations to focus solely on a review of some fundamental theses.

Whatever might be Foucault's hesitations with respect to structuralism properly speaking and the distinction to be maintained between science-generated (Saussure and Lévi-Strauss) and philosophical points of view, it is obvious that, in his work, structuralism has been raised to the level of a general formulation of knowledge of the systematic type. A rehabilitation[2] of sophistry against Plato's dialectic is also proposed, and this is to pledge allegiance to the 'sovereignty of the signifier', since sophistry consists wholly in defining the meaning of a discourse as a pure product of language, in which it is actually only a game, the operating of a semiotic chain. Such an idea (and we will come back to this) is quite simply contradictory and betrays the industrialist or 'engineering' ideology that inspires it: everything can be manufactured, even meaning; just let the signifiers do their work—within the sophist is a man of pure and blind power, the preeminent form of absolute tyranny, violence as such, and not the victim of the 'will to truth', as if the worst 'will to truth' were not precisely the 'truth by willing' of the sophist.

As we know, Foucault's reflections start with his taking into consideration an event that occurred in 1656 and through which the 'asylum system'[3] was established. We have already mentioned this, as well as its corollary: reasoning formed by its structural relationship, its 'contrastive features', with the 'confinement'. "To say that this rationalism is pure," writes Michel Serres, "is to say from what it has been purified by

1. We are foregoing any examination of Lacan's ideas on symbolism here. They are of interest to psychoanalytic therapy, but do not differ philosophically from the general approach of Structuralism.

2. *Discourse on Language*, in *The Archaeology of Knowledge and the Discourse on Language*, trans. A.M. Sheridan Smith (New York: Pantheon Books, 1982), 218, 232.

3. *Madness and Civilization* (New York: Random House Books, 1965), 39.

exclusion, scorn and denial. What seemed to be only image, copy, reciprocity, becomes then a basis."[1] We have remarked that this 'great confinement' of madness is contemporary with a great confinement of the sacred, of which it is only a reflection, a confinement far more important than the first and for which we have not stopped paying the price. (Quite remarkably, Foucault says nothing about it.) However this may be, what he basically wants to show is that rationalism rests on an unthought, unconscious structure which, as soon as it appeared, deprived reason of its axiological primacy, since it is the irrational that gives it its value. However unconscious it might be, this structure should not be seen as a 'depth' in the field of knowledge. The whole error of eighteenth- and, above all, nineteenth-century 'historical' theories is here: they do identify a certain number of 'signs', that is, they get closer to a semiotic conception of cultural determinisms, but these are still "visible sign[s] directing us towards a buried depth."[2] Here we find then, straightforwardly expressed, what we call surfacialism. Foucault's position in this respect is quite clear: depth is an illusion. The analysis of discourse that a society maintains about itself, and by which it expresses itself, does not consist in "rediscovering, beyond the utterances themselves, the intention of the speaking subject, his conscious activity, what he means, or again the unconscious play which comes to light despite him in what he has said.... One does not seek, *beneath what is manifest*, the semi-silent babbling of another discourse; one has to show why it could not be other than it was, in what it is exclusive of all else, how it assumes, in the midst of others and with respect to them, a place which nothing else could occupy. The right question for the analysis of discourse might be formulated in this way: what is then this irregular existence that comes to light in what is said—and *nowhere else*?"[3] Thus, similar in this to Velasquez' painting *Las Meninas*, the commentary of which opens *The Order of Things*,[4] the culture of an era is not, for Foucault, a faithful or adulterous expression of the truth, a certain grasp of the *meaning* of things, a revelation (or an obscuring) of their hidden essence, but the structural organization of an *episteme*, a perfectly 'surfacial' display. Everything is given, 'indicated': "*the order behind is, today, an order without a within*."[5]

1. *La communication*, Editions de Minuit, 1968, 178.

2. M. Foucault, *The Order of Things* (London: Tavistock Publications, 1970), 229.

3. M. Foucault, "Réponse au Cercle d'Epistémologie," published in *Cahiers pour l'Analyse*, Editions du Seuil, num. 9, 1968, 17. Our italics.

4. *The Order of Things*, 3–16 and 307–312.

5. F. Wahl, *Qu'est-ce que la structuralisme?* Philosophie, 79.

Surfacialism refuses to discover, then, the raison d'être of culture, nature, or man in a founding (divine, human, cosmic, or historical) *Logos*. It situates it in a certain spread-out arrangement of contrastive substitutable unities, with the variations produced being only a structural reshuffling. As a result there is, from what Foucault calls a 'displacement', a total subversion of the notions of truth and meaning, which are no longer originating and legitimizing causes, foundational as expressions of knowledge, but provisional and contingent *effects* relative to the epistemic structure of a moment: "The episteme is not a form of knowledge or type of rationality which, crossing the boundaries of the most varied sciences, manifests the sovereign unity of a subject, a spirit, or a period; it is the totality of relations that can be discovered, for a given period, between the sciences when one analyses them at the level of discursive regularities."[1]

But we think that it is with Derrida that the philosophy of structure dares go right to its end, at least as far as conceived by this philosophy. Less sensitive than his 'master' to the sparkling diversity of scholarship, his goal is a more radical conceptualization of the structural process. A text like *Structure, Sign and Play in the Discourse of the Human Sciences*[2] in this respect constitutes a model of its kind, but begs for a word-by-word discussion that cannot be undertaken here. In our eyes its importance is, as with Foucault, to in some manner 'give away the game', in other words to say aloud what structuralism is 'thinking' to itself, but is often hesitant to formulate because it recoils at the consequences. Now these consequences are such that they introduce into the speculative field a complete upheaval (one more) which makes the Nietzschean uproar seem like a metaphysical popular tune.[3] This time these are instances of the critique itself that are being called into question, the very ones alleged to accomplish its purifying work: pure nature, pure reason, and even, in a certain manner, pure culture. To be perfectly frank, Derrida's conclusions are especially valuable, because the truth of the semantic principle is established in *opposition* to them, and because, having deduced them himself with all his characteristic rigor and authority, he spares us from having to do so less skillfully and much

1. *The Archaeology of Knowledge*, trans. A.M.S. Smith (New York: Pantheon Books, 1972), 191.

2. *Writing and Difference*, trans. A. Bass (Chicago: University of Chicago Press, 1978), 278–293.

3. "Nietzsche, Freud, and Heidegger, for example, worked within the inherited concepts of metaphysics." Ibid., 281.

more suspiciously. And it is of extreme importance that these conclusions are presented, not as the advent of a new philosophical system, but as a coming-to-awareness of the 'structurality of structure' when it was clearly necessary 'to begin thinking' about it,[1] in other words as the finishing of the whole history of western metaphysics, or, more exactly, of the critique of metaphysics, of what this critique has not stopped thinking—without knowing it—and which should now be explained.

As already noted and as Derrida himself declares, until now structure was thought about in the manner of a center present to everything that it controlled, a presence that was contradictory because at once inside and outside of everything. "Successively and in a regulated fashion, the center receives different forms or names. The history of metaphysics, like the history of the West, is the history of these metaphors and metonymies. Its matrix . . . is the determination of Being as presence in all senses of this word. It could be shown that all the names related to fundamentals, to principles, or to the center have always designated an invariable presence—eidos, archè, telos, energia, ousia (essence, existence, substance, subject), alethia, transcendentality, consciousness, God, man, and so forth."[2] But to definitively call into question the metaphysics of center and presence, it is necessary to pass through a critique of the symbolic sign (and through its reduction to pure semiotics), because it is essentially this that renders the center present as the transcendent and invisible referent that it designates, which explains why it can be at once there and not there. Moreover, this is what Derrida recognizes. The sign is essentially a sign of presence; however, "the metaphysics of presence is shaken with the help of the concept of sign. But . . . as soon as one seeks to demonstrate in this way that there is no transcendental or privileged signified and that the domain or play of signification henceforth has no limit, one must reject even the concept and word 'sign' itself—which is precisely what cannot be done. For the signification 'sign' has always been understood and determined, in its meaning, as sign-of, a signifier referring to a signified, a signifier different from its signified."[3] How can we not notice the contradiction that there is in continually speaking of a signifier which is, in reality, from the structural perspective, in-significant? Here we have, in fact, a rather enormous paradox, and it would be at least appropriate to point out its

1. Ibid., 280.

2. Ibid., 279–280; as for the Greek terms, *eidos* = essence, *arche* = principle, *telos* = end, *energia* = in act, *ousia* is defined by the terms in parentheses, *alethia* = truth.

3. Ibid., 281.

existence. This contradiction is, moreover, only tenable at the cost of an artifice: on the one hand, sign is reduced to signifier by eliminating the two other poles of its definition (referent and meaning), while, on the other, the word 'signifier' continues to function in the discourse with its ordinary semantic value—'that which signifies'. Derrida is therefore undoubtedly right: to think of the structurality of structure, in all its rigor, necessarily leads to the successive rejection of a transcendent referent, of the logocratic meaning, and even of the signifier as an entity pointing towards whatever the signified might be. Nothing is left of the symbolic sign.

Therefore it is necessary to make "the beginnings of a step outside of philosophy. The step 'outside philosophy' is much more difficult to conceive than is generally imagined by those who think they made it long ago with cavalier ease, and who in general are swallowed up in metaphysics in the entire body of discourse which they claim to have disengaged from it."[1] Such is indeed the case with Saussure, and even to some extent with Lévi-Strauss. In Derrida's eyes this is attested to by the linguistic 'phonocentrism' of the *Cours de linguistique générale*, which has been solely interested in spoken language, in language as structure-of-speech and not in writing. Now, if this is so, Derrida thinks, it is because there is a great difference between their respective signifiers. The linguistic signifier is in some manner more transparent, is better at letting the *Logos* appear, while the written signifier is much more opaque.[2] The true reason for Saussurian phonocentrism is revealed in this way: it is an unconscious logocentrism, a primacy granted to the signified, that is to a Meaning prior to every signifying inscription. This is the affirmation of a Truth in itself metaphysical, a Truth which transcends every formulation, and yet contradicts the most essential demands of structuralist analysis, since this analysis tends in fact to restore the power of the sign. But then it must be admitted that "to restore the original and non-derivative character of signs, in opposition to classical metaphysics, is, by an apparent paradox, at the same time to eliminate a concept of signs whose whole history and meaning belong to the adventure of the metaphysics of presence."[3] And so it is necessary to purge the concept of sign in structuralism itself of its metaphysical double, and set the play of

1. Ibid., 284.

2. On this whole question, cf. *Of Grammatology*, trans. G.C. Spivak (Baltimore and London: Johns Hopkins University Press, 1976), Part I, chaps. 1 and 2.

3. *Speech and Phenomena*, trans. D.B. Allison (Evanston, Illinois: Northwestern University Press, 1973), 51.

signifiers in their *différance*, that is, basically, in their complete otherness. Language is not only a system of differences, but the signifier is, in its very being, dare we say, pure '*difference*', not the *other* of some *same*, the otherness of some identity, which at final reckoning would be Truth, Meaning, *or Logos*, for which the sign would be the exteriorization, representation, or substitute: "The substitute does not substitute itself for anything which has somehow existed before it."[1] Meaning, Truth, and *Logos* do not overarch the order of dispersed signifiers. To the contrary, they are actually its product: "there is no absolute origin of sense in general."[2] "What is written as *différance*, then, will be the playing movement that 'produces'—by means of something that is not simply an activity—these differences, these effects of difference. This does not mean that the *différance* that produces differences is somehow before them, in a simple and unmodified—in-different—present. *Différance* is the non-full, non-simple, structured and differentiating origin of differences. Thus, the name 'origin' no longer suits it."[3] Thus 'différance' is a *game* that indefinitely and interminably haunts the thinking process, a game that should not be conceptualized: "there is no *name* for it at all, not even the name of essence or of Being, not even that of '*différance*.'"[4]

Such are then the conclusions that follow from Structuralism and lead the evolution of philosophic thought to its end. That untenable *theses* are involved, theses that one cannot posit without contradicting oneself: this is what we are going to establish; and, besides, anyone can see this for themselves. That, nevertheless, they are a consequence of the viewpoint of structure and the endeavor to think about structure in itself should be clear. But the structural point of view itself arises after taking into consideration the *signs* of the Transcendent that appear in the mirror of consciousness and must indeed be taken into account, and upon which, quite precisely, the magic of alienation runs aground. Also, if these signs have to be accounted for, this is precisely because the Transcendent either does not exist or should not be signified. Its modes of cosmic presence—realized by symbolic signs—are quite simply impossible: nature is only matter spread out in space. And, if nature is only

1. Ibid., 280.

2. *Of Grammatology*, 65.

3. *Margins of Philosophy*, trans. A. Bass (Chicago: University of Chicago Press, 1982), 11. Derrida says again: "Since language, which Saussure says is a classification, has not fallen from the sky, its differences have been produced, are produced effects, but they are effects which do not find their cause in a subject or a substance, in a thing in general, a being that is somewhere present, thereby eluding the play of *différance*." (Ibid.)

4. Ibid., 26.

geometry, this is by virtue of an ontology of individual substance by which Aristotle has definitively chained down western metaphysics, a conception of being which limits and encloses it in its presence *here and now*, as if being could be wholly *there*, even when a being is just a mode of being and a certain degree of reality, as Plato teaches. We need not be surprised then to see today this ontological substantialism rejected as a constantly threatening enemy to be warded off. The error—and failure—of the critics of this substantialism, whether Nietzsche, Heidegger, or Derrida, is to identify this ontology with metaphysics, proving by this that, in reality, they are truly incapable of escaping it and exhaust themselves denouncing it. Basically, they commit the same error as Aristotle, but with much more serious consequences, because the Aristotelian 'error' is possible and even justified *in view of the requirements of a simple cosmology*, while theirs is impossible and spells intellectual death. The conception of Aristotelian ontology corresponds to a reduction of the real, not its destruction; it is a temporarily acceptable limitation insofar as it involves the founding of a science of sublunar realities, while the semiotic reduction propounded by structuralism corresponds to a disappearance of the real and meaning through sheer scattering, that is through the exploding of conceptuality as such. So we again find—and there is nothing surprising in this—the space of the Galilean cosmos as pure exteriority, that nearly realized explosion we attempted to describe at the end of chapter 2. One is the image of the other and each finds in the other its truth and its model. The exploding of the *Logos* and its complete dispersal into the radical uncenteredness of *différance* is correlative to the exploding of a world whose *center is nowhere and outer limit everywhere*; the one is inscribed in the 'logic' of the other and could not but be manifested some day. As we see, the end of the crisis of symbolism is in perfect agreement with its beginning, thus proving by its very coherence the truth of our analysis. This truth and coherence were unsought: they came of themselves as we unwound the thread of the critique of sacred symbolism. No, the only surprising thing that we could make a case for is that the Nietzsches, Heideggers, and Derridas have never found the answer to their question in Plato's own text. Simply glancing over the *Parmenides* or the *Sophist* should have been enough, though, to make them understand that there can be *comprehension* (in every sense of this term) of being only from the point of view, which is not a point of view, of the super-ontological. For want of establishing oneself in the super-ontological (one either is or is not there), one can never speak *of* being, but only *from* being, even though speech itself is then quite simply impossible. Here is to be found the answer to the question Derrida poses to Foucault: is there an 'other' of the *Logos*, and what is it? Or is there none?[1] And this answer is the following: the *Logos* itself is the other (than being), contrary to what Parmenides asserts, who conceives of it as word-of-being (*ontologos*); otherwise, how would it be possible to say what is not? *how would it be possible to be a sophist*? The *Logos* must indeed be other than *ontologos*: it is the reflection of Above-Being in least-being. This is a truth hidden from those who reject sacred symbolism, that is, the radical immanence of the radically Transcendent. This is because the essential 'other', which first awakens us to an awareness of otherness, is the experience of the sacred. To have an experience of the sacred is to accept it and receive it as a word from the beyond coming to pass here-below. Far from madness or *ubris* being the preeminent other of right reason, of the *logos*, they represent only its shadow or falsification, its inverted image perhaps, but not insofar as they are hand-in-glove and share a common lot. Awareness of this bi-polar opposition implies a prior and underlying one, which is conferred through an encounter with the religious element, a difference not horizontal but vertical, of which the previous one is only an image. The fool, the true fool, is the one who says in his heart: there is no God. In other words, insanity is contrary to reason only *in* the rejection, by reason, of the Transcendent. For want of recognizing the transcendent and vertical otherness of Above-Reason that it experiences in the sacred, reason, in order to situate itself, is constrained to set itself up against its horizontal other, irrationality. But, while from human reason to Divine Logos there is the mediation of the symbol in which vertical otherness is transformed into deifying identity, between reason and its inverted double there is nothing. Just the impossibility of rejoining oneself and abandoning oneself.

1. This question is posed in the important study that Derrida has devoted to Foucault's *Madness and Civilization*, reprinted and completed in *Writing and Difference* (32–63) under the title "Cogito and the History of Madness." The question is posed and developed 39–42.

8

Culture, Mediatrix Between Man and the World

The Excesses of Cultural Relativism

We have just seen that the true metaphysics of symbolism, far from rejecting the challenges of structural surfacialism, is alone able to give them a meaning and to situate them in their true light. But this is not just concerned with the concept of sign and the ontology of substance. In closing, we would also like to stress how it is possible to give a positive meaning to certain assertions of a 'super-culturalism' which, on the basis of a reduction of culture to a structural organization, bestows on culture an absolute power over man and the world. We have encountered this thesis with Lévi-Strauss. It is found among many Saussurians, who, however, hardly pride themselves on their knowledge of philosophy. This is so for Martinet, who, with regard to the concept of language nomenclature, writes: "The notion of language as a catalogue is based on the naïve idea that the whole world is ordered, prior to its perception by man, into perfectly distinct categories of objects, each of which necessarily receives its appropriate designation in each language. This is true up to a certain point, for instance in the case of different species of living creatures; but it does not hold good in other domains." And he lists a whole series of examples proving that languages categorize the real in extremely varied ways, colors being the standard example. This is why he concludes: "to each language there corresponds a particular organization of the data of experience."[1] But, as we know, this thesis was

1. *Elements of General Linguistics*, trans. E. Palmer (London: Faber and Faber, 1964), 20–21.

already partly formulated by Karl Wilhelm von Humboldt, and above all in our time by B.L. Whorf, who was perhaps unaware of Humboldt. This doctrine basically represents a linguistic Kantianism:[1] the categories that give order to the world and make it into a system of objects are no longer categories of the understanding, but ones of language, defined by B.L. Whorf as "the linguistic relativity principle." "Formulation of ideas is not an independent process, strictly rational in the old sense, but is part of a particular grammar, and differs, from slightly to greatly, between different grammars. We dissect nature along lines laid down by our native languages. The categories and types that we isolate from the world of phenomena we do not find there because they stare every observer in the face; on the contrary, the world is presented in a kaleidoscopic flux of impressions which has to be organized by our minds—and this means largely by the linguistic system in our minds. We cut nature up, organize it into concepts, and ascribe significances as we do, largely because we are parties to an agreement to organize it in this way."[2] Whorf follows the statement of his principle with a certain number of examples copied from a comparison between European and Indian languages, Hopi and Eskimo in particular. For example, French has only one word to designate all kinds of snow, while in Eskimo this is a matter of different things designated by different words. And likewise for the categories of time and space.

We can see how such a principle fits in with the semiological theses of structuralism: a word's meaning does not depend on the referent it designates, since, to the contrary, it is the linguistic sign that forms the objective referent by designating it. Meaning is therefore only a reflection of the structural organization of language and culture in general, and, consequently, the world itself. As such this doctrine is quite simply false, if not even absurd, at least in the form it assumes with certain linguists. As for Whorf's writings, they pose much more interesting problems, especially in what concerns the notions of time, matter, speed, event, etc.[3] Clearly this doctrine, precisely and yet with a surprising unawareness,[4] is guilty of the very fault that it denounces, namely

1. Kant's influence on Humboldt moreover seems undeniable (R.H. Rabbins, *Brève histoire de la linguistique de Platon à Chomsky* [Paris: Seuil, 1976], 185).

2. *Language, Thought, and Reality* (Cambridge, MA: The MIT Press, 1956), 213.

3. Thus Whorf informs us that the two basic categories of the real among the Hopi are the Manifesting and the Unmanifest (*Language, Thought, and Reality*, 59). Now, in a certain manner, these are also two categories of Hindu metaphysics (*vyakta* and *avyakta*).

4. Mounin does not lapse altogether into this error; however, he hastens to write: "In the proper sense of the term, it can be thought that what is named does not exist distinctly" (*Clefs pour la linguistique*, 73), which rightly understood makes no sense. The

nomenclature language. We are told that the diversity of linguistic categorizations proves that a word does not designate a thing. Very well. But this principle should operate in both directions: not only from thing to word, but also from word to thing, in such a way that nothing proves that different words designate different things, since, precisely, there is no nomenclature language. And hence by no means can it be inferred, from the variation in terminology for colors or the different kinds of snow, that this involves different (or non-distinct) *objects* according to their designations. In other words, language nomenclature is denied in one direction and made to operate in another. Besides, this is inevitable; otherwise the irrelevance of language nomenclature could by no means be brought to light. Actually, if the linguists did a little more philosophy, for which they have only the most excessive scorn, they would recognize that, if language determines both the categories of thought and those of the real, and not the objective organization of the world, we have precisely no means to perceive this. But, by an inconsistency we have already observed on many occasions and will soon analyze in its own right, the linguist asserts that there is nothing beyond language, except itself, which through an inexplicable privilege can cast an extra-linguistic glance at the world so to ascertain precisely, with respect to this cosmic invariant, cultural variations.

Besides, it would mean pursuing very careful analyses to verify many of the assertions cited as examples. But, be that as it may, it is obviously not the absence of a corresponding vocabulary that prevents "the Sango woman from distinguishing the colors of everything she calls *vuko*," a term that encompasses "violet, indigo, blue, black, grey, and dark maroon,"[1] no more than the non-existence of a linguistic marker (in French) to designate the male swallow implies that, for the French, all swallows are female. The demonstration would be much more difficult if it dealt with 'syntactical' categories, ones concerned with the general conditions of existence. And yet clearly it is the same for these categories as well—there are common and universal categories behind the varied forms of our mentalities. Every other supposition is purely and simply devoid of significance, since, if the basic categories of the intelligence were completely heterogeneous, we absolutely could not account for

exact opposite could even be upheld: only that which truly exists can be absolutely named.

1. Mounin, *Clefs pour la linguistique*, 72. Sango is a language of the Ubangi. A Canadian correspondent of ours has confirmed that the diverse labels for different kinds of snow by no means prevent the Eskimo from having the concept of *the* snow as one and the same reality.

them. Contrary to what some think they can advance, there is difference only at the core of identity, which by no means signifies that this identity excludes difference.[1] We cannot postulate then anything but a *Logos transcendent to all particular mental forms and yet present in each of them.*

In fact, as we have already indicated above, this structural culturalism is a prisoner of the 'metaphysics' it wants to reject and to which it claims to reduce all metaphysics. For our thinking is inevitably conditioned by the experiences that it has (we are not saying that it is confined to them); and this is why our judgments judge us and betray our conception of the real. Structuralism, a western thing, has had the experience of a certain kind of reason, that of classical rationality such as it is expressed, more or less ideologically, certainly not in the greatest works, but across the '*koine*' of western intellectuality, a conception more or less identifying reason with positivist scientism—and is still today, whatever may be said about it, that of the scientific majority. The famous Lévy-Bruhlian distinction between prelogical and logical mentalities is, in all its naïvety, a good witness to this. But the diversity of mental forms, however vast, does not go so far as to make understanding and translation impossible. A true culture is always 'total' in its way and reflects, according to its own perspective, the Spirit's totality. It will offer analogies then for effecting possible translations. Surely no culture does so without losing something, since the raison d'être of a cultural perspective is precisely its unicity, a unicity non-transposable by definition. What remains to be grasped, however, is whether this is where the essential lies. One translates, not to repeat—this is impossible—the unicity of a culture, but to understand it and be nourished by its truth, which definitively leads us back to *the universality of the Spirit.* When we read B.L. Whorf, what surprises us, much more than Hopi conceptions about time and space, is that he can declare as if self-evident: "The metaphysics underlying our own language, thinking, and modern culture (I speak not of the recent and quite different relativity metaphysics of modern science) imposes upon the universe two grand cosmic forms, space and time; static, three-dimensional infinite space, and kinetic one-dimensional uniformly and perpetually flowing time—two utterly separate and unconnected aspects of reality (according to this familiar way of thinking)."[2] Surprising, yes; because it is untrue that this is the meta-

1. We have explained the conception of an identity including difference (or alterity) in our article "Le Zeuxis ou de l'analogie" (*Revue de Métaphysique et de Morale*, 1968, num. 3, 180–193, an article reprinted in *Penser l'analogie* [Geneva: Ad Solem, 2000], 136–192).

2. *Language, Thought, and Reality*, 59.

physics underlying our language. Quite the contrary, we have shown that we conceive of space (and this is also true for time[1]) as being in a qualitative and symbolic form. Spatio-temporal indefiniteness, as a neutral and empty container, is a late and superficial concept, which enters more or less consciously into conflict with the much more basic and spontaneous conceptions of the mythocosm.

Culture as Agent Intellect—Outline of a Metaphysics of Culture

The previous considerations, however, should not let us forget a no-less-certain fact: the diversity of cultures, and the diversity of conceptions of the world that spring from the former. In this sense, how can we call into question the importance of the role played by cultural structures in the forming of our mental categories? But the recognition of this fact by no means entails the adoption of structuralist theses. Returning to analyses already developed in an earlier work,[2] we will only recall the doctrine of culture as an 'agent intellect' that is collective in its chief characteristics.

This doctrine is based on a certain conception of human intelligence that might be summarized in this way: *intelligence is sense of being*. This formula signifies three things: 1) being makes sense only for the intellect—just as sugar makes sense only for taste and red only for sight—and not for the will or imagination; 2) being is sense for the intellect, that is to say what gives it its meaning, orients it, specifically determines it, and actualizes it; 3) intellect perceives being only through sense. *The first point* defines the intellect as a sense, a perception, *of itself* direct and intuitive, whatever the modes assumed in other respects. *The second point* defines being, the real, as that which founds the intellective act and therefore, in a certain manner, as the hereafter of its aims, or its absolute end. From this point of view, the intellect is the sense of the Absolute 'as such', and this is what accounts for its intrinsic quality of objectivity,[3] as

1. Linear time is an abstract and mathematical time, a real transcendental illusion; that is to say we are considering, not time itself, but the formal rule for representing time. This is therefore a boundary concept, just as the straight line is the curve's limit when the curve's radius increases. Real time is always cyclical in its qualitative rhythms and linear in their successiveness.

2. *Amour et Vérité*, 54–60; cf. likewise, *supra*, chap. 3, 2.

3. The intellect is objective by definition, or we would otherwise have no idea about objectivity. Let us recall that an animal does not posit the world as an objective reality in itself: only its *Umwelt* exists.

well as a discrimination between the real and the illusory and therefore its sense of the relative. *The third point* indicates that, from nothingness to the Absolute, the degrees of reality are perceived as degrees of intelligibility (or again: perfection). We are not actually saying that the intellect perceives being only 'as' sense, but just that it perceives it only 'by means of' sense, with the help of sense. The intellect requires intelligibility, like the eye requires light, and this intelligibility is being's *revealer.* This means that we do not perceive being as such, naked being, but 'such' being, being that is 'just so'. If we perceived being as such, we would know only absolute Reality, and, in a certain manner, this is indeed the case: knowledge of the Absolute constitutes most profoundly the essence of intelligence, but it is an essence-limit, a 'remembrance' our intelligence carries within itself as a secret, as its transcendent heart. It can only be identified with it through a radical conversion of its gaze, by passing through the ordeal of its own death,[1] in other words by becoming 'infinite ignorance': this is the crucifixion of the *Logos* transpierced through the heart.[2] Then, in Unireality, it will know the innumerable multiplicity of beings, fused but not confused. Meanwhile, under its present form, the intellect is actualized by sense or the intelligible.

Of the three significations identified, the first is concerned with the *nature* of the intellect; the second is concerned with the *metaphysics* of the intellect, what is ultimately implied by its possibility; the third and last is concerned with the '*physics*' of the intellect, what is implied by its functioning under a properly human form. This third aspect is more especially involved with culture. We should then clarify this.

To say that human intelligence requires intelligibility, or again, that it is actualized by the senses; that is, it is in itself intelligence in potency with respect to all possible senses, with respect to the multiplicity of the intelligibles (although in its ultimate depths it is in act with respect to the pure Reality known to be unchangingly within it). This is why Ruyer can say that man is a universal reader, or Aristotle that the intellect is *in*

1. "Fortunate the intellects that know how to shut their eyes," says Dionysius the Areopagite in *Mystical Theology*, 1 [Trans.—in conformity with Gandillac's French translation used by the author]. And likewise the *Vedanta* teaches that the unconditioned Self can be attained only by passing through the state of "deep sleep" (*Mandukya Upanishad*, *shruti* 5). Or again, from another point of view, the *Mundaka Upanishad* declares: "The Self is accessible neither to lack of strength, or illusion, or to knowledge without renunciation" (*Mundaka Upanishad* 3, sec. 2, 4).

2. On infinite ignorance (an expression borrowed from Evagrios of Pontus), cf. *Amour et Vérité*, 350–352; on the 'gnostic' significance of the Crucifixion, cf. ibid., 207–209 and 332–334.

some manner everything. This means that we should also learn about the multiple. The question is then of knowing: who will teach us this?

It is precisely this potentiality of the intellect with respect to the multiple that, in our opinion, accounts for the 'mental' mode it has assumed in our intelligence, which is not the case for animal intelligence, and even less for plant or mineral intelligence, to the extent that it is possible to speak about the intelligence of stone or plant. Non-human intelligence is in fact essentially direct and is distinguished by the extent of its field rather than by its modes: thus the root of a plant directly knows whatever substances it needs for its nourishment; its knowledge is, so to say, only one with its being. Angelic intelligence is of this order, although its field is incomparably more extensive. But the field of human intelligence is not only more extensive, it is potentially unlimited (this is why we have the idea of totality), unlimited to the very extent that it is, in itself (that is, as a field of intelligibility), detached in some manner from the knowing being as well as the known being. Knowledge is the common act of the knower and the known, yes, but only inasmuch as it is knowledge and not inasmuch as the being of the knower and the being of the known are but one, at least according to the ordinary mode of knowing. Thus human intellective knowledge is *indirect*, meaning that it operates with the help of a mediation in which subject and object are conjoined in intelligible mode; this mediation is called a concept or a thought, and this is what defines its mental nature. But how is this concept (or thought) induced? *We think it an effect of culture, and suggest it be seen as a true agent intellect under the action of which our potential intelligence is informed.* So, how does culture do this? Always through the mode of linguistic or symbolic signs, through semantic expressions in which the intelligible form of the natural being with which it enters into relation is taught to the human intellect. Such is at least, we think, the ordinary mode of the intellect's actualization. Culture (symbolism and language) is clearly then a mediator between nature and man that reveals to man the world (as objective totality) and himself (as subjective unity).[1] From this point of view, the concept is

1. The role of culture (language and symbol) is not then to cut up the real experienced in a particular fashion to distinguish and isolate objects from each other. That is the role of sensory knowledge, and the Gestaltists have shown the great importance of the proper organization of perceived reality and 'good form'. And, besides, there are no absolute objects, and the degree of the individuality of objects varies considerably according to the nature of a being. The role of culture is to teach us the meaning of things, their profound significance, their essence. On the notion of agent intellect, its culturological interpretation, and traditionalism, see Appendix 6 at the end of the present chapter, 305.

only a mental sign,[1] just as the sign or symbol is only a cultural 'concept'. The concept (or the thought, to avoid using too precise a term) is the form with which the intellect clothes the cultural sign when it is accepted by it and with which it is expressed, speaks to itself about it. Culture teaches the world to the intellect, including man insofar as he is a being of the world; it tells of the meaning of beings and things, that is, it presents them to the intellect as significations, essences, intelligibles—not that it uselessly duplicates the function of empirical knowledge, which puts us in direct contact with existence, but it 'sensifies'[2] what we experience. This is why mental knowledge, which is then a knowledge through signs, is basically a *reflexive* knowledge. By this term we are indicating that human intelligence is *able* to turn back upon itself, but does not do so necessarily, which would be properly designated by the term *reflected* knowledge. This capacity is only possessed in permanent mode, that is: it can be objectivized in itself, precisely because it operates with the help of signs that are already themselves mediatized objects, *to* which, therefore, it is possible to return—in other words, which are not dependent on (and indistinguishable from) the momentary current interests of the intellect that thinks about them, but which possess, from their cultural origin, a certain characteristic permanence. This mode of information must not be conceived of, however, in an exclusively linguistic manner. Far from it, and this is why the term concept, more directly tied to language, might be misleading. It would be better to speak of 'thought', for through symbols (of all kinds: dances, rites, costumes, legends, songs, etc.) culture teaches us so much 'sense', so much of the intelligible that is hard to conceptualize (that is to say express in language) and is just the same extremely *precise* and perfectly *positive*.

Now we see how this idea of culture as a 'collective' agent intellect reverts to the previously indicated notion of a linguistic and ultimately cultural Kantianism. However, this idea has an entirely different meaning in structuralism. Here the categories are neither those inexplicably *a priori* for the Kantian understanding (why those and not some other ones?) nor the merely differential categorization of a cultural unconscious, but should be seen as particular determinations emanating from the *principial Logos*. A cultural system can be secure in its mediating function only if, within itself, the intelligibles contained in the *Logos* are

1. This definition is standard for St. Thomas (H.D. Gardeil, *Initiation à la philosophie de saint Thomas d'Aquin*, t. I, 66).

2. That is, it gives a meaning to, makes sense of empirical data. This neologism is due to Ruyer.

reflected. In other words, and if we would view matters from the most general vantage point, we should recognize a triple emanation of the *Logos*: in beings and cosmic relations, in human intelligence, and in the cultural system; in objective beings, as their own essence; in human intelligence, under the form of a forgotten memory—for the intellect in potency can only know what it recognizes, and all the cultural information in the world would not actualize the least intelligibility unless it inwardly gave its assent (in other words, what culture teaches human intelligence must 'tell it something'); and lastly in the cultural system, and these are specifically symbols which, as we have already said, should be seen as transcendent, semantic rays.[1] This triple emanation of the *Logos* safeguards an extreme possibility that cannot be discarded *a priori*, as strange as it may seem to a modern westerner: we mean the possibility of a direct knowledge of things through contemplating their archetype, either in nature itself as revealed by sensible knowledge,[2] or in the 'unconscious' substance of the mind by sheer concentration on itself.[3] On the other hand, where cultural emanation is involved, this direct knowledge points to a kind of essential discontinuity between the order of the world and the order of the mind, a discontinuity which is that of the order of symbolic signs as a whole. We have already too sufficiently stressed this aspect in *Histoire et théorie du symbole* for there to be any need to go over it again at length. But we will just reiterate that the symbolic order manifests in this way its *sui generis* reality, its disruptive nature, its transcendence, and its essential independence with respect to the field of man-world relationships, the splitting-apart and semantic uprooting produced specifically enabling each of its terms to be formed into its own ontological order. Besides, this is why symbolic forms, although always borrowed from nature, in the final analysis and as a function of their participative analogy by which they symbolize are not however direct emanations of natural entities. There is always something in them not of the 'arbitrary', which is excluded, but of the instituted and therefore irreducible to sheer naturalness, something that might even appear artificial or conventional. Culture does not speak solely of nature, it tells 'a bit more' and 'a bit otherwise' about semantic

1. *Histoire et théorie du symbole*, chap. VI, art. III, sect. 4.

2. This case would seem to correspond to the experience described by C.M. Erdsman in "'Le Buisson ardent': Contribution nordique à la mystique de la lumière," published in *Mélanges d'histoire des religions offerts à H.C. Puech* (Paris: PUF, 1974), 591–600. We are likewise thinking of the mode of knowledge spoken of by Fernand Brunner in *Science et réalité* (Aubier, 1954, 128–129 and 202–227).

3. Proven, in our opinion, by the intuitive manifestations we come across in scientific history.

surplus and semantic difference which are absolutely inherent to the symbol itself.[1] This is why Cratylism[2] and Hermogenism are equally true. The 'matter' of every (symbolic and linguistic) sign is Cratylian, but its 'form' is Hermogenian. The symbol receives this form with its introduction into a fixed cultural system. If symbols were actually mere emanations of nature, there would only be a single culture, or even none at all. But as a matter of fact each culture, since it is other than the nature that it teaches, reveals only a certain synthesis; each culture is only a fixed constellation among all other possible constellations, none of them exhausting the totality of nature's intelligible aspects. This is because its finality is in any case not just making the world known to man, but also teaching him how to be rescued from it as from that which cannot but imprison him within its ontological limits.

Such is the conception of mediatory culture identified by our analyses, and this in some fashion brings together all our conclusions: it ends the critique of the symbolic sign by revealing the functioning of the metaphysical pyramid.

Our doctrine still implies, though, a last thesis concerning the divine origin of culture, a thesis we cannot imagine developing at present. We will just say that, from the philosophical point of view, we only have a choice between two hypotheses: either admit, along with Lévi-Strauss's materialism, that linguistic and symbolic structures are, in the last analysis, produced and determined by the macromolecular structures of DNA—an unintelligible theory—or admit that they emanate from the Divine Logos which they reflect according to a certain point of view. Both solutions imply in fact a common thesis: the 'pre-existence' of culture, or at least its unengendered character. In a genetic fashion, it is actually impossible to account for the origin of language or the symbolic order as such. This is at least one point on which we are in agreement with the structuralists. Just as Minerva sprang fully armed out of Jupiter's forehead, so the least word or the least symbol supposes the totality of the linguistic and symbolic structure. On the other hand, the term 'revelation' in no way prejudges the particular modes under which it might be accomplished and none are excluded *a priori*: it only indicates the transcendence of the origin of cultures. After all, the structure

1. These considerations are not unrelated to the appearance, in human societies, of a Justice that is institutional and no longer immanent (the recourse to Athena in the *Oresteia* of Aeschylus).

2. In the *Cratylus* Plato has Cratylus, upholding the thesis of the conformity of words to the things and ideas that they express, hold a dialogue with Hermogenes, who upholds to the contrary that language-based forms, as with Saussure, have no relation to their meaning.

of a living being or a DNA macromolecule also has a transcendent origin. There is not necessarily any radical heterogeneity between the transmission mode of a cultural tradition and the reproductive mode of genetic structures—(biological) life is also a tradition—and it is obviously the first alone that can give us some idea of the second and not the reverse, whatever the differences might be in other respects.[1]

The critique of the symbolic sign is at an end. It has led us to a philosophic impasse and by that very fact compels us to make a speculative conversion. Clearly we have already anticipated this conversion, this reversal, and this is inevitable because the term cannot be achieved without reaching beyond the term. Recall however: *this lengthy history of western thought is basically one of a rejection and exclusion of the religious.* We know this assertion will seem exorbitant in attributing, some will say, so much importance to what is not so important. And yet we think we have shown just the opposite. We think that if western thought casts an objective glance over the secret wish that supports it and the lasting desire that guides it, it should recognize that for three centuries it has only sought to erase from its heart the divine traces left there by sacred symbols, even at the price of its own death.

Appendix 6

Intellect, Culture, and Traditionalism (cf. 301, n1)

The distinction in the mind between an agent and a patient (or possible) intellect arises, in Aristotle, from the following consideration: our intellect is in potency for all the intelligibles, that is, everything is *a priori* thinkable, intelligibly graspable, which implies that, by itself, it is in act for no intelligible in particular. Aristotle compares this 'patient intellect' to a writing-tablet upon which everything can be inscribed precisely because it is free of any inscription.[2] However, the intellect is in contact with known objects only through the intermediary of the senses which convey to the soul's imaginative faculty only images of individual things: a particular flower or stone, etc., while the intelligible (the flower-idea or stone-idea) is in itself a universal, abstracted from its individualizing material conditions. Where then does the actuation of

1. This is the whole meaning of Ruyer's work.
2. *On the Soul*, III, 430a1.

this intelligible come from, it being certainly present in a thing (it is its essence) as well as in the mental image of a thing, but, as an intelligible in act, only existing in the intellect? The existence of a faculty by whose light the mental image lets the intelligible form of this image show through must be supposed in the human soul. This faculty, this power of our spirit, is what Aristotelian tradition (not Aristotle) calls the agent intellect. For St. Thomas Aquinas this agent intellect, belonging to each human being, is a certain participation in the light of the Divine Intellect.[1] The intelligible, abstracted from the mental image by the revelatory light of the agent intellect, is impressed onto the possible (or patient) intellect, which, by reflecting on this 'impression', expresses it, tells it to itself (this is the 'interior word') under the form of a concept.

Considerations of language and culture do not intervene in this cognitive process (at least in keeping to the concise presentation just given). Everything transpires between knowing subject and known object, between thought (reason, if preferred) and the world. Now, if we grant that the doctrine of abstracting the intelligible by the light of the agent intellect actually describes the reality of the cognitive process (as one might logically presume), we also think that it fails to take into account the concrete conditions for its realization. This scheme only takes into account a philosophically constructed cognitive apparatus. This is not a question of doubting its validity, but simply of noting that it functions somehow 'all alone'; we are told nothing about what sets it working, like an internal combustion engine that functions without a starter. Or rather, the role of starter is played here by 'sensible experience' alone. Perception alone would be sufficient, given the cognitive apparatus, to set everything going and for the intellect to produce a concept.

Now this thesis, that might be allowed as an initial rough estimate, seems to ignore an undeniable fact: sensible experience does not put us in contact with sensible realities of only one kind (as if the category 'sensible object' were unique), but with those of two: things and signs (sonic or visual). And there lies the whole question. *As to their sensoriality, nothing distinguishes a sign from a thing*: both are 'perceptual events' among a limitless number of others, and confusions between the two are always possible. Confining ourselves to their effect on our senses, both only serve to inform us of the existence of 'something'. The sign is not perceived sensorially, then, as a sign. To perceive it as a sign requires that, by an act of the mind, its sensible presence be understood as being there not to point to itself, as is the case with perceived objects, but *for something else*, for something 'invisible': a meaning.

1. *Summa Theologiae*, I, ques. 79, ans. 4.

We have analyzed this 'experience of the sign' at length in *Histoire et théorie du symbole* in connection with the case of Helen Keller. We see in it the birth of the human being to spiritual awareness. In the discovery of language the human being is born to a distinctive awareness of self and world, because he grasps—this is the miracle of the spirit—that there is in the world something that is the visible face of an invisible world, one of meaning, essence, the intelligible, and the concept. And whoever speaks of language speaks of culture.

But this is not all. The experience of language and culture not only awakens our thinking to an awareness that there is something 'conceivable'. By naming the things of the world, by inserting them into conversations, dances, songs, gestures, behavior, and myths, their archetypal significance is given a setting: we are taught what they are. It is also in this sense we say that culture is as if a 'collective' agent intellect. On the one hand, as experience of the sign, it gives rise to the activity of our agent intellect; on the other, by its teaching, it guides us towards the essences.

This culture- and language-function is not perhaps totally absent from Aristotle's philosophy. In the *Nicomachean Ethics* we read: "Intellectual virtue in the main owes both its birth and its growth to teaching...."[1] In any case, it is related to what was implemented at the beginning of the nineteenth century by Joseph de Maistre, Louis de Bonald, Félicité de Lamennais, and the whole current called traditionalism. This current, the object of several ecclesiastical condemnations because it seemingly denied that reason could, by itself, attain to a sure knowledge of God's existence, His chief attributes, and the spirituality of the soul, represents in fact, on the part of profoundly religious thinkers, a reaction to the ideology of the French Revolution, fruit of eighteenth-century philosophic rationalism. However, opposed as it may be to 'Enlightenment' ideology, traditionalism is also a child of its time and retains one of its major lessons, namely recognizing the role of culture and language in the make-up of human beings. As previously shown, 'culture' represents one of the three poles which, with 'nature' and 'man', govern the whole thought of the eighteenth century. All truly speculative works fit into this triangulation, and, of the three, the newest item is 'culture'; new, yes, because almost absent from the philosophic problematic of prior times, with some notable exceptions, one being Leibniz. Contributing to the appearance of a 'culturological awareness' were those ethnographic discoveries speaking of peoples unknown to the Bible, of oriental languages ('Sanskrit' foremost), but also the interest in

1. II, 1, 1103a 15–16 (W.D. Ross translation).

the intellectual life of blind and deaf people, and therefore in the functioning of signs, in the origin of our ideas and language, to which must be added the development of research into grammar, focused on explaining language, showing its logic, and not just a codifying of good usage. Obviously, the culturological awareness of the traditionalists is at the antipodes of the one promoted by the 'Enlightenment' which tended to reduce the life of the mind to something physiological, whereas Maistre and Bonald attributed its source to the existence of a primitive Revelation. Nevertheless, this is clearly an advance in culture, in the signs and history of languages, within the context of a major body of knowledge for a science of the mind. The Roman Church was certainly right, in 1832 and 1834, to reiterate the rights and powers of 'natural' reason, even if it might be thought that the consequences of this condemnation were not always happy. However, this 'natural' reason (that of Aristotle, for example, demonstrating the existence of God as prime mover) is only so with respect to supernatural revelation, while it is in itself given form by all the currents of its time. This is all too obvious to insist upon it. To this same traditionalist school could be linked, in some respects, the symbolist movement made famous in France, in the second half of the nineteenth century, by Cardinal Pitra, Msgr. Devoucoux, the Abbé Lacuria, and some others, among whom were René Guénon and his school, at least in some ways.

Our own reflections were not formed by contacts with these authors, whom we hardly knew or seldom read, with the exception of Guénon. But at many points we had encountered them, especially in our concern to lay the foundation for a metaphysics and, even more, a *theology of culture*. However, traditionalist doctrines did not seem to include, as we maintain, the integration of culture with the spiritual life at the time of the sign's discovery. Surely the semantic experience as an act of giving birth to the conscious life of the spirit presupposes the existence of signs and language. About the origin of this we confess our ignorance, except to refer to a divine or angelic revelation, as we believe Scripture teaches more or less explicitly. But far from denying the role of native reason, thinking, and intelligence, the semantic experience equally presupposes their existence: the discovery of language does not 'produce' the spirit. What is left is to maintain that reason only operates if an object is given to it, and that, among these objects, this sensory data, some are things while others are signs.

PART IV

The Semantic Principle or the Primary Evidence of the Logos

Introduction

The Last Question is One of Sense

The question posed to us is extremely clear and may be summarized in the following manner. When taking into account the certainties of Galilean science, *the existence of an array of sacred symbols* in the human cultural field appears to be basically anti-rational, in other words irrational. Clearly, this irrationality must be rationalized; it must be explained how human intelligence has been (or still is) capable of such behavior, the insane character of which, objectively incontestable, goes subjectively unperceived. That hundreds of millions of people, for millennia, have been able to think, live, and act according to this symbolic universe (the theses for which are held to be, with respect to scientific truth, only so many aberrations) without having the least awareness of it, actually poses a rather formidable problem for critical thinking. How can reason produce the irrational? What is its genesis?

The first type of response consists in seeking this genesis within the hidden folds of our consciousness. Two things must in fact be explained at the same time and with a single principle: that there is insanity, and that there is no awareness of it, that is to say this insanity coexists without the least conflict with a behavior perfectly adapted to the real. The concept of alienated consciousness responds well to these two requirements: an alienated consciousness is obviously unaware of this alienation, and on the other hand the *object* by which consciousness is alienated is a necessarily unreal representation (produced by consciousness itself), and therefore unable to be directly contradicted by reality either. The solution would consist then, not in proving the unreal character of the alienating representations (for example: to prove that God does not exist), but in appealing to consciousness itself to explain how it is its own victim and its own 'evil genius'. Consciousness is thus set up as a fabricating power unconscious of its illusions. Nonsense has a sense and, even more, derives from sense.

Yet the difficulties with this response are obvious and amount to knowing 'which came first'. Since this is about a genesis, it is actually about time and therefore about a beginning. Must consciousness already be alienated to produce alienating representations? Should not representations necessarily precede the consciousness that takes them for an object? And are not these representations formed specifically by the sum total of sacred semiologies?

It must be concluded that in reality the first alienation of consciousness consists in positing itself as subjectivity in the secrecy of which lies hidden the genetic principle of its alienation. We are in this case led to situate the unconscious on a much more radical level, that of the structural organization of the signifiers that determines consciousness and its discourse. Sense is thus defined as a production of a chain of signifiers that are themselves devoid of it. There is no sense 'in itself', and symbolic behavior is not 'aberrant' with respect to a reputedly normal rationality; there is no metaphysical difference between 'primitive' and positivist mentalities, and the scandal of sacred symbolism is finally erased by the 'indifference' of structuring differences.

In short: Galilean science, hence religious aberration; religious aberration, hence alienation of consciousness; alienation of consciousness, hence supremacy of the signifiers.

And this does not involve 'erasing' or 'destroying' meaning, but "rather, it is a question of determining the possibility of *meaning* on the basis of a 'formal' organization which in itself has no meaning."[1] The only remaining question is precisely the following: does this undertaking make sense? *Is it really possible to fabricate sense out of nonsense*? Such is the last inquiry posed by western thought today. Now, by a not-at-all surprising coincidence, this question clearly seems to have arisen only from forgetting something: the primary truth that philosophy, at the dawn of its long history, had understood under the form of a paradox; this authentic initiatic (because initial) test guarded the entrance to the speculative Temple, letting through only those who knew how to understand its significance. As for the others, the sophists of all kinds, past and present, let them rest content: they have never entered this Temple out of which they imagine it so hard to 'take the slightest step'. The anti-metaphysical heroes, those who hide nothing of their appalling efforts to tear off the mantel of philosophy, who have rejected everything except for the spirit of seriousness, are imaginary heroes: they were never bound by the shackles they thought held them prisoner.

This is then the paradox that we will now strive to understand. What

1. Derrida, *Margins of Philosophy*, trans. A. Bass (Chicago: University of Chicago Press, 1982), 134.

it has to say to us is quite commonplace. The first truth, the semantic principle, that we would like to formulate is known to everyone, obvious to everyone, and forgotten by many. Our whole ambition is simply to help with its remembrance.[1]

1. In particular, it has been formulated most lucidly by R. Ruyer in *Néo-finalisme* (Paris: PUF, 1952), 1–7, under the title of 'axiological cogito', meaning that every proposition implies a previous affirmation about its 'truth value', or its meaning. We prefer the term 'semantic principle'. Besides, contrary to what Ruyer says, at least in this chapter, this principle equally has an ontological bearing and is moreover, as we will see, identical to the argument characterized in this way by Kant.

9

The Paradox of Epimenides

Approaching the Semantic Principle

'Semantic', from the Greek *semantikos*, is the scholarly adjective for the Latin *sensus*, 'meaning'. Generally it is used in linguistics to designate everything having to do with the meaning of words (problems of origin, evolution, variation, etc.). Here, it is understood as "what makes sense for the intellect," and therefore as the objective quality of that which is intelligible. Insofar as primary metaphysical principle, the semantic principle affirms that the absolute of Meaning is presupposed with every judgment, so that every doctrine that posits a complete absence of meaning, and for which, as a result, the apparent meaning is only the random product of actual nonsense, is philosophically nil. The semantic principle could be worded then: "Meaning is with the Principle," that is: "In the Principle was the *Logos*."

PARADOXES AND ANTI-PARADOXES

We propose to show that the essential function of the paradox of Epimenides is to establish the truth of what we call the semantic principle. Doing this, we turn aside from the exclusively logical point of view under which this paradox is ordinarily viewed. Not that logic has nothing to teach on this subject—and moreover we will gather together its teachings. But we think that the most profound significance of paradox is precisely to reveal the insufficiency of the logical point of view and to teach us how to surpass it; in other words, this paradox is a symbol.

Besides, in a general manner, to see a paradox as a logical difficulty intended to exercise the shrewdness of the learned or dazzle the simplicity of the ignorant is a misinterpretation. Rather a paradox should be regarded as the 'visible face' of a truth in itself invisible, or again, too obvious to be perceived: like light, truth is hidden in its very transparency. Paradox is only the *verso* of a *recto* that could be named 'anti-paradox'.

§1 — *Principles do not condition intelligence, but constitute its nature.* Among the anti-paradoxes should be counted first and foremost what are called 'principles of reason'. The formulation under which they are ordinarily presented should not make us forget that their obviousness is only revealed by forgetting them and the consequences that follow from their negation. This means that they should not be proven (they would no longer be principles) and appear to be demanded by reason; otherwise this demand would simply be a constraint imposed on reason from without. Reason, at its very core, acquiesces to this demand through a real intellectual intuition, the necessity for which is rightly 'irrefutable'.[1] Also these principles are in truth metalogical, or metarational, in the sense that logic and reason designate the order of a purely discursive knowledge, that is, a purely mediate (and therefore demonstrable) knowledge. They are similar to the light without which the eye will not see, but yet never sees, and discovers only indirectly by its nocturnal absence.

The principles of reason are therefore principles for reasoning, that is they govern the good use of reason and are norms for this art of conclusions that comprises logic. Such is the case for the first among them, the principle of non-contradiction. But, being metalogical in their essence, they are grasped in their true nature only by philosophy itself, which therefore transcends logic (without contradicting it). They are superior to logic insofar as they set norms to which it is subordinate. Better put: logic is nothing but the sum-total of intellectual operations by which the human mind subordinates itself to principles in its cognitive activity. But philosophy is not subordinate to principles: it thinks about them, or rather they are thought about in it and are connatural to it ("to think," says Lagneau, "is to think about order"). In other words, the intellect knows them implicitly by knowing itself. They are, however, posited and defined as such only by the intellect reflecting on its own cognitive activity, when it becomes aware of the natural structures brought into play by this activity. Knowledge of them therefore implies a initial intellectual act which, as stated in the previous chapter, is essentially an intuition of the real as such, and which means for us an awareness that something is real (or again: being is meaningful for the intellect). This is why St. Thomas can write: "Our intellect, therefore, knows being naturally, and whatever essentially belongs to a being as such; and upon this knowledge is founded the knowledge of first principles, such as the *impossibility of*

1. On reason as submission of the mind to intellective principles, cf. *Amour et Vérité*, 108–113.

simultaneously affirming and denying, and the like. Thus, only these principles are known naturally by our intellect."[1]

And so these natural structures of the intellect should not be viewed in a Kantian manner, as if they were conditioning the intellect *a priori*. They are to the contrary perfectly transparent to it, they are intellect itself, they are the *Logos*; the intellect, 'obeying' them in its cognitive operations, is simply free and follows its 'nature' totally: it is itself because, by definition, it cannot be 'other than itself', which implies first that whatever belongs to the order of intelligence is necessarily in itself intelligible, that is: everything of the intellect is necessarily *meaning*, *Logos*. Kantian (or post-Kantian) structuralism is in fact equivalent to a disguised naturalism. Far from answering Hume,[2] transcendentalism has only justified him philosophically by rectifying his excesses, but by itself falling, as we will see, into the same contradiction. In reality, no intellect can say: "Those are the unintelligible principles that condition me, I obey them without understanding them." And *what intellect* could ever decide that its understanding of them is purely illusory? Principles are clearly then the reflection of the intellect's structures in objective knowledge, granted, but these principles are also those of the *Logos* in itself, necessary and purely intelligible, for that is the sole proposition that the intellect can intelligibly hold—which implies that there is no *essential* heterogeneity between our intellect and the *Logos*.

§2 — *Intellect, will, reason*. This is not to say, however, that the intellect only operates in identity. Certainly, like light (its primary symbol) revealing in the objects that it illuminates its inherent colors, the intellect reveals the intelligible nature of all that it touches. But this revelation is possible only on the condition that the intellect is open within itself to the otherness of its object, being; for the intellect *is* 'relation

1. *Summa Contra Gentiles*, bk. 2, 83, [31], trans. J.F. Anderson (Garden City, NY: Doubleday & Company, 1956), 281–282. There is clearly for St. Thomas, then, a knowledge that does not come from the senses: "certain seeds of knowledge pre-exist in us" (*De Veritate*, XI, 1, *ad. resp.*). "Our intellect knows some things naturally; thus the first principles of the intelligibles, whose intelligible conceptions ... naturally exist in the intellect and proceed from it" (*Summa Contra Gentiles*, bk. 4, 11, trans. C.J. O'Neil [Garden City, NY: Doubleday & Company, 1957], 87–88).

2. Hume's radical skepticism is known to have had the effect of 'awakening' Kant 'from his dogmatic sleep'. He agrees with skepticism that we cannot know principles in themselves. But he refuses to make of them simple habits of thinking and sees in them the *a priori* conditions for all possible knowledge. These conditions are *transcendental* to all experience, but we cannot posit them as *transcendent* objects.

to'.... There is within it then as if an essential otherness: it fulfills its own nature only in its openness and submission to something other than itself, it receives its completion only as it becomes present to something else, its own 'other side' first and foremost. In other words, the intellect is only light in an illuminating relationship with the reality by which it is *informed*, and which, in some manner, actualizes and fertilizes it; this is true for sensible as well as metaphysical knowledge. And this is even a major theme of our reflections, since sacred symbols are precisely the semantic objects under the action of which metaphysical understanding awakens and develops, as will be seen in the following chapter, to the extent that it accepts subjecting itself to the cultural contingency of their presence. Thus "nothing is in the intellect that was not first in the senses," according to the scholastic adage; to which must be added the Leibnizian correction: "except the intellect itself," for, we repeat, by submitting itself to its objective other, it is itself that the intellect discovers and rediscovers, it is by itself that it remembers itself, and such is the meaning of Platonic reminiscence.[1]

The intellect is therefore ordained to being, just as the eye is ordained to light. But if it should allow itself to be invested by being, if this involves a consent to be open to the object of its aim, this is because the intellect, in itself purely a relation to being, is not altogether itself in its mode of human actuation. In other words, its eye is not always open, or always oriented towards the only object worthy of it, that is to say capable of actualizing it according to the perfection of its essence.[2] In short, if the intellect is ordained to being as to its 'other side', this is because it is existentially rooted in a 'this side' of itself upon which it depends for its actuation. In itself, the intellect is a mirror which cannot but reflect its object, provided that it is oriented towards it. This dependence, upon which is stamped the *potential* condition of the intellect; this existential rootedness defined by one's 'point of view', one's cosmic 'point of anchorage' out of which the intellect tends towards its investing by being—

1. Platonic doctrine is often criticized because it implies the innateness of Ideas to the mind. This is to forget that the mind does not know these Ideas by simple inspection: it should be reminded of them, it should give birth to them; precisely the sensible world, the symbolism of myths, and the spiritual master are of use here. The spirit finds everything in it, but through the mediation of the sensible and the symbol. —The formula of Leibniz is taken from *New Essays on Human Understanding*, II, chap. 1, §2, 111 (trans. P. Remnant and J. Bennett [Cambridge: Cambridge University Press, 1996]).

2. Not only openness but also orientation is required, for it could be objected that the symbolism of hearing [*ouïe*] (and hence understanding [*entendement* = both 'hearing' and 'understanding']) is not 'closable' and therefore not 'openable', but yet 'orientable'.

and which is then also its 'blind spot'—we designate by the term *will*, the consequence and 'expression' in the order of life of our metaphysical state of freedom. In other words, the intellective mirror is not purely transparent (otherwise it would not reflect), but it possesses an *obscure face*, something 'this side' (of its aim) that 'situates' it ontologically, and by which it 'escapes' the luminous flux continually radiating from being. This means that, however pure intellection may be, it requires at its root the will's acquiescence.

By what mystery, now, can the will (blind by definition) acquiesce to the intellect's openness and its orientation toward the light of being? How can it obey an unknown order? Must we not suppose a capacity for intuitive knowledge in the will itself that draws us into an indefinite regression: intellect supposes will which supposes intellect, etc.? This is a formidable question, one which, if we unwound its thread, would cover the entire field of philosophy. We will simply say, first, that will and intellect are not two distinct 'things', but two modes of being of the same spiritual entity: the person. If there is then a will for the intellect, this is because there is an intellect for the will, or because the will is at least capable of some cognitive perception.[1] Next we will say that this agreement, humanly inexplicable—but not impossible—is precisely the work of grace. An intervention from 'On High' is required to incite the will, on the one hand to allow the intellect to be open to the light, and on the other to be itself obedient to the reality perceived by the intellect. This intervention from 'On High', this grace, presents itself under many forms. Thus, faith is a grace, and every intellection requires the grace of this faith. But it also assumes the form of a revelation, a culture, an education, that trains the will and teaches it how to be conformed to the perception of the true. Conversely, the intellect itself, in its potential dimension, that is insofar as it is not in act, is able to be educated (otherwise it is necessarily autonomous in its pure act: the intellective grasp[2] cannot learn to understand or command). This education of the intellect, this habit or *habitus* that it is inclined to receive

1. With reference to the far eastern symbol of *yin-yang*, we will say that the will is *yin* and includes a luminous point, whereas the intellect is *yang* and includes a dark point.

2. Simone Weil has shown better than anyone that the intellect, in its act of intellection, is perfectly free, and no authority, no will, even our own, has any power over it: one cannot be forced to understand what is not understood. But she has also admirably shed light on the importance of *attention*, the intellect's sole voluntary dimension. Likewise, the Cartesian method, although 'willed', is basically summed up as 'paying attention'; as for proof, it happens or does not. Every other doctrine destroys the very idea of truth, namely: whatever is self-imposed on the spirit.

is, as for its active structuring, reason.[1] Reason is therefore the acquired submission of the spirit to norms. It is identical to rationality's will, to that *willed* part of the intellect which, without any real intuition of the true, swears an oath to conduct itself according to its principles, perceived as demands. Hence the possible definition of reason as the '*habitus* of principles', and therefore also as rational *behavior*. But this oath can be broken, precisely because it depends, in its very being, on an implicit and almost unconscious act of will. The will, tired of being obedient to something that is in a certain manner totally elusive, can rebel and refuse the gift: such is irrationality or insanity. Insanity (mental deficiency excluded) is a breaking of the pact that binds us to the *Logos* at the very moment when the soul discovers that reason is rightly only a pact, or at least seems no more than that. With this the soul experiences its own unbridled freedom, and so the principles, the cognitive expressions of reality, have no other power than that of our obedience. This is not the case with the intellect, which, at reason's core and in an almost subconscious fashion, recognizes the obviousness of principles. Within the intellect the fulfilling of its free nature prevails over the sense of its determination by being. What remains, however, is this: the intellective eye has to give its consent to be opened. Before it opens there is only obscurity, and it is into this very obscurity that the grace of a presenti-

1. According to a thesis we have already developed elsewhere, in every intellectual act we ought to distinguish pure (unteachable, unengendered, autonomous) intellect, which is however ordained to being, and the categorical (either conceptual or mental) intellect. As a capacity for acculturation, the latter consists in the grasp and use of various languages through which information or knowledge is communicated, as well as the conceptual forms extricated therefrom and which the mind elaborates for itself. Being an intelligible grasp, the categorical intellect is therefore not only memory, the simple possibility of being informed, nor just a concept-building activity: since we are only informed by something when we grasp its *meaning*, it also participates in pure intellect. Conversely, the former participates in a certain categoriality: either directly, insofar as it is 'informed' by the 'category' of being (which, to tell the truth, transcends all categories), or indirectly because it only functions, in the human state, with the help of instruments (bodies of knowledge and concepts) provided by the categorical intellect. Without a minimum of intellectual equipment—which may be possibly confined to just the mother tongue—can there be any actual intellect? It is in this sense that we have spoken of culture as a veritable agent intellect. Let us note however that a weak categorical capacity can, for a particular individual, go hand-in-hand with a strongly intuitive intelligence: a primary and deep acculturation can make it hard to acquire other languages. The opposite is no less true: the ability to acquire and handle all languages can betray a basic inability for intellective understanding. Now, in the categorical intellect itself, one can distinguish among its informative contents bodies of knowledge and an orderly structure: reason; as for concepts, they consist in both at once. See, *supra*, chap. 8, 2 and *infra*, chap. 11, 3.

ment of light comes to touch it: in this sense, as we have stated, intellection presupposes a kind of faith. But, submissive to its investing by being, the intellect realizes its true nature, and the more it submits to this, the more it knows the joy of its freedom. And so, in its very depths, the intellect requires what is intelligible, it is nourished by meaning and lives only for this; in short, the law that constitutes and defines it in its very essence is the *semantic principle.*

FROM THE LOGIC OF BEING TO BEING AS SYMBOL

We can now attempt to describe the semantic need in its full breadth and show just where it leads. For this we will follow the path indicated by Plato in the *Sophist*, for it is clearly in obedience to the semantic need that he is compelled to the 'parricide' of Parmenides,[1] to this need that presses the intellect to find meaning itself in what seems to contradict the most basic rule of intelligibility.

§1 — *The logical principle of non-contradiction.* Let us recall then that Parmenidean philosophy appears to be based on an equivalence of Being and *Logos*. The Logos means only Being: "It is the same to think and the thought that [the object of thought] exists, for without Being, in what has been expressed, you will not find thought."[2] How in fact do we speak of something other than being? How could the *Logos* be anything but *onto-logos*, since "For never shall this be forced: that things that are not exist; but do you hold back your thought from this way of inquiry."[3] To speak of something other than being is to speak of non-being, which is nothing; this would therefore be to say nothing. Here we see where a strict application of the principle of non-contradiction leads, and how, in fact, it reduces reality to whatever logic can measure. For the experiences of error and, even more, sophistical *illusion* are also 'realities' that consist in saying what is not so and making it pass for being. It is the sophist's very being that contradicts Parmenides because, if he were right, his existence would not be possible. Parmenidean logic, then, does violence to being. But it also does violence to logic. The principle of non-contradiction is found to be contradicted, undermined by

1. *Sophist*, 241 b. In the analyses to follow, the Parmenidean philosophy is viewed in its standard sense and without prejudice to its real nature.

2. Fragment VIII, 34–36; trans. L. Tarán, *Parmenides* (Princeton, NJ: Princeton University Press, 1965), 86. Another translation: A.H. Coxon (*The Fragments of Parmenides* [Assen/Maastricht, The Netherlands and Wolfeboro, NH: Van Gorcum, 1986], 70): "The same thing is for conceiving as is cause of the thought conceived; for not without Being, when one thing has been said of another, will you find conceiving."

3. Fragment VII, 1–2; trans. Taràn, op. cit., 73.

its very definition. And this is precisely what the intellect, insofar as it wants to be faithful to its own nature, should not accept. The intellect establishes that there is error and illusion, and this is what it should take into account, that for which it should find a meaning, that is: find how this non-reality, or apparent reality, can indeed be integrated into universal reality. For the intellect is the sense of the real, as we have said again and again: it is submission to the real (how are we to deny the reality of the experience of illusion?), but a submission that demands meaning. The intellect is open to the real, but this openness is an intellection, a semantic act. The parricide of Parmenides is demanded by the semantic principle, and therefore by that very thing that logic calls the principle of non-contradiction.[1] In his introductory note to the *Sophist*, A. Diès speaks on this subject and in an analogous way about a "*fundamental postulate*: that without which logical thought would not subsist is true."[2] This is not a restricting of being to whatever limits the principle *seems* to set for its boundaries. We must, to the contrary, under the very demands of reality, see the principle in its most absolute universality, and such that Non-Being itself can no longer contradict Being, that is, accede to a conception of absolute Reality such that it includes in itself its own contradiction, or again to the conception of an Identity non-exclusive to its Otherness, in short a strictly non-contradictory Reality.

§2 — *Sophistry, like Structuralism, is Parmenidean*. Only from this point of view does the odd kinship that joins Parmenidean onto-logic, sophistry, and Lacan's or Derrida's structurality come to light. It is in fact clear that the equivalence of being and *logos*, the absolute determination of the second by the first, makes of every discourse a pure product of being, that is to say of what is, without it being able to have, in this case, a true or false value. Discourse participates in the same positivity as

1. We reject here every restriction of this principle's validity, contrary to some of its modern detractors (Lupasco for example) who claim to have worked out a 'non-Aristotelian' logic, under the pretext that the physically real would present us with instances of realized contradictions. But it is still in the name of the non-contradiction of thought, in other words out of a need for rigor, that they wish to 'over-step' this principle, which is . . . contradictory.

2. Collection 'Budé', 285. In short, to speak of what is not is to make non-being be, which seems to violate the principle of non-contradiction. But we can actually speak of what is not. This is then because, in a certain manner, non-being does not exclude being: otherwise this possibility would be unintelligible, extra-semantic. The possibility of stating the principle of non-contradiction implies the reality of non-being. On the *Sophist* and the sophistic, see *Penser l'analogie* (Genève: Ad Solem, 2000), 135–161.

being. Being is what is as what it is—this is its 'factuality'; likewise all that can be said about discourse is that it is as it is, that it is 'just so'—this is its 'textuality'. Moreover, by virtue of this positivity, being is reduced to what discourse states about it. Even though being determines discourse, discourse necessarily determines in return: textuality equals factuality. One sees that we are in a 'state of determinateness', and it is this that conditions Parmenidean onto-logic; for, after all, if Parmenidean being were nothing, one could no longer say anything about it. Parmenidean being is then that 'something' as such, the 'determinateness' in itself for which every *logos* is only the determined effect. But as this is in any case about discourse, and also Parmenides, it is the *logos* or rather the sum total of *logoi* that constitutes the determinations of being. Now the sophist is in exactly the same situation: the meaning of his discourse is a product of the discourse itself, since the *logos* is fabricator of the true as well as the false. Giving an existence to what is not: this is the whole art of magician and sophist that wants to prove that the true, what is, is just as fabricated as what is not, that there is no difference between the two, between true and false. And so, in reality, the sophist concurs with Parmenides, is a Parmenidean without knowing it. The theses that everything is true (as with Parmenides) and that everything is false (as with the sophists) amount to exactly the same thing: the absolute naturalism of the first coincides with the absolute artificialism of the second, but of course the one is unaware that it is the other, and can only be unaware of this because the one like the other, in stating his thesis, *wants to speak the truth*. Each of their theses is enveloped in a signifying intention that masks their limits and contradictions, and each renders the other impossible, but both are based in the same 'determinateness'. As for structuralism, the proof is even more simple, because structuralism is actually the truth of Parmenidism and therefore manifests its limits in an exemplary fashion; here, being as factuality proves to be textuality: identified purely and simply with the structures of a *logos* devoid of meaning, entirely reducible to them. Surely Parmenides' radical substantialism can appear the opposite of structuralism's no-less-radical 'relationism': here being (with a lower case 'b') entirely determines discourse, while there discourse entirely determines being. But, by reversing Parmenidism, structuralism is only repeating its text.

§3 — *Of Non-Being* (*or Above-Being*) *as 'possibility' of Being and* Logos. This is why the semantic principle requires that we surpass the viewpoint of ontological 'determinateness', of being as source and root of all determination, as pure and absolute positivity, as the general formality of all forms. Super-ontological Reality, the absolutely and infinitely Real, is that for which pure Being is itself *only* the symbol, the principial self-determination; it is the metaphysical All-Nothing, the absolutely Non-Posited, the Beyond of every affirmation (and therefore every negation), the supreme Non-Contradiction, that which can be 'defined' as infinite Possibility, to the very extent that the possible is that which comprises no contradiction.[1] By this we clearly see that the logic of being leads us beyond being. Being is only the first 'form', the primary 'image', the initial 'revelation' of supreme Reality, its principial self-affirmation. Certainly, without this form, without this self-affirmation, the *logos* would have no knowledge of supreme Reality, which is only 'manifested' through it. And besides, neither would there be any *logos*, since it is essentially a sense of being. However, the *logos* would be just as impossible if supreme Reality, the super-ontological Absolute, were purely and simply identical with its affirmation, in other words if this principial ontological 'manifestation' 'exhausted' all of Reality's possibilities. For then there would no longer be any 'place' for *logos*. Such is Parmenidean being, the full and solid reality of which excludes every other possibility, that of the super-ontological as well as the infra-ontological. With such a conception being has no 'meaning', nothing can be said about it, since it itself *says nothing*, and therefore the word of being is, precisely, no longer possible. Truly, *logos* can be revelation or principial utterance of the sense of being only if being makes sense, and being makes sense only on condition of it being seen, not as the ultimate sphere of Reality, but as the supreme and absolute affirmation of Reality: that is, dare we say, its principial function. This is moreover why being, at this level, is opposed to its negation: the very negation of being clearly proves that being is an affirmation, the first affirmation of Supreme Reality. Thus, in contemplating being, the *logos* is implicitly contemplating Above-Being, as the depthless Depth that 'gives' to the *logos* its meaning under the form of being, that is, under the form of Its ontological self-revelation. To 'say' (*logos*) being is to say Above-Being, for, without the Above-Being (or Non-Being) *in which* it is uttered, there would be no *logos*. With respect to the Sun of Being, everything that is not seems to it a dark nonexistence, and surely, for the human *logos*, there is no other luminous experience than the one coming to it from this Divine Sun. However, it also happens that this Sun sets. That

1. Such a 'definition' constitutes the upper limit of every possible conceptualization; cf. René Guénon, *The Multiple States of the Being*, trans. H.D. Fohr (Ghent, NY: Sophia Perennis, 2001), 10–11. It comprises the minimum of determination beyond which there is only the ineffable.

which hides it and impedes its shining is the motion of the Earth that revolves around it; in other words, this is the human being when turned towards him- or herself, and no longer oriented towards receiving the ontological light. By the same stroke is revealed then both the sparkling of the super-solar stars and the more-than-luminous Darkness of nocturnal Silence that the daystar obliterates with its splendor. This is to say that the creature, in its autonomy and by its own existence, proves, as the hither side of Being, the super-ontological Reality for which the Sun is itself only the blinding revelation.[1]

CONCLUSION: THE NEED FOR INTELLIGIBILITY AND ASCENSIONAL REALIZATION

We are beginning to glimpse just what the semantic principle is, an expression by which we wish to designate nothing but the need for intelligibility (or meaning—'semantic' being the adjective for it) proper to our intellect. We have just shown, following Plato's inspiration, just where this need leads. We should make clear, however, that even considered on a lower level this need remains identical to itself. The least intellectual act as well as the loftiest intellection manifest their nature. But by no means does this signify that the intellect does not include degrees, but only that each of these degrees is in continuity with those preceding

1. The distinction between Non-Being and nothingness was formulated by Georges Vallin: *Lumière du non-dualisme*, Presses universitaires de Nancy, 1987 (Vallin's account seems, however, rather systematic, misconstruing in particular the transcendent and apophatic notion of 'Divine Person'). Schelling had already insisted on not confusing *me on* (non-being) with *ouk on*, "nothing at all" (*Philosophie de la mythologie*, Aubier, tome II, 13e leçon, 63–65 and Appendice 361). The doctrine of Non-Being or Above-Being is explained by R. Guénon at the beginning of *The Multiple States of the Being*. This doctrine, although not expressly represented there, is considered an essential teaching of Shankara's nondualist *vedanta*. This is likewise the best translation of the Chinese *wu*, which literally means 'not to have' (Max Kaltenmark, *Lao Tzu and Taoism*, trans. R. Greaves [Stanford, CA: Stanford University Press, 1969], 34). But clearly it also belongs to western traditions, especially the Platonic and Neoplatonic. Thus Proclus states that "there is in us a sort of seed of that Non-Being" (*Commentary on Plato's Parmenides*, 1082, trans. G.R. Morrow and J.M. Dillon [Princeton, NJ: Princeton University Press, 1987], 432). And from there it passed to St. Dionysius the Areopagite and Meister Eckhart. Actually, it is encountered throughout the Middle Ages and with one thirteenth-century author in particular, Thomas Gallus, who writes: God is "beyond every substance and every being," and again: "above all... above one and unity, and being and entity" (*Mystical Theology: The Glosses by Thomas Gallus and the Commentary of Robert Grosseteste on De mystica theologia*, ed. and trans. J.J. McEvoy [Leuven, Belgium: Peters, 2003], pages 33 and 27 respectively). We have dealt with this doctrine in two articles for the review *Connaissance des Religions*: "Du Non-Être et du Séraphin de l'âme" (I, 1), and "Connaissance et réalisation" (III, 2–3).

and following without a break in continuity anywhere, each superior degree realizing the conversion of the inferior degree to its own essence and truth. The intellect is thus like a ray, uniting all degrees of reality among themselves: it is identical, in 'subjective' mode, to the semantic ray that is the symbol in 'objective' mode. And just as the symbol is a call, an invitation to anagogic ascension, so each intellection is a spiritual injunction to realize and become what is known. For, we repeat, if the intellect is a 'universal reader', this is because it is nothing of what it knows, since the known object is only received within it apart from its particular conditions of reality. The human intellect, viewed in this way in its rudimentary exercise, is therefore in some fashion the 'universal nothing', the hither side of every determined being. It is out of this potentially universal nothing that the intellect points toward being and opens itself to its intuition as to what grounds and specifies it. Intellect is therefore, by definition, desire for being, orientation and aspiration to the real. From this 'nothing', from this hither side of being, the shadow and reflection of the beyond of being, the human *logos* points toward being, and, under the effect of the desire and love that bears it along, enters into the process of realizing itself by an ascending conversion into its own essence: "we who reflect as in a mirror the glory of the Lord, we are transformed into this same image, going from glory to glory."[1] The intellect, being in fact nothing, can only become what is reflected in it. Its ascensional realization is therefore a function of the degrees of being that it knows and of which it becomes actually conscious, that is as a *reality* and not just as a speculative need. It is this path and this conversion of the intellect that Plato teaches in the symbolism of the Cave: it teaches us that true philosophy is not at all a conceptual game or a simple exercise of thinking activity, since it involves the whole being in an ascent towards truly supernatural realities. Such are the true stakes of the semantic principle.

Epimenides and His Paradox

"Epimenides the Cretan affirmed that all Cretans were always liars." This proposition is the most common and surely the best formulation of this famous paradox. However, it will be helpful to inquire about the origin and traditional attribution of this sentence, as well as about the rather enigmatic figure of Epimenides, for the historical data and the legend are especially illuminating.

1. 2 Cor. 3:18.

As for its attribution to Epimenides, first of all it creates a problem at least under its usual form. Some, in fact, see in this formulation only a particular expression of the more general paradox of the *Liar*: when I say I lie, is this because I am lying or telling the truth? Now, as we know, this argument was attributed by Diogenes Laertius to Eubulides of Miletus, who lived in the second half of the fourth century before Christ, and who was disciple and successor of Euclides of Megara. "The Liar" is moreover cited among seven other arguments which, almost all of them, bring in the theme of the 'veil' or 'veiling'.[1] Must we then withdraw his authorship from the sentence, as do some of today's writers? We think not, and this is why.

There is a rather well-attested tradition that attributes to Epimenides another sentence, akin to the previous one but not a paradox: "Cretans, always liars." This declaration would have been from the *Theogony* mentioned by Diogenes Laertius.[2] St. Paul, who relates this sentence, cites it as coming from "one of them, a *prophet* of their own,"[3] in which Clement of Alexandria and St. Jerome recognize Epimenides.[4] Must we suppose then that, starting with the Christian period, the paradox of the "Liar" and the sentence of Epimenides were brought together to end up with the formula "Epimenides the Cretan says that all Cretans are always liars"? This is possible, but no one is able to determine at what date this connection took place. We are faced then with a sentence of unknown historical origin, which an immemorial tradition attributes to Epimenides, but in which specialists see an alteration of a Megaric argument invented by Eubulides of Miletus. So why not consider this sentence as itself having two sides, two meanings, one exoteric, that of Eubulides, and the other esoteric, that of Epimenides? Would not the skeptical use made by Eubulides constitute an exteriorization of what, in itself, should be viewed in an entirely different frame of mind, so that, in this perspective, it is Eubulides' argument that should actually go back to Epimenides and would be a kind of vulgarization?[5]

1. *Lives of Eminent Philosophers*, Book II, 108; vol. I, trans. R.D. Hicks (Cambridge, MA: Harvard University Press and London: William Heinemann Ltd., 1925), 237.

2. Ibid., Book I, 111; vol. I, 117.

3. *Titus* 1:12.

4. *Stromata*, I, 59, 2. We cite St. Jerome according to *La sainte Bible*, of Pirot and Clamer, Letouzey et Ané, tome XII, 251.

5. It is not always easy to determine the true import of Greek skepticism. Is this 'modern' doubt that purely and simply denies access to truth, or rather a highlighting of the relative and contradictory character of formulations (language-based and mental)

What inclines us towards this thesis are Epimenides' basic character traits as revealed by legend and which seem to express the idea of 'veiling-unveiling'. If we are right, this means that our meditation's essential themes are rediscovered as a synthesis in the figure of Epimenides: the theme of symbolism which reveals by veiling, the theme of esoteric meaning; that is, the theme of a secret meaning and initiatic testing, the theme of a critique of the signs of the Invisible, and lastly the theme of the semantic question or the principle of metaphysical understanding. Who then was Epimenides?

To tell the truth, we know very little about him. Diogenes Laertius informs us that he was considered to be the manifestation of Aeacus, judge of Hades along with Rhadamanthus in a tribunal presided over by Minos.[1] This is then a semi-divine being whose function as judge makes him a master of discrimination, of meting out, and of 'initiatic' passage. It is he who puts people to the 'test' to differentiate those who have an understanding of signs from those who do not. Moreover, Plato himself speaks of the 'divine Epimenides' and introduces him as at once a prophet, the establisher of a sacrifice, a purifier of hallowed places, a benefactor of the city of Athens, a wise man endowed with amazing powers, and someone able to be nourished in a miraculous manner thanks to a mixture of mallow and asphodel.[2] For Aristotle, however, this prophet reveals the past rather than predicts the future,[3] which seems to indicate that we have here a divinatory power directed towards knowledge and not towards action. Modern scholars had long thought that this personage was legendary, until the moment when Diels succeeded in attributing to him a certain number of fragments[4] and dating them. The majority of scholars agree that he lived in the sixth century before Christ, with Cnossos as his birthplace, and see in him a possible eminent representative, the most important after Pythagoras, of the

---

with a view to surpassing all form, and of a direct seizure of the intellect by Truth itself? Are we not dealing with a kind of 'negative theology' that nails the mind to its own contradictions so to, by this sacrifice, effect its necessary divesting? Is there not something here quite close to certain Zen methods? By definition, appearances are deceiving. But the similarity between these two methods is incontestable.

1. Ibid., Book I, 114; vol. I, 121.

2. *Laws*, I, 642d–e, III, 677e. Obviously this involves, as Jean Biès suggests (*Empédocle d'Agrigente* [Paris: Éditions Traditionnelles, 1969], 39), the use of a sacred potency contained in certain plants and mentioned by all traditions.

3. *Rhetoric*, 1418 a 24.

4. E.R. Dodds, *The Greeks and the Irrational* (Berkley, CA and London: University of California Press, 1951), 141–146, and 166, n. 40.

shamanic tradition, which, from remote Asia, would have penetrated as far as Greece through Thrace.[1]

But there are basically three characteristic 'signatures' of the person Epimenides that seem to converge on the question of sacred symbolism and are, from this point of view, mutually illuminating.[2] First signature: Epimenides, having gone off in pursuit of a sheep, withdrew to a cave on Mount Ida sacred to Zeus, where he found himself plunged into a state of deep sleep for fifty-seven years[3]; after which he awakened without retaining any memory of the time elapsed. Second signature: Epimenides was tattooed, they say, over his entire body, which was only discovered after his death, so that 'Epimenidean skin' became a proverbial expression "to designate whatever is hidden."[4] Lastly, the third signature: the attribution of the paradox of the *Liar* to Epimenides. As we see, each one of these traits is connected to the theme of veiling-unveiling, of hiddenness and discovery, the esoteric and the exoteric. The first is concerned with the initiatic Cave, the one where, as Plato teaches, the knower effects a conversion of seeing and being, discriminates between the real and the illusory, and, aware of the secret, is freed from the temporal condition: what appears as a death outwardly is actually an awakening and a rebirth. The second seems to refer, according to historians, to a sign of dedication as a 'servant of God', the Greeks using "the tattoo-needle only to brand slaves."[5] And yet how can we not see in this the symbol of . . . the symbolic function of the body and, more generally, of the corporeal form as theophany? The very fact that it was only discovered with his death shows that a living symbol is transparent and that the symbolic form reveals its opaqueness only when it can no longer exercise its presentifying function. Conversely, this also means that

1. Ibid. On this subject do not forget that Plato, in the *Gorgias*, assigns to Rhadamanthus the function of judging eastern souls and to Aeacus that of judging western souls, which might point to the European function of Epimenides, the manifestation of the Aeacian principle. If one recalls that, as Guénon remarks (*The Reign of Quantity and the Signs of the Times*, 4th rev. ed., trans. Lord Northbourne [Hillsdale, NY: Sophia Perennis, 2001], 131), the sixth century is an 'historic barrier' where a traditional readaptation for the major cultures takes place (Pythagoras, Lao Tzu, Buddha, the reconstruction of the Jerusalem Temple), the 'prophetic' function of Epimenides is perhaps something similar.

2. The bringing-together of these three traits is due to Enrico Castelli, in *Mythe et foi* (Paris: Aubier, 1966), 13–14.

3. Diogenes Laertius, op. cit., I, 109; vol., 115. Other traditions say fifty-one years.

4. Dodds, op. cit., 163, note 43, citing Suidas, that is, the fictitious author of the *Suda*, a Greek encyclopedia of the tenth century.

5. Ibid., 142. Sacred tattooing existed moreover in Thrace, and is in fact encountered in all religions (Hinduism and Native American traditions in particular).

appearances, insofar as appearances, are invisible and are perceived as appearance and veiling only with respect to a higher reality. Finally, the paradox of the *Liar* also speaks without telling and tells without speaking, for it is silent about what it wishes to signify and denounces the signifier with the help of the signifier itself. Such is how this enigmatic figure venerated by Pythagoras himself is presented.[1]

Concerning its Logical Significance

First of all we will mention that, from the psychological point of view, the solution to the paradox of Epimenides[2] poses no difficulty. Or rather that, psychologically speaking, there is no paradox. We can just as well allow that Cretans betray a strong propensity to lie, that is to deceive those with whom they speak. This would involve an almost national character trait. We could just as well say: the French are frivolous, the Germans serious, the Italians talkative, which does not mean that they are always so.

We find, however, that this judgment is the work of a Cretan. But there is not for all that any subsequent contradiction. To the contrary, one might suppose that, being himself a Cretan, Epimenides knew Cretans better than anyone and that his judgment has more of a chance of being true. Yes—but, some will say, a lie is involved and not frivolity or volubility; that is, the tendency to utter deceptive judgments. To which it is necessary to reply that a lie implies, precisely as a condition of its possibility, knowledge of the truth. To say what is false without knowing it is not to lie. This is to deceive oneself, not others. But anyone not knowing that he is speaking falsely is equally unaware when he tells the truth, as Socrates shows in the *Lesser Hippias*. Awareness of the true supposes that one does not impulsively say everything that comes to mind, and therefore that one is aware of the difference between thought and speech, and between reality and thinking. And this awareness implies the possibility of a lie, since a lie is precisely the realization of this difference: the liar says what he does not think is true. The possibility of a lie is therefore the criterion for true discourse.

The lie expressed introduces a remarkable point because, in fact, the

1. Francis Vian, "Grèce archaïque et classique," in *Histoire des religions*, *Encyclopédie de la Pléiade*, 1970, tome I, 560.

2. On this topic we have mainly consulted: A. Koyré, *Épiménide le Menteur* (*Ensemble et Catégorie*), Hermann et Cie éditeurs, collection 'Histoire de la Pensée', IV, 1947; Stephen C. Kleene, *Mathematical Logic* (New York: John Wiley & Sons, Inc., 1967), especially §35; Bertrand Saint-Sernin, "Les Paradoxes," article published in *Revue de l'Enseignement philosophique*, 25e année, no 1, October–November 1974, 31–43.

meaning of a lying discourse has no other reality than that of its expression. Here meaning is truly reduced to the signifiers and the use that is made of them. Meaning is then truly an effect or product of discourse, or again, if preferred, of a chain of signifiers. But what we have here is only an exterior or apparent sense, for a lying discourse is itself produced with a view to hiding the interior and real sense. The sense of a lying sense, its raison d'être, is to replace, to disguise, to occupy the 'place' of a true sense.

If we wish to find a 'point' to this sentence, we should therefore see it from a logical perspective, that is, from the perspective of its own structure, and no longer from the perspective of its psychological possibility. From this perspective the Epimenides version seems to be a contradiction or an absurdity. Logicians call it a "semantical paradox."[1] It is worked out in this way: Epimenides being a Cretan, by virtue of his own declaration about Cretans, lies by saying they are liars. And therefore Cretans are not liars and so neither is Epimenides. And therefore he is telling the truth. And therefore Cretans are liars, etc.

This paradox is more deserving of being called a paradox than is the argument of the "Liar" (if someone says 'I lie', is he lying or telling the truth?[2]). The "Liar" is not actually a true paradox, for, as Russell notes, it does not constitute a true judgment, a true assertion. Nothing is asserted by saying that one is lying and by wishing to signify by this that the proposition 'I lie' is a lie,[3] whereas the *Epimenides* version does indeed entail a judgment, an affirmative act: that of Epimenides asserting that Cretans are liars. Logicians have produced moreover other paradoxes of the same type: such as that of the village barber who shaves all the village men who do not shave themselves.[4] Again, we will cite Richard's paradox, according to which "a non-definable number in a finite number of words" is exactly defined by that very fact. All this is well-known and may be reduced to the contradiction between the content and form of a proposition. Thus, for a statement to assume the *form* of a true judgment it is necessary that it is not impossible to tell the truth; but the *content* of the Epimenidean proposition specifically denies that this condition is realized.

1. The expression is due to the English logician Ramsay, who in 1926 distinguished between logical and semantical paradoxes; Kleene, op. cit., 189.

2. Cicero (*Academica*, II, 29) writes: "If you say that you are lying and that you are telling the truth, you lie. Thus you tell the truth and lie at the same time."

3. Bertrand Russell, *Principia mathematica* I, 30 *seq.*; cited by A. Koyré, *Épiménide le menteur*, 12.

4. Koyré, ibid., 21. We owe this paradox to Russell.

These paradoxes have given rise, on the part of specialists, to extremely varied and often very critical appraisals. This is the case with Koyré and some authors whom he cites in his account: "how did it happen that simple sophisms, which would not have puzzled a disciple of Aristotle or a student of the Faculté des Arts of the University of Paris for one second, could be taken so seriously by minds as eminent as Russell, Frege, etc.? The answer seems abundantly clear. The cause of this curious blindness lies in the formalism of logistic reasoning and, first of all, in the *interpretation of judgment in extension*."[1] It is in fact certain that the paradox, properly speaking, subsists only if we place ourselves in the formalist perspective of set theory, which regards a proposition's definition of terms only in a purely 'extensional' manner. Historically, these are the contradictions to which set theory leads and which restores to the Epimenides version its full importance. And, as concerns the validity of these paradoxes, the critics have come to refuse to situate themselves in set-theory logic.

Disregarding the understanding of terms,[2] an understanding that seems too metaphysical, too intuitive, and not very rigorous, since it consists in the sum of determinations 'comprising' the *nature* of the defined, modern logic wants to be extensivist. The extension of a term in fact defines it in an *invariable* manner, since it only designates the defined set appropriate to this term. In a way it comes down to a relationship of properties. Whether the extension increases or decreases, the term does not vary. For example, the term 'man', understood according to its extension, does not vary, whether applied to a specific person or to everyone of a specific nationality. The idea of the set actualizes in this case the logical formalization of the idea of extension. All men to whom

1. Ibid., 24. A concept is defined in extension and in comprehension. Extension designates the set of objects to which a concept is applied; comprehension designates the set of characters comprised in the definition of the concept. When comprehension increases, extension decreases conversely. 'Animal-reasonable' comprises two characters and is extended to all mankind; 'animal-reasonable-masculine-French' comprises four and is extended to just male human beings born in France. A detailed and exhaustive critique will be found, as far as modern mathematics and logic are concerned—a critique made from the Aristotelian-Thomist point of view—in a special issue of the review *Itinéraires* (n° 156, september–october 1971) especially in the articles by Paul Bouscaren: "La sophistique des ensembles mathématiques" (34–92) and Father Guérard des Lauriers: "La mathématique. Les mathématiques. La mathématique moderne," 95–252. (Actually the latter is more a book than an article.)

2. Cf. on this subject F. Chenique, *Éléments de Logique classique*, Éditions Dunod, 1975, tome I, 62–64, et tome II, 341. As regards modern logic, by the same author: *Comprendre la Logique moderne*, Dunod, 1974, tome I, 16–17, and tome II, 416–419.

the property 'French nationality' is appropriate belong to the set of Frenchmen, and, conversely, the items of the set are *entirely defined* by their belonging to that set.[1] And so a rigorous and universal method for defining the terms of a statement is at our disposal. Abandoning intuitive representations based on the random and undefinable grasp of a nature or meaning, the notion of a set is raised to a norm for every definition. To define any item whatsoever is to name the set to which this item belongs. Such is, it seems, the condition for a rigorous discourse.

It is then that we stumble upon the paradox of the set of sets brought up by Russell in 1905. A normal set is naturally distinct from its components; in other words, it is not a component of itself. Consider then set E of all the normal sets, that is, sets that do not contain themselves. Each of these sets is a component of this set E. Now, is E itself a normal set? It is impossible to answer this question. If it contains *all* the normal sets, as it is obliged to do by its definition, it should contain itself, but then it is no longer normal. And if it does not contain itself, it is no longer in conformity with its definition, that is, it does not contain *all* the normal sets.

We will halt our logical considerations here. Clearly the idea of a set of sets, as abstract and formalized as it is, simply puts us in the presence of the *concept* of set as such, of its significance. But, defined in extension and according to the rigor of the most formal logic, the *concept* of set is no longer anything but the *set* of all the sets to which such a concept can be applied. And since we have here the only logically valid definition, the result is that logic's most important idea is contradictory. It is the same for the village barber who cannot either shave himself or not be shaved (for then he would no longer be barber *of* the village or else he would not shave all of the village's men); this contradiction comes in fact solely from the component 'village barber' being considered as *entirely defined* by the relationship 'shaving all men who do not shave themselves'. If this were not so, if the barber were also a man, there would be no contradiction with the barber as a man shaving himself.[2]

1. An 'in-comprehension' definition of Frenchmen would be along the lines of: light-hearted, analytical, scoffing, individualistic, etc. Its difficulties and lack of rigor are obvious.

2. To eliminate this kind of paradox, Russell worked out his famous theory of the hierarchy of logical types, according to which defining the belonging of any object whatsoever to a class makes sense only on condition that this class is of a *type* superior to the object under consideration. *Type zero*: individuals; *type one*: classes of individuals; *type two*: classes of classes of individuals; etc. Rule: every type-*n* object belongs to a class of

And likewise for Epimenides, who falls into contradiction only if he is considered to be entirely defined by the 'liar' characteristic.

Concerning its Metaphysical Significance

As we see, considered from the vantage point of logic, the importance of the paradox is rather limited. But it is no longer the same if one decides that the aim of this paradox is to make *plain* the contradiction inherent to the scientific project of a total closure of the concept.

First of all, notice how—something hardly mentioned by logicians—the Epimenides version realizes better than any other one the basic contradiction of every paradox, because it is concerned with lies, that is the falsifying of statements, while the barber paradox only indirectly calls into question the logic of statements. We could also imagine 'overly strong' paradoxes bordering on outright absurdity, which would be canceled out then as paradoxes: for example, "I am dead," or even "I am mute, he cried out," etc. Paradoxes of logic or mathematics are rather 'overly weak'; that is, their paradoxical character is, for the uninitiated, only detectable with the help of painstaking analysis. The paradox of Epimenides, to the contrary, exhibits a contradiction, but this contradiction, although easily noticed, remains however as if veiled beneath the perfectly natural and normal character of the sentence. It is not the complicated invention of an extremely subtle logician; it can occur in quite ordinary speech. This is moreover why Tarski concludes that it is impossible to escape the risk of such a paradox in natural language, but that artificial languages (like logic), by being impoverished, can be shielded from this risk.[1] Being impoverished means here eliminating all terms by which a language speaks about itself: in other words, a language cannot, without contradiction, speak about itself; to speak about a language, a metalanguage needs to be used, a language, then, with a semantic power stronger than that of the language it takes as object. There is thus, say the logicians, irreflexivity between the formal system of language and that of metalanguage. And by this we meet again the corollary of Gödel's theorem, a corollary "according to which it is

type-$n + 1$; L. Vax, *Lexique: Logique* (Paris: PUF, 1982), 155. By observing this rule, and therefore by excluding every proposition such as *a* belonging to *a*, the paradox is in fact avoided. This solution, appreciated in varying degrees (cf. R. Blanché, *Introduction à la logique contemporaine* [Paris: Armand Colin, 1957], 146–166), is simply a rule of syntax and presupposes that there is a fault of logic only against syntax—which constitutes a perfectly contradictory proposition.

1. Saint-Sernin, "Les paradoxes," *Revue de l'enseignement philosophique*, 25e année, nº I, 32 and 41–42.

impossible to represent, in a formal system corresponding to the hypotheses of the theorem, a demonstration of the non-contradiction of this system."[1]

And so it seems that the philosophy of logic still remains a prisoner to the circle in which Epimenides enclosed it. What logical thinking considers a rigorous and therefore scientific language, and gauge of truth, is one in which discourse never escapes from itself, but to the contrary entirely possesses itself: that is, a language in which the terms and the relationships they support are entirely defined. Now, when is a term entirely defined if not in the instance when the *definiendum* is totally reduced to the *definiens*, where there is nothing more in the defined than is put there by the definition? In such a rigorous discourse the whole being of the defined comes from definition. This is why we are led to consider the definition as the rule for the producing of the defined. Philosophically this means that terms evade discourse by their origin, but, by becoming their sole origin, discourse claims to be in possession of terms. To the contrary, terms, concepts, ideas, thoughts, or whatever name by which they might be designated, to the extent that *they have a meaning* that is their own property and with which in a certain manner they are identified, and therefore to the extent that they do not receive this meaning from discourse itself, evade discourse. Only their formal translation is to be found there. Like the umbilical mark on the human body, which witnesses to our origin, the semantic reality of a concept, which constitutes its linguistic further side, is the point by which it is connected to being. It is the invisible side *of* language to which intellect alone can gain access. Such a concept, viewed according to its semantic reality, has never been closed upon itself, and therefore never defined linguistically; and we can never *just make use of it*[2] in language. To utilize it, it is also necessary to contemplate it, know it in itself

1. Jean Ladrière, "Les limites de la formalisation," in *Logique et connaissance scientifique, Encyclopédie de la Pléiade*, 317–322. Kurt Gödel, "the greatest logician since Aristotle" (von Neumann), by stating his 'incompleteness theorem' has revived the question of the basis of mathematics and has not ceased to stimulate philosophical reflection. See: Ernest Nagel and James R. Newman, *Gödel's Proof*, rev. ed. (New York: New York University Press, 2001).

2. We especially object to Wittgenstein's famous proposition: "For a *large* class of cases of the employment of the word 'meaning' [*Bedeutung*]—though not for *all*—this word can be explained in this way: the meaning of a word is its use in the language" (*Philosophische Untersuchungen*, §43 [Frankfurt-am-Main: Suhrkamp Verlag, 1967], 35; we are using the G.E.M. Anscombe, P.M.S. Hacker, and J. Schulte translation [*Philosophical Investigations*, rev. 4th ed. (The Atrium, Southern Gate, Chichester, UK: John Wiley & Sons, 2009), 25].

and recognize it. And even the use we make of it will ever be but an approximation and a limiting of its semantic reality, which is in itself informal and inexhaustible. Such a concept or such a term is therefore open in itself, at its core which is also its origin, its original openness. It is open to being as to that which gives it meaning and reality, and towards which it is a way among all the other ways offered by the innumerable multitude of intelligible constellations.

Now the paradox of Epimenides offers precisely the case of a discourse entirely the producer of its own terms, of an entirely *artificial* [*artificiel*] discourse because entirely *'crafty'* [*artificieux*]. Logical thought is fascinated by the apparent contradiction of the paradox, to such a point that it no longer perceives the reason why the contradiction arose. It clearly distinguishes the contradiction's abstract structure, which enables it to manufacture, as Russell says, "a strictly infinite number of contradictions,"[1] but, doing this, it forgets what makes the contradiction inevitable. It claims that this contradiction stems from the transgression of the irreflexivity of statements, since it is produced when a statement takes itself for object. Surely this transgression is inevitable, says Tarski, since the natural language includes terms about itself, either denoting linguistic objects (for example: noun, verb, adjective, etc.) or involved with "the relations between linguistic objects . . . and what is expressed by these objects" (for example: 'truth', 'designation'), so that "for every sentence formulated in the common language, we can form in the same language another sentence to the effect that the first sentence is true or that it is false."[2] Hence the necessity in logic to make such a contradiction impossible, to eliminate from artificial language all metalinguistic statements or terms. But this conclusion, which is perhaps imperative for logic—because of our incompetence in this matter we cannot make a determination here—seems deadly for philosophy. As we have shown in *Histoire et théorie du symbole*, it is not language that is metalinguistic, it is human thought that is reflexive by nature. The reflexiveness of language is only mirroring the basic reflexiveness of thought, that is, its capacity to take itself for an object, or again to *surpass itself*: no human thinking is outside the possibility of this 'self-transcendence' or this 'self-transitiveness'. Obviously the linguistic sign plays a part in the actualization of this reflexivity, since, as we have seen, it is by its mediation, by the entry into meaning it elicits, that the mind

1. *My Philosophical Development*, rev. ed. (London and New York: Routledge, 1995), 59.

2. Cited by Saint-Sernin, "Les paradoxes," 42, who makes use of two Tarski texts: *The Semantic Conception of Truth* and *Truth and Proof*.

becomes aware that it is able to think its thoughts. Not *by itself* then does language 'denote' itself. It is by thought using language metalinguistically, a metalinguistic usage in conformity with the primary nature of the linguistic sign, since it is not the creator of but the catalyst to the awareness of thought's reflexivity, to the very extent that, as we have stated, a sign is first a sign of itself. A sign indicates first that it is a sign, and the entry into meaning is nothing but that.

There is only contradiction when this reflexivity is *rendered* impossible, that is when thought is reduced, in one manner or another, to language's materiality, which basically amounts to *using language contradictorily*, to making it no longer the means of expression for thought, but the producer of thought. This is precisely what the formalized languages of logic and science do, their ideal being, ultimately, to dispense with logician and scholar and function all alone, like a machine. With such languages contradiction is only eliminated in a formal and apparent manner; that is, it suddenly appears when a statement is applied to itself, whereas the imprecision of common language makes the presence of contradiction less perceptible. But this is at the price of a real and hidden, and, we say, universal contradiction. The contradictions of common language are circumstantial and local; the contradiction of formalized languages is permanent and underlying, since it consists in an act by which the logician says that it is not he who is speaking. Besides, it must be remarked that contradiction is . . . contradictory: to say that it is impossible is to posit it; to say that it is possible is to deny it, since it is impossible. If common language tolerates contradiction (it is even, we are told, inevitable), this is perhaps because it is not really a contradiction. And, if formalized languages exclude it and do not tolerate it, this is perhaps because in that case it is a true contradiction. Hence it simply follows that there is only contradiction of and in diction, not of and in thought. Intelligent thought can no more contradict itself than light endure darkness. Only when the intellect allows words to think in its stead (unfortunately this is almost inevitable) does it happen to contradict itself. This rightly proves that no language, by itself, excludes contradiction. Only the human mind can impose on itself the task of avoiding contradiction; it cannot entrust care for this to any logical machine. Quite the contrary, it is by making the logical machine—formalized language—work that contradiction is produced, which is in that case so only '*in terminis*.' Anyone can say: "the circle is square," but nobody can actually imagine it. This can be said because words have a lexical meaning, that is, based on this meaning they are mediators between a signifying intention and a signified intention, and therefore because they apparently continue to function and produce meaning even though they express no signifying intention. If words were not contractual bearers of meaning, it would be otherwise: each signifying expression would be a unique event, with its meaning *indistinguishable* in other respects; in short, there would be no signs, that is distinct entities endowed with a set meaning. In other words, the sign is a *sign* of itself, posits itself as *sign*, and, thanks to this metasemiotic property, it can specifically 'pretend' its purely semiotic functioning is possible. In the proposition "the circle is square" nothing is said and nothing is meant, or rather it means only what it says (a statement) and is nothing real; the semiotic entities are left to function 'all alone', but this is only an apparent, purely mimetic functioning. One is unduly taking advantage of the language's lexical properties, one is breaking the contract. And if somebody asks: but then, what are you thinking about when you say: the circle is square? since it must be thought about in order to be said, we will reply that we are *thinking about words and not ideas*, and this rightly proves, in an irrefutable manner, that *to think truly is to think about ideas and not words.*

Such is the meaning of the paradox of Epimenides, and our conclusions are in some manner the opposite of logic's. We readily agree that, formally speaking, the contradiction of the Epimenides version is a result of the statement being applied to itself. But this application, as such, is not responsible for the contradiction. We can, moreover, conceive of propositions applied to themselves that are by no means contradictory.[1] There is no circle here to be mentioned: Epimenides the Cretan asserts that Cretans always tell the truth, even though this statement is obviously false. The proposition's reflexiveness is only contradictory when it is impossible, that is when it is at once required and excluded by the proposition. What Epimenides the Cretan says of Cretans necessarily applies to himself and, at the same time, cannot be applied. That is the contradiction. Now, when is reflexivity at once required and impossible? Each time a discourse states that the meaning of a discourse is simply a product of the discourse, in other words, each time a discourse claims to apply to itself the thesis that no discourse ever applies to anything. For the meaning of a discourse is what the object designates to which it is applied, the goal of the signifying intention for which the discourse is only a mediator. It is therefore extra-linguistic and passes right through the sign. The sign is precisely a sign only on this condition: that of being traversed and able to be traversed by the signifying intention. But if the sign claims to signify that it is opaque,

1. Besides, modern logic admits relationships of reflexivity; cf. F. Chenique, *Comprendre la logique moderne*, Dunod, tome II, 356–367.

that it itself produces meaning instead of being only its means of expression, at that very moment it stops being a real sign and having a meaning, and therefore would not signify. *Thus, there is contradiction each time a sign intends to actually signify that meanings are only apparent.* Now this is exactly what the permanent lie realizes: it defines a condition such that, without calling for a formalized language or structural theories, meaning becomes merely a product of the apparent functioning of the signs of discourse alone.

Our conclusions, however, must be pressed somewhat further. The Epimenides paradox does not in fact tell us only that Cretans are liars. It also tells us that they are *always* so, in other words that they cannot help but be so or, again, cannot tell the truth. This second condition of the paradox is as important as the previous one, and is moreover necessarily bound up with it since, precisely, the previous condition depends on it: a lie is not the work of a single semiotic chain, but the result of an intention to deceive. This time we are then at the level of signifying intention, which we have said is bound to be deceiving. Now here is how: this intent to deceive is either voluntary or involuntary. If involuntary, either it is unconscious, or imposed on consciousness from without by a "malignant demon."[1] But these cases amount to the same thing *from the viewpoint of the meaning of discourse*. What is in fact the opposite of the will to deceive if not the will to tell the truth? And what is the will to tell the truth if not the will to say what is, in other words to renounce one's own will to speak in order to submit to what being says, to allow being to speak? Thus, the will to truth implies the capacity for thought to allow itself to be informed by being, the freedom to acquiesce to this speculatively. We have just said that there are signs only when traversed by signifying intentions. Now we say that there is only a signifying intention when traversed by a knowledge of being, in such a fashion that to 'signify', to 'say a meaning' is, in the final analysis, to imply that being makes sense and speaks through us. To suppose the contrary is to imply that meaning is produced entirely by the signifying intention, is entirely a work of the will, is simply artificial. And whether

1. 'Malignant demon' is a translation of the Latin expression *genius malignus*, used by Descartes at the end of the first of his *Meditations on First Philosophy*. This involves a supposition, an hypothesis that enables us to understand those very doubts about the most obvious and least doubtful truths (mathematical truths, for instance), as if there existed a universal power of deception which, from without and unknown to us, falsified every approach of our reason. If there is a truth that withstands the ordeal of this hyperbolic doubt, then this truth would be established with an absolute certainty, a metaphysical certainty, which, Descartes states, is quite superior to mathematical certainty. This truth is that of my thinking: if I doubt, this is because I think.

this artificiality is voluntary or involuntary changes nothing as to the origin of meaning. However the contradiction, no longer now only 'in terms', is as flagrant as before because meaning would not be meaning if it were not meaning—that is to say intelligibility—of what is. And this is valid in every case. But if the deception is the fruit of a constant will, then the very possibility of the sign disappears, since there would no longer be any linguistic contract. Or if it is involuntary and produced by the ineluctable workings of consciousness, whether internal or external, then there would be no one to perceive it and therefore neither would there be any possibility of deceit. The one is moreover equivalent to the other. There is no difference between a constant will to never tell the truth and an involuntary inability to do so, since a free will is, by definition, a will that can change the direction of its willing. This second conclusion of the Epimenidean paradox therefore establishes that there is no proposition that does not imply these two conditions: an intellect capable of knowing what is, and a will capable of acquiescing to this knowledge.

We have discovered in this way the two essential dimensions of the semantic principle: intellect and will, which are also the two poles of the human being, the union of will and intellect being realized through love. The Epimenidean paradox is not a demonstration, but a 'test' of this. It is only an occasion for the mind, to the extent that it submits itself to this test, to become aware of it, that is, to understand that it has never ceased thinking about it: under the shock of this paradox the native evidence of the intellect is reactivated. By no means does this involve a novelty or a subtle and hard-to-understand truth. Purely and simply, it establishes in a universal and absolute manner that *it is impossible to reduce any proposition whatsoever to the conditions* (internal-speculative or external-semiotic) *that produced it.* Yes, 'impossible'. For, if human thought were reduced to the conditions that produced it, there would never be anyone to speak of it. The very possibility of stating such a deduction proves its non-reducibility. And this is why the contradiction is valuable. A language in which it would be entirely banished (made impossible by the very manner in which this language would be constructed) would be a language that could never become aware of itself and, as a result, would not be the primary locus for the discovery, by thought, of its own nature and the world's existence, since, let us recall once again, it is in the experience of the sign (namely: that the sign is a sign) that the distinction between knowing subject and known object is realized.[1] Just as St. Augustine, in connection with orig-

1. Cf. *Histoire et théorie du symbole*, chap. V.

inal sin, could cry out: blessed fault that has won for us such a Savior, so we could say: blessed contradiction, since through it there descends on us the salvific proof of the *Logos*.

However, through a mystery which, in our eyes, is the most astonishing of the modern West's whole philosophic history, everything transpires as if the Epimenidean contradiction did not exist, and for two centuries we see the flourishing of reductionist theses, theses too eager to destroy in us the signs of the divine to grasp their own impossibility. There is then a *critical sleep*, and only paradox can awaken us from it.

Two Applications of the Semantic Principle

THE THREE REDUCTIONISMS

It is impossible to reduce any proposition whatever to its conditions of production. This is quite precisely what the paradox of Epimenides demonstrates. Positively, it establishes the proof of the semantic principle: *the truth of meaning derives solely from the meaning of truth*. Now the semantic principle, or founding principle of all metaphysical doctrine, is ignored by the reductionists. Their critique always consists in claiming to demolish the dogmatic propositions by disclosing the unconscious conditions that produced them, and therefore sparing such critics a direct refutation of any particular dogmatic truth. This is why, to the 'dogmatic sleep' (Kant's expression), we oppose the 'critical sleep' revealed by the semantic principle, and about which we would now like to say a word.

Every statement is conditioned: 1) by language or, more generally, by the semiotic system with the help of which it is expressed; 2) by the nature (generic, social, individual) of the person expressing it, and 3) by the form of the intellect conceiving it. Hence we have a triple conditioning: semiotic, anthropological, and cognitive, and three kinds of critical reductionism: structuralist, hermeneutic, and philosophical. The third kind, illustrated by Kantianism, is also the most important: to this we will devote a chapter. In light of the semantic principle the first two, which we have dealt with at length, clearly betray their inherent faultiness.

CONCERNING SEMIOTIC (OR STRUCTURALIST) REDUCTIONISM

Our line of argument is simple: it comes down to observing that the discourse of structural surfacialism (which denies any transcending of discourse) always sets itself in a privileged position with respect to every other discourse; nothing is more blinding than a critical intention.

This is so for Lévi-Strauss, who declares: "meaning always results from a combination of factors that are not meaningful themselves....

In my perspective meaning is never a primary phenomenon: meaning is always reducible. In other words, behind every meaning there is a non-sense, and the converse is not true."[1] Since moreover, as the author declares, this involves "understanding how the human mind functions,"[2] we should surely consider Lévi-Strauss's texts as an example of the human mind's functioning. We are also informed: "if, in this realm (of mythology where the greatest freedom seems to reign) the human mind is enchained and determined in all its operations, *a fortiori* it must be so everywhere."[3] As a result, in good Epimenidean logic, the meaning of Lévi-Strauss's texts is, too, entirely determined and enchained, and therefore purely "phenomenal,"[4] and consequently has, by itself, no truth value. Lévi-Strauss's discourse is thus juxtaposable with *Zuni*, *Hopi* or *Tereno* discourse and, like them, is nothing but the effect of a surfacial structure. But then, if this is indeed so, this discourse, having rid itself of any right to the truth, must also lose any ability to give a verdict on the human mind, savage or non-savage. Far from it…

As for Foucault's philosophy, its definition of *episteme*, not as "a form of knowledge or type of rationality which, crossing the boundaries of the most varied sciences, manifests the sovereign unity of a subject, a spirit, or a period," but as "the totality of relations that can be discovered, for a given period, between the sciences when one analyzes them at the level of discursive regularities,"[5] presupposes for this *episteme* itself the privilege of constituting the transverse unity of all bodies of knowledge and holding out to them the transcendent viewpoint of simple truth. Certainly Foucault intends to speak the truth, and even the most hidden and imperceptible truth. However, if he is right, then Foucault's *episteme* is itself only a determined structural arrangement which, as such, enjoys no transcendence with respect to the others, nor any depth, and which is therefore perfectly determined and so clearly required by the details of this structural situation that it becomes altogether contingent, unless, by a most unlikely exception, it is to be considered the only absolute thing in an ocean of relative concretions.

As for Derrida's doctrine, we have already stated that, in our opinion, it may be the most remarkable effort in contemporary philosophy to follow thinking about structure *through to the end*. This is because it

1. *Discussion avec C. Lévi-Strauss* (with Paul Ricœur, Mikel Dufrenne, etc.), published in the review *Esprit*, November 1963, 632, and 637.

2. Ibid., 648.

3. Ibid., 630.

4. Ibid., 637.

5. *The Archaeology of Knowledge*, 191.

continually touches on the Epimenidean contradiction, even more because it is occupied from within by this contradiction and shaped by it. It is clearly conscious of being, in a certain manner, an outer limit of philosophy, a frontier philosophy, and yet it does not truly know this; otherwise it would have broken through and passed to the other side, and only then would it have discovered its impossibility, its contradiction, if not its lie. And so, as we have noted, Derrida clearly perceives that there is a contradiction to overturning metaphysics and transcendence with the help of the concept of sign, which is inseparable from its referential aim, and therefore from the assertion of a transcendence. But he overreaches, more desirous of finally destroying metaphysics than respecting the demands of his own course; "...*there is no absolute origin* of sense in general,"[1] he declares, without seeming to suspect that this assertion, to be pertinent, demands the exact opposite. Likewise, when Derrida, meditating on the mythopoetic function of "ethnographic *bricolage*,"[2] asserts that this function "makes the philosophical or epistemological requirement of a center *appear* as mythological, that is to say, as a historical illusion,"[3] we cannot help but ask ourselves: indeed, with what absence of center can all that *appear*? Continuing, in the same text (*Structure, Sign, and Play*), his commentary on Lévi-Strauss's method, Derrida explains that although the cultural field can be considered as a play of 'infinite substitutions' enclosed within a finite totality, this is not because it is infinite, but because "there is something missing from it: a center which arrests and grounds the play of substitutions."[4] What takes the place of the empty center is the sign which, precisely because it substitutes for an emptiness, a lack, signifies nothing determinate, and thus manifests a radically *supplementary* or, as Sartre would say, 'too much' significance, which would not be related to anything since the sign is *set*

1. *On Grammatology*, 65. Derrida himself declares quite rightly that "coherence in contradiction expresses the force of a desire, (*Writing and Difference*, 279).

2. A Lévi-Strauss thesis according to which myths are constructed out of the debris of no-longer-useful knowledge, in the same way that patching-together consists in using pieces of material or objects to fit into a new construction.

3. Ibid., 287. Our italics.

4. Ibid., 289. Already Edmond Ortigues (*Le discours et le symbole*, Aubier, 1962), but from a very different perspective, compared language to a 'Loyd's puzzle' where numbers are arranged in a set order on a tablet by sliding and not lifting them, which requires an empty spot. The human subject can only enter into language, that is to say play with it, thanks to this empty spot which, for Derrida, is nothing but the illusory subject. Now, if it is true that the subject's presence cannot be grasped at the level of 'mundane' things and entities, this is not because it is illusory, but because it is 'absent' through transcendence.

in place as a 'nothing'. And Derrida concludes: "The *overabundance* of the signifier, its supplementary character, is thus the result of a finitude, that is to say, the result of a lack which must be supplemented."[1] True. But why precisely *must* this lack be supplemented and how might this be done? We agree that the sign is 'too much'. Out of the notion of 'semantic surplus' we have even fashioned one of the essential ways for characterizing a symbol: the symbol is a symbol, that is to say a sacred sign, only because there is in it a semantic power that exceeds every mundane or cosmic referent. More radically still, this is true for every sign (but is fully realized only in the sacred symbol), because every sign is first *something other* than an object of the world or human subject. It is a *diacrisis*, an otherness (it is in this sense that it is first indicated as a sign), a disruptive experience through the mediation of which man and the world are posited in their correlative and inseparable distinctness. We acknowledge these themes as our own and they underlie our entire meditation.[2] They are quite simply true. But, as a matter of fact, there would be no possibility of being in the least aware of them or speaking about them if the center, the subject, and the transcendent were only *sheer* absence. Simply put: it is not the lack or absence that accounts for the surplus, but the surplus or transcendence that accounts for the lack and finitude. Derrida's strategy (and Lévi-Strauss's, but with less philosophic rigor) is clear. Seeing that the 'supplementary' nature of the sign is recognized—how can we avoid it?—the only non-metaphysical solution capable of accounting for it is to relate this 'supplementarity' to 'nothingness', because in fact, with respect to nothing, emptiness, absence, everything is radically 'too much'. But it should be admitted, at least *a priori*, that there is another possible solution, a solution according to which the semantic surplus that gives scope to semiotic entities and enables them to be used is the effect, the manifestation of the transcendence of the *Logos* with respect to a determinate degree of being, and, in the final analysis, of the transcendence of Above-Being relative to its ontological self-determination, Above-Being, which is basically absolute Beyond-More.

Now, that this 'solution' is the only one possible is what must be recognized, if Derrida's solution is to be stated and conceived of without contradiction: every word, every thought is a word and a thought *about* something, that is transcendent to what it is aware of and about which it speaks. All words and thoughts come from 'beyond' and present them-

1. Ibid., 290.
2. We have explained them in *Histoire et théorie du symbole*, chap. V.

selves at first as such. The emptiness, or lack, or finiteness *about which* Derrida speaks should also be *that out of which* he speaks; otherwise he would not, as a matter of fact, be speaking about them. And besides, Derrida clearly intends to reveal the true sense, the true vision of things, that with respect to which there is no longer any surplus, because it integrates into its concept all signifiers and all significations. In other words, if Derrida is right in all that, then truly there would be a 'too much' discourse, *his own*. Derrida could be right only if his discourse erased itself in its very manifestation.[1] But, in this case, we would never have any knowledge of it and it would never say anything, and the content of the discourse would be abolished along with its form. Meanwhile, Derrida has thought and spoken, necessarily considering himself as somebody 'somewhere outside the world', implying by the very existence of the Derrida *logos* the transcendence that he claims to deny. And this transcendence is undoubtedly also *in* the world (and by definition), as is the one in whom it is manifest, the man, but it is there in the manner of a transcendence, that is to say as the presence of an absence. For everything is dialectic.

Absence of a center is not exactly just any absence, but the absence of the unique and absolute Center from which all centrality derives and which no determinate or finite reality can contain, unless as its own lack and its own finiteness. And never will a pure and simple absence be able to be known as the absence of a center and a transcendence. Moreover, it is only by reason of transcendence that we will be able to account for the dual character of finiteness, which is not only negation, but also positivity. If a determinate reality were not finite, if it were everything no matter what, it would quite simply not exist. To be and to be *such* are inseparable. Its limits are what give existence to a being, and, from this point of view, this is a positive fullness. Negative finiteness only appears then out of radical infinitude; it is only its inverted reflection, the projected shadow, and therefore also, in a certain manner, manifestation. For, as Descartes states so irrefutably: "my perception of the infinite . . . is in some way prior to my perception of the finite,"[2] which in objective mode is tantamount to saying that universal manifestation is only the sum of all the *possible* limitations of infinite Possibility. And it is precisely because we cannot 'leave the infinite' that we perceive the finiteness of everything that exists, *a priori*—*a priori* and not after examination, for we can also bury ourselves in the positivity of things without encounter-

1. Nothing can actually make Derrida's philosophy be itself a 'sign' which, in order to be rigorously conceived, requires its own disappearance.

2. *Meditations on First Philosophy*, Third Meditation, 45, 31.

ing finiteness, emptiness, or absence, and for a very good reason: we cannot experience that which is not.[1] This is only then a primary and founding 'knowledge' presupposed for every recognition of finiteness, a cognitive or existential knowledge like that of suffering and misfortune. The infinite, the unlimited is the 'unsurpassable' and inaccessible horizon, as well as the transcendent heart of all knowledge. And so, once again, our favorite axiom is verified: *only the More can do the less.*

Far then from reproaching Derrida for quitting the strictly positive terrain of 'scientific' structuralism, we think that his approach is an effort to isolate those philosophical conditions thanks to which thinking about structure is possible. This effort leads him to place in relative 'prominence' the Epimenidean contradiction that occupies surfacialism from end to end, and which only a certain speculative laziness, reinforced by a hatred of metaphysics, can explain. Logically, the step that Derrida's approach wants to take outside philosophy should lead him towards a metaphysics of the infinite and the limitless as the only possible 'milieu' out of which thinking originates. But for that it would be necessary to break with thinking about structure and with the scientistic ideology that this expresses.[2]

CONCERNING ANTHROPOLOGICAL (OR HERMENEUTIC) REDUCTIONISM

Semiotic reductionism, because more radical, manifests more clearly than any other variety its contradictory nature. But, although the con-

1. Just as we cannot experience corporeally the absence of space or visually the absence of light, or see 'the edge of our vision'.

2. The metaphysics of the infinite or the limitless signifies that there is no point of view about Reality; It is Itself its own knowledge. This absolute and infinite Reality can indeed be called the 'One', or 'supreme Unity'; still this is so only in a manner of speaking. Likewise for the *En sof* of the Kabbalah, literally the 'Endless', which can be reduced to *Ain*, literally the 'Nothing'. And likewise for other traditions, Neoplatonic as much as eastern or western. But it is also necessary to add that the metaphysics of absolute Surplus, of the Infinite, by no means excludes principial affirmation or ontological self-determination, since to the contrary it implies it as the consequence of its own infinitude. In no way does this involve our positing an impersonal Principle beyond the 'personal God'. By this it is believed that one can escape from religious anthropomorphism, whereas one is only positing, above the thing 'Being', a super-thing that will be called Non-Being, Above-Being, or Brahman. Surely, an 'existential' anthropomorphism is thus avoided, but only to fall into a speculative anthropomorphism worse than the previous one. For the mystery of the human person very really opens us to what is beyond every thought, whereas our thinking always encloses us in a 'thing': to think is always to think about something. The only way to avert the almost inevitable risk of speculative reification is to enter resolutely into the infinite transcendence of the Divine Person.

tradiction of anthropological reductionism is less patent, it is more dangerous: it does not erase the human subject in a structure like a piece in a puzzle; it dispossesses it of itself and, occupying it from within, it thinks and acts in its stead, in imitation of a true diabolic possession.

Marxism and Freudism see in all the discourses of consciousness an unconscious interpretation of an alienating situation for which they themselves constitute the sole truly disalienating hermeneutic. Obviously this explanation is at once impossible and reversible. Impossible, for either this false-consciousness manufacturing alienation acts on every consciousness, and in that case just as well on Marx and Freud, who then produce only alienated discourse; or else it does not. But if they themselves are miraculously unscathed by it, why not others? And from that moment all the vigor of anthropological reductionism disappears, since alienation ceases being necessary and formative, and becomes contingent and avoidable. Reversible, for it should be asked if, in this case, Marxism and Freudism are not simply the ideological expression of socio-economic or psychic determinisms. Marx would be nothing but a petit-bourgeois intellectual marginalized by the industrial revolution, suffering from his own powerlessness and replacing the economic power that he was deprived of with hope for a socio-political dictatorship. Freud would be nothing but an unhappy child, and the vast psychoanalytical edifice would be then no more than the expression of his own obsessional neurosis, an over-compensatory construction by which Freud alone will succeed in subjecting his desires to the order of reality, all while excusing himself from experiencing them.

If however we notice that these two ideologies have, so to say, invaded the entire world and the consciousness of hundreds of millions of people, making them act and think like veritable automatons, it will become clear that their respective founders could be only 'prophets' of quite a dark 'revelation'.

As for Feuerbach, it would be useless to recount what we have said on this topic. His thinking too is at once impossible and reversible, he who is the first to claim to make religion speak all while beginning to silence it, he who above all would set man in the place of God: but can man continue to be a god for man when there is no longer any God?

10

Theophany of the Intelligence

Introduction: The Commander of the Philosophers

"*Alla laloumen Theou sophian en musterio*"—"But we speak the wisdom of God in a mystery"—theosophy under the veil of the symbol, a theosophy that declares "that eye has not seen, nor ear heard: neither has it entered into the heart of man, what things God hath prepared for them that love him."[1] These words of St. Paul describing the content of theosophy 'in a mystery' also perfectly define the content of religious discourse and, consequently, sacred symbolism: namely, that which surpasses all natural human experience. That the human mind can nevertheless pursue such a discourse puts into question cognitive structures as a whole and, ultimately, all anthropology. By degrees, as we have just verified, western culture, in accounting for this scandal, has arrived at the destruction of reason itself, at the radical denial of semanticity, that is, at the transgression of the semantic principle. As surprising as this may seem, the conclusion thrust upon us leads to a complete reversal of the speculative attitude: *since we cannot deny sacred symbolism the supernatural referent presentified in it, without being constrained to affirm (contradictorily) the basic non-semanticity of all discourse, we find ourselves needing to admit then the possible truth of this discourse.* By no means are we saying that all symbolism is true, that any statement with a supernatural referent requires our belief, but only this: it is strictly impossible that *every* affirmation of the Transcendent, either intentionally or unintentionally, is a lie. Consequently, some are necessarily true, even if others are only imitations: every imitation supposes a model. Moreover, and contrary to what is ordinarily assumed, nothing is

1. 1 Cor. 2:9.

harder to invent than a religion, and the logic of symbolism is second to none in coherence and persuasive power. But lastly, what remains is that the necessity for metaphysical knowledge should not be denied *a priori*—or, therefore, in the final analysis, for an experience of the supernatural which will assume, in that case, the form of a revelation. *Because the fact of sacred symbolism is irreducible, metaphysical knowledge is inevitable.* As surprising as such a conclusion might seem to western rationalism, the evidence for this is nonetheless compelling.

However, a further objection can be raised, not actually against the conclusion itself—for we do not think that anything can overturn it—but against its possibility. Even while admitting that what Simone Weil would call 'supernatural knowledge' is necessary, we would still be obliged, so not to find ourselves confronted with an insoluble dilemma, to sanction it for the human intellect. That the semantic principle is required by our mind is not enough to show that it is able to satisfy it. In short, what remains is to reverse the source of every critique and every reductionism—Kantianism—and recall how it is caught in the most irreducible of Epimenidean contradictions. Thus our going back across the various reaches of the critique of sacred symbolism ultimately leads to the redoubtable fortress that controls entry to it. For two hundred years the one unhesitatingly called 'the greatest of philosophers' has jealously watched over the door to metaphysics, forbidding anyone to cross the threshold. He himself has bolted the entry as tight as he could, undoubtedly for want of finding its key. Others have come, enlisting themselves more or less under his banner, people from here and there, who believed in Heaven or did not, all of them busy with re-enforcing the wall or raising it higher. But could they prevent the wind from bringing to our ears the miraculous joy of Reality's song?

The Epimenidean Contradiction of Criticism

METAPHYSICAL ILLUSION ACCORDING TO KANT

Kantian philosophy differs considerably from the reductive etiologies of religious symbolism in that the latter deny the referent's existence, while Kant affirms it; but that changes nothing about their identity from the viewpoint of knowledge. The affirmation (the postulate) of the existence of God or the immortal soul is in fact only possible from the viewpoint of moral action, in such a way that the *ideas* of reason—world, self, and God—are found to be in the same situation as sacred symbols in Freud, Marx, Feuerbach, or Lévi-Strauss: these are illusory transcendentals. We know nothing about these ideas because they represent nothing of what we think they represent. These ideas are only transcendent in appearance, an inevitable appearance, but not the reality. It must be explained then why this is so.

The basic reason for this situation consists of two causes: first, there is no other means to know about the existence of something than by receiving certainty of it through direct experience, and so we need a receptivity capable of being directly affected by something else; second, although we indeed have a sensory receptivity, we do not have an intellective receptivity—our thinking is always active, never passive, and cannot be directly affected by an intelligible reality. In short, we have no intellectual intuition, "the possibility of which we cannot understand."[1] As a result the senses are the only source of knowledge, and an idea or a concept[2] which is not applied to some sensory datum is an empty form that does not cause us to know anything. Now the ideas of reason, those of world, self and God, by definition surpass every possible experience: no one has ever seen the world, a self, or God. To what then do they correspond? These are in fact the conditions or requirements for the functioning of reason. To think, for reason, is necessarily to strive to bring everything back to unity: unity of phenomena (the world), unity of thinking subject (the self), and unity of all beings (God). Reason is only satisfied if it can lead this requirement to its goal. But this condition, this need for reason, should not be transformed into objective knowledge. This is however what we do: we take "the subjective conditions of our thinking for objective conditions of things themselves and . . . a hypothesis that is necessary for the satisfaction of our reason for a dogma."[3] Such is the secret of the transcendental illusion: the ultimate conditions of our knowledge are objectivized and are themselves considered to be things because reason, in constructing its object, unconsciously projects them onto this object and then imagines itself grasping them as determinations of reality itself. The source of illusion is then in the objectivizing or, in other words, 'speculative' attitude of human thought, an attitude that is itself, moreover, perfectly natural. But such is also the source of metaphysics and the history of metaphysics. Why is there a history of philosophy if not because there is a succession of opposed metaphysical theses? In other words, this history is one of an

1. *Critique of Pure Reason*, trans. P. Guyer and A.W. Wood (Cambridge: Cambridge University Press, 1998), 361; AK, III, 210.

2. Kant specifically makes it a rule to reserve the term 'idea' for concepts of reason which can be applied to no experience, while concepts properly so-called only arise from the understanding insofar as the latter is applied to sense data in order to think about them (ibid., 399; AK, III, 250).

3. *Prolegomena to Any Future Metaphysics*, §55; trans. G. Hatfield (Cambridge: Cambridge University Press, 2004), 100; AK, IV, 348.

endless conflict, each one thinking itself in reality contrary to another one. Now, if the opposition were really an opposition according to contradictories, there would be a solution to the conflict, for two contradictories are mutually exclusive: if one is true, the other one is false. We would be able to know whether the world is finite or infinite, whether God is or is not. But in fact this involves a 'simply dialectical'[1] opposition, for these propositions include the common term *world* or *God*, which is thought apt to be taken as a subject of attribution, whereas we can place no determinate knowledge under these terms. When we say *God* or *world*, we quite simply do not know what we are saying. Thus these metaphysical oppositions only occur between contradictories, but, as everyone knows, two contradictories can be equally false.[2] Dialectics is therefore the science of transcendent appearances, and this is the proper and altogether new work for 'transcendental logic'. This involves setting up "a court of justice" which denounces the appearance of knowledge, and "this court is none other than the *critique of pure reason* itself." And Kant specifies: "By this I do not understand a critique of books and systems, but a critique of the faculty of reason in general, in respect of all the cognitions after which reason might strive *independently of all experience*.... It is on this path, the only one left, that I have set forth, and I flatter myself that in following it I have succeeded in removing all those errors that have so far put reason into dissension with itself in its nonexperiential use."[3]

All that is, moreover, well known, and is a 'major site' for western philosophical reflection. But it needed to be recalled after all our preceding analyses, to identify its true nature. Clearly, what is stated here is the concept of *philosophy as a general hermeneutic of the metaphysical illusion*. It is here that this attitude, which we have encountered with all the *hermeneuts of suspicion* from Feuerbach to Freud, finds its model and primary support. Kant does not struggle against a particular metaphysical thesis, no more than Nietzsche or Marx. He does not oppose one affirmation with a contrary one; but, rejecting the contents of all theses, he suspects them and deprives them of their truth value in the name of their basic speculative unconsciousness. What good is it to argue with the dogmaticians? They are sincere, but they do not know what they are saying. And so the thinker is divested of his most precious possession: the consciousness of what he is thinking about, not in the

1. *Critique of Pure Reason*, 517–519; AK, III, 347.

2. Here are two equally false contradictory propositions: "everybody smells good" and "everybody smells bad."

3. Ibid., 100–101; AK, IV, 9.

name of a loftier and assumptive truth, but in the name of empirical and critical truth. Kant proceeds with a veritable 'psychoanalysis of theoretical reason', bringing to light its metaphysical unconscious (the dogmatic sleep), and reducing all the theses of prior philosophers to *their conditions of appearance and production.* Such is the authentic significance of Kantianism, and such is the unique task left to pure reason, for the "resting-place for human reason . . . can only be found in a complete certainty, whether it be one of the cognition of the objects themselves or of the boundaries within which all of our cognition of objects is enclosed."[1] And since the cognition of intelligible objects, the noumena, is not given to us, there remains no other certainty than one of boundaries, in other words, of its own conditioning. But to reduce knowledge to its conditions of production . . . it is precisely in this assertion that the Epimenidean contradiction resides, and this is why Kantianism merits the label of reductionism.

THE CRITICAL ILLUSION

In order to bring to light the contradiction of this speculative reductionism, it is enough to pose the pre-eminent Kantian question. Under what conditions is a critique of pure reason possible? Our question in no way contradicts the Kantian project since, with criticism, reason takes itself as object: "Pure reason is in fact concerned with nothing but itself, and it can have no other concern."[2] Now, and according to Kant's own terms: "That which *bounds* must be distinguished from *that which is bounded* by it."[3] Consequently, reason can limit the faculty of reason, in which the work of criticism consists, only on condition of being superior to this faculty, that is to say only if it is, in its essence, participatively enlightened by divine and absolute Reason. But precisely in that case criticism is useless and dogmatic metaphysics is in the right. To the

1. Ibid., 654; AK, III, 497. This is a question of knowing whether the *Critique of Pure Reason* should be seen as a propaedeutic for the new metaphysics, or else as constituting the entirety of philosophy. Kant writes: "the assumption that I have intended to publish only a propaedeutic to transcendental philosophy and not the actual system of this philosophy is incomprehensible to me. Such an intention could never have occurred to me, since I took the completeness of pure philosophy within the *Critique of Pure Reason* to be the best indication of the truth of my work." ("Open letter on Fichte's *Wissenschaftslehre*, August 7, 1799" in *Philosophical Correspondence*, 1759–99, trans. A. Zweig [Chicago: University of Chicago Press, 1967], 254; AK XII, 396–397. Cf. R. Verneaux, *Le Vocabulaire de Kant* [Paris: Aubier, 1967], 37–38.)

2. *Critique of Pure Reason*, "Appendix to the Transcendental Dialectic," 610; AK, III, 448.

3. Ibid., "Antinomies, Section Eight," in fine, 523; AK, III, 448.

contrary, if reason is no more than reason, if there is, at the core of reason, no receptivity of the intellect enlightened by the *Logos* "who enlightens every man coming into this world," reason will never be able to be aware of its limits, and any critique of reason by reason will be impossible. Unless we concede that, by some unheard-of privilege, since the time that philosophy and thinking began, Kant is the first to escape the transcendental illusion and to know how to see through it. But this is absolutely denied by the *Critique* itself: intellectual intuition "is not our own" (*die unsrige*).[1] As a result of every cognition being conditioned and constructed, the cognition that reason apprehends of itself, by virtue of the principles of criticism, should be equally conditioned, and the object of critical reason (which is reason itself) is necessarily, too, a 'constructed' object. Or else it would be in that case the sole object with the ability to be known *in itself*. But if this indeed involves a 'constructed' object, by the same stroke the critique loses all value for truth and certainty (surely not as to its analyses of detail,[2] but as to its general thesis called a "Copernican revolution"[3] by Kant himself).

In other words, the notion of an essential limitation[4] of cognition by itself is quite simply contradictory, that is to say impossible. Just as the eye cannot see the edge of its vision, knowledge cannot trace out the boundaries beyond which it cannot know. What knowledge can do—and this is what the doctrine of intellect as sense *of* being sufficiently accounts for—is to become aware of the boundaries imposed on it by the existential conditioning of its human realization. But it is an extrin-

1. Ibid., 361; AK, III, 210.

2. We in no way deny the pertinence of this.

3. In connection with the Copernican revolution, it is amusing to see how, since its first edition in 1905, the Trémesaygues and Pacaut French translation (Paris: P.U.F., 1950) has Kant speaking cosmological nonsense: "... he (Copernicus) wondered if he might have more success in making the observer himself turn about the fixed stars" (19). If Copernicus had sought to construct such an hypothesis, his scientific reputation would not have surpassed that of Cyrano de Bergerac. Regrettably, numerous collections of selected passages have reproduced as is this blunder which Kant never committed (he who had taught cosmology), but this does not seem to have upset the philosophers unduly. Barni, another translator of Kant into French, does not make this mistake.

4. Kant himself declares that a transcendental appearance is essential to reason: "what we have to do with here is a natural and unavoidable illusion which itself rests on subjective principles and passes them off as objective." The conflicts resulting from this illusion constitute a dialectic, itself also natural and unavoidable, "one that irremediably attaches to human reason, so that even after we have exposed the mirage it will not cease to lead our reason on with false hopes, continually propelling it into momentary aberrations that always need to be removed" (ibid., 'Transcendental Dialectic," Introduction I, 386–387; AK, III, 237).

sic and not an intrinsic boundary, and one the awareness of which is only possible precisely by virtue of the internal limitlessness of knowledge. This conditioning basically determines on the one hand the forms with which knowledge is endowed, and on the other its internal structure. As to the forms, they are defined as the sensible and abstractive or conceptual modalities that the mediations of the senses and the mentality imprint on knowledge. As for the internal structure, it characterizes knowledge as an intentionality, a dynamic orientation, an aiming of subject towards object, which stems from the fact that the cognitive act is accomplished in an individual and finite being. But, however it may be with these boundaries, they would be in no way conceivable if, at its heart, the intellect (or knowledge: they are one and the same) were not *more* than what it knows and *more* than the knowing subject. Surely this involves a participation in, an illumination by the infinite light of the *Logos*. And yet this intellect-knowledge is infinite only in itself, but this internal limitlessness is truly the transcendental, the ultimate and necessary condition of every determinate cognitive act. This is why Meister Eckhart can say that "the intellect is uncreatable insofar as such,"[1] an assertion that earned him condemnation and gave rise to many difficulties and various interpretations, but which comes down to the following question: is the light that infuses a crystal produced by the crystal? And if not, how do we distinguish one from the other? Thus, insofar as such, that is to say in act, *the intellect, in its superhuman essence, is uncreated and uncreatable.*[2]

The question "what can I know?" cannot then be posited *a priori*. It can possibly have an *a posteriori* sense when someone reflects on their own knowledge and ascertains its errors and limits by the fact of their existential finitude. But all that can be said of knowledge *a priori* is that it is self-assuming, that it is itself the non-conditioned condition of every determinate cognitive act, and therefore that it is quasi-absolute and *unengendered*: such is knowledge; this is what we have concluded, and this is another manner of stating the semantic principle. Knowledge understands itself as the 'universal witness' presupposed in all its acts. Such is the certainty of the Self spoken of by Shankara: "The Self... because it is the basis for the operations of the norms and the functions of knowledge, attests to its own existence even before they operate. And

1. *Quaestiones Parisienses. Quaestio Gonsalvi. Rationes Equardi*, 6; Magistri Eckardi, *Opera latina*, Auspiciis Instituti Sanctae Sabinae, ad codicum fidem edita, edidit Antonius Dondaine O.P., Lipsiae in aedibus Felicis Meiner, 1936, 17.

2. J. Ancelet-Hustache has summarized what is essential to this question in volume I of his translation of the *Sermons* (German), Seuil, 1974, 27–30.

it is impossible to refute such an existence, for what is adventitious to you can be repudiated, not what is essential."[1]

THE PHILOSOPHIC THEATER

From this we must conclude then that criticism, as such, is an illusion and a 'trompe-l'œil' since, if it were right, it itself would not be possible. In other words, Kantian philosophy has by no means 'gotten away' from dogmatic metaphysics (it is even often hyper-dogmatic), for the simple reason that it is impossible to get away: to think (and speak) is to think about something and speak about being. How then can we account for this illusion?

It seems that its secret resides in the most Kantian notion of all those worked out by the philosopher, the one that constitutes his own discovery and is also the key to criticism: we mean to speak of the notion of the transcendental. Like the notion of reflection for the Marxists, this one functions in two opposite directions: "I call all cognition transcendental that is occupied not so much with objects but rather with our mode of cognition of objects insofar as this is to be possible *a priori*."[2] This is how the 'transcendental critique' delivers us from the 'transcendental illusion'. And since it delivers us from it, but without abolishing it, given its 'natural and unavoidable' character, it enables us to abandon ourselves, *without risk*, to the delights of illusory transcendence. For such is truly the 'trompe-l'œil' of Kantianism. It is enough to break with a dogmatic *attitude* and adopt a critical *attitude*, the transcendental gaze, to enable criticism to do everything forbidden to dogmatics, that is to speak about God, the world, the thinking subject, causality, substance, time, space, etc.: in short, about everything that would be the object of dogmatic metaphysics, provided only that a 'transcendental' sign is allocated to these objects. Once this agreement is made there is nothing to limit the field of objects dealt with by reason. This is precisely because, thanks to this agreement, they can, at will, escape from

1. *Brahma Sutra Bhashya*, II, 3, 7: cited by O. Lacombe (*L'Absolu selon le Védânta*, Librairie Orientaliste Paul Geuthner, 1966, 232) adds that the certainty of the Self is like "a central and essential lightning-flash, which we have not sought, which we cannot seek because it is absolutely first and as if prior and interior to the gift that it makes of ourselves to ourselves in that awareness immanent to the understanding of an object, which we should no longer flee, which we can neither escape nor renounce, which we cannot deny, repudiate, or *transcend by a critical reflection*, because it constitutes our basic inmost depths, the most basic possible without any real splitting possible" (232–233); our italics.

2. *Critique of Pure Knowledge*, Introduction VII, 149; AK, III, 43.

the illusion of objectivity, all while exhibiting sufficient consistency to be objects of a philosophic discourse:

> I am bird, see my wings.
> I am mouse, long live the rats!
> — LaFontaine, "The Bat and the Two Weasels"

The term 'transcendental' actually belongs at once to the order of subjectivity insofar as it is applied to cognitive conditions of the thinking subject, and to the order of objectivity insofar as by these conditions there is an object, since it is they that construct it. The Kantians are then free to devote themselves to all the metaphysical objects snatched from dogmatics, seeing that they have severed them from their ontological root. Anything can be spoken about since we have taken the precaution, once and for all, of declaring that we would be saying nothing, that our words would not call into question any noumenal reality. Truly, if Kantianism were right, we would never know it: either we would remain caught in a dogmatic illusion that is 'natural and unavoidable' and 'irremediably attached to human reason', or else, we would be endowed by nature with a critical lucidity: space, time, substance, causality, world, self, and God could never become the *objects* of our discourse. But Kantianism, like all criticism, is a parasitic philosophy. It is fed, objectively, by the illusion of objectivity that it denounces among the dogmatists, and so claims to have its cake and eat it too. If Plato, Aristotle, Descartes, and Leibniz had not spoken about God, the world or being, by definition criticism would have nothing to say. But it can now steal from them their own words, since it has discovered they do not know what they are talking about. This is how critical philosophy transforms metaphysics into theater. It has clothed the ideas of reason with a transcendental livery, and then, attired in this way, has made them do battle upon the speculative stage. Does not its program consist in showing scenes of "the divisions and dissensions occasioned by this contradiction in the laws . . . of pure reason?"[1] Western intellectuals have been endlessly fascinated with the spectacle offered by Kantian theater. They have truly believed that the 'trick' was finally found enabling them to continue thinking, they for whom this was a craft, but without for all that running the risk of a commitment to the real and to being, which they abhorred, being above all fearful of the ridicule scientistic positivism would heap on them. They could roar with laughter at the debates that would unfold on the transcendental stage between grotesque shadows proudly draped in the mantle of their dogmatism, thesis clashing against thesis, full of serious-

1. Ibid., 460; AK, III, 282.

ness and importance, while the clever director discretely shows the strings by which he controls these unconscious marionettes.

Thus we discover that reason has only ever dealt with mental forms (reason, understanding, sensibility), never with realities. And so we are invited to enter into a *symbolic* philosophy that only works with signs without referent, 'symbols' that reveal nothing of the noumenon that they nevertheless require. Thus the reason of Immanuel Kant unfolds the amazing scene of pure reason's image-world in its shadow theater, isolating the human mind from a forever-unknowable Reality, and bewitching it with the spectacle of its own game. And so this manifold game, where the subtlest abstractions intermingle at the core of a most rigorous discourse and which excels in giving the impression of a perfectly conscious self-mastery, presents itself, whether intentionally or not, as a substitute for Reality. Having opened the eyes of dogmatic reason, Kant reveals to it that what it took to be real objects are only mere appearances. But the performance continues. It is precisely these appearances that will now *show* us this and act as objects of reason. "Reason is occupied only with itself." What a terrible condemnation! Occupied with these empty forms that transcendental usage has ontologically emasculated, it no longer runs any risk. Metaphysical conflicts are abolished, peace is finally established—and Kant is not a little proud of this—but at what price! At the price of the deadening unreality of all its cognitive acts.

Clearly, Kantian theater is the exact opposite of the Platonic Cave. Whereas for Plato mankind is shut away in a cave, the symbolism of which *reveals* that it is only a portion of a great cosmic theater, so that here it is the theater and illusion that give us access to the true nature of things and save us, for Kant the Cave and the theater are within man, *are* each person's reason, and the revelation of this illusion effected by criticism inexplicably, far from delivering us, definitively enchains us there because it establishes its natural and unavoidable character. Here there is no opening onto the beyond, the doors are closed, no one has come back down from up there to speak to us about the Sun. We are in the century of the Enlightenment—only the fires along the ramp glow, illuminating a performance whose conclusion is perpetually known to us.

The Theophanic Intellect

GOD, THE HUNDRED THALERS, AND ST. ANSELM

We come now to what is the heart of Kantian criticism, namely its *rejection of the ontological argument*. This is a key moment in our reflection, for here symbolism and metaphysics are reunited and mutually engendered, as will be discovered. Besides, *the ontological argument is itself only an extreme form of the semantic principle.* It is important then, in order to establish its universal validity, that it brave this ultimate and decisive test. The semantic principle, implicit in every intellectual act, becomes 'apparent' only with its own negation. And the latter happens only in connection with transcendental statements, that is, statements in which a knowledge of realities surpassing the world of common experience is expressed. It is then that the 'intrinsic logic' of the intellect is suspected, and the semantic illusion within which it is found is denounced: according to its 'logical consciousness', according to the law experienced by its *logos*, the intellect believes itself to be still moving in the region of meaning, whereas it no longer knows what it is talking about. This illusion is only explicable if one admits to a radical fallibility of intellective consciousness itself, of the intuitive and direct knowledge that it has of its own semanticity, and therefore if the intellect is only, in the end, a 'functioning' entirely determined by its conditions of production. But, among all the transcendental statements, none are more radical than those dealing with transcendent Reality itself: God. And, among theological statements, none are more 'potent' than the ontological argument, since the intellect claims, in this argument, to be able to *conclude that God exists simply by the knowledge that it has of an idea or concept of Him.* Here the illusion of intellective consciousness is at its maximum, or else, conversely, its irreducible semanticity is affirmed most victoriously. This is why it should be obvious that this argument is actually only the loftiest form of the semantic principle. Likewise, one perhaps begins to understand that it is not so much the demonstration of God's existence—which is out of the question—that is at stake as knowing whether the intellect loses its senses when daring to think about this supreme and absolute Reality, or is to the contrary fulfilling its most profound nature. It is precisely the *sense-less* who deny God!

According to its ordinary presentation, the ontological argument[1] claims to demonstrably attribute the predicate 'existence' to the subject 'God'. The Kantian refutation will consist in showing that existence is not a predicate that would define a perfection of the being under consideration, adding to the other perfections contained in its essence. Thus the concept of the hundred thalers is the same, whether or not the hundred thalers are in Kant's pocket. "Otherwise what would exist would not be the same as what I had thought in my concept, but more

1. It was Kant himself who designated this argument in this way, first in 1763 (*Œuvres philosophiques*, Pléiade, tome I, 432; AK, II, 160). This term has been commonly accepted since then, although it is not without ambiguity.

than that, and I could not say that the very object of my concept exists."[1] In other words, if one makes existence a quality that can be attributed to a subject, there is a choice to be made. Either one affirms that the concept of God implies his existence, that is, existence is one element necessarily added to other elements to form the concept of God. But then this would not be the concept of God that we would be thinking of when we thought about this attribute independently in order to specifically demonstrate it, and therefore it is not the existence of God that is being proven. Or else, it is actually the concept of God that is being thought of when considered without the predicate 'existence', and once again it is no longer the concept of God's existence that is being demonstrated, since the addition of a new attribute (existence) has necessarily transformed it. In fact, (sensible) experience alone can be the basis for a judgment about existence: "If, on the contrary, we tried to think existence through the pure category alone, then it is no wonder that we cannot assign any *mark* distinguishing it from mere possibility."[2]

As a result of this analysis, concerning what we think about God as necessary being, it does not follow that God really exists. This critique is along the same lines as that of St. Thomas Aquinas with respect to St. Anselm of Canterbury, or that of Caterus with respect to Descartes:[3] real existence should not be a predicate that one thinks attributable or not to the subject. Now we are specifically proposing to show that the offending argument does not imply a similar assertion, and that reality is not to be seen there as one attribute among others.

That the problem for Kant is posed only in terms of the attribution of a predicate to a concept we understand without difficulty by virtue of the distinction of analytical and synthetic judgments: either a predicate is related to the subject analytically (this would be the case here where the existence of the concept of God is deduced by analysis)—but then our knowledge does not grow out of the experience of a real existence,

1. *Critique of Pure Reason*, 567–568; AK, III, 402.

2. Ibid., 568; AK, III, 403.

3. *Summa Theologiae* I, q.2, a.1; and Descartes, *Œuvres Philosophiques*, Éditions Garnier, ed. F. Alquié, 1967, tome II, 513–518. This passage was written after a single reading of the *Proslogion* (Koyré-Vrin). Since then we have encountered readings similar to ours (in particular: Karl Barth, *Anselm: Fides quaerens intellectum: Anselm's Proof of the Existence of God in the Context of His Theological Scheme*, trans. I.W. Robertson [Richmond, VA: John Knox Press, 1960]). On the other hand, from now on it will be best to refer to *L'œuvre de S. Anselme de Cantorbery* (Cerf, 1986), an edition under the direction of the great Anselm specialist, Michel Corbin, S.J., that is as remarkable for its [French] translations (with Latin on the facing page) as for its introductions and notes. [trans. note: for cited passages the Deane translation will be used.]

and we learn nothing that we did not already know—or else the predicate is related to the subject synthetically, either empirically (experience alone establishes a judgment of existence) or *a priori* (*a priori* syntheses of sensibility and understanding). But the ideas of reason (the idea of God being one of them) do not establish such connections: being related to no experience (we have no intellectual intuition), they are not formative principles like the categories, but only regulatory principles that urge the mind towards the unconditioned without giving it, obviously, the possibility of reaching it. The necessary existence of God is clearly then only a concept.

Now this does not agree with St. Anselm's or even Descartes' actual approach.[1] We will go back to the first so to consider it as it is actually presented, for, without claiming to innovate, it nevertheless seems that, despite so many philosophical or logical commentaries, Anselm's real approach is still not understood. Let us just recall that, by definition, God should exist in the manner of a hundred thalers, and it is out of the question whether He is or is not to be found in Kant's pocket. The idea of God's existence has then only an analogical relationship with the idea of the existence of a hundred thalers. On the other hand, St. Anselm defines the idea of God as the idea of a being such that I can conceive of nothing greater (*aliquid quo nihil majus cogitari potest*).[2] There is the strength of his argument and any substitution of another definition for that one renders it inoperative. So, to shed a clearer light on the rather surprising character of this 'definition', we will refer to the procedures of the infinitesimal calculus of Leibniz, especially integral calculus, which will serve here as illustration.

1. Hegel spoke in this connection of sophism by *ignoratio elenchi*, that is, by ignorance of what is in question: cf. *Hegel's Science of Logic*, trans. A.V. Miller (Atlantic Highlands, NJ: Humanities Press International, 1969), 86–89.

2. Literally: something in relation to which nothing greater can be thought. The designation of the divine Essence as "what is greater" will surely seem somewhat 'naïve', not metaphysical enough, lacking in complete transcendence, which, in the eyes of some, shows the intrinsic superiority of eastern perspectives. Now, when we consult one of the loftiest Asiatic authorities, the *Tao Te Ching* ("Treatise": *ching*; "of the Way": *Tao*; and "of Virtue": *te*) we read this declaration of Lao Tsu: "It is a being, undifferentiated and perfect, born before heaven and earth.... We can consider it to be the mother of this world, but I do not know its name; I will call it Tao, and if a name must be given to it this would be: immense (*ta*)" (XXV, Kaltenmark translation, [*Lao-Tseu et le taoïsme*, Seuil, 39]). The translator adds: "No name could be suitable for the absolute. But when Lao Tsu nevertheless declares that, if it were absolutely necessary to choose a name for the Tao it would be *Great* (*Ta*), clearly he understands this word in an absolute sense: the *Immense*, the *Incommensurable*" (40). Commentators point out that, instead of simply *Ta*, *yi* (unity, identity) is added, giving *Tayi*, which is translated as 'supreme Unity'. We think it

Suppose that someone proposes to think of a regular polygon, inscribed in a circle, having the greatest number of possible sides. Whatever the number of these sides, a higher number can always be thought of, unless, *by a passage to the extreme*, the circle is thought of as the extreme that integrates all possible sides within itself. Is the circle one polygon among others? Does it define a certain number of sides that might be attributed to a polygon? No, it is the synthesis or integral of all possible sides. Likewise for existence, or rather Divine Being (*esse*), which is the infinite perfection of being, whereas all creatures have a lesser, or else nil, being.[1] By declaring that it is *greater* to exist *in re* than only *in intellectu*, does St. Anselm transform the *esse* into one predicate among other predicates? This would in fact be the case if a comparative were involved. But then we would no longer understand why it might not be still added to other predicates so to make the subject even greater. Now, in reality, an absolute superlative is involved here, that is something that clearly proves the impossibility of adding a predicate to the subject to make it greater. In other words, if we wish to follow the impulse proper to Anselm's approach, this would not be to ask if existence can be legitimately deduced from a concept, but rather to focus our attention on the concept of a *quo majus nihil*, and ask under what condition the *esse*, about which St. Anselm asserts that he is replying to the question posed, can actually put an end to the conceptual increase of the greatest possible. For such is indeed what is required here. God being, by definition, the being such that its concept should verify the *quo nihil majus* property, under what condition can this *esse* fulfill this requirement? We reply: on the sole condition that this *esse* is not viewed as one perfection among others, or as one *esse* among others, but as the basis and synthesis, that is to say as the *integral* of all possible perfections, precisely because it is considered to be pure and infinite Reality. Thus conceived, *esse* indeed puts an end to the progress of perfections

more exact to translate it as 'All-One', the 'infinite One', or even 'infinite Identity'. The term *yi* also designates the state of an artist who has attained a total identification with the object to be represented, in such a way that the brush 'paints by itself' on the canvas. We will likewise point out this very 'Anselmian' text of Shankara in his commentary on the *Taittiriya Upanishad* (II, 1) in connection with the verse: "The knower of Brahman attains the highest," where he declares: "Brahman is so named because he is *greatness* brought to its ultimate point.... The highest is what nothing can surpass.... The highest, that is Brahman" (cited by O. Lacombe, *L'Absolu selon le Védânta*, 215).

1. St. Anselm, *Proslogion*, chap. III, 55; also *Monologion*, chap. XXVII, 135: relative being scarcely exists or is "almost non-existent—this assuredly may be rightly said to be in some sort non-existent" (St. Anselm, *Basic Writings*, trans. S.N. Deane, 2nd ed. [La Salle, IL: Open Court, 1962]).

since it is their unsurpassable limit, just as it is impossible to imagine a polygon whose number of sides surpasses the circle within which it is inscribed. But we have here an 'infinite integral', a limitless limit, or rather a (conceptual) limitation by essential unlimitedness. Such is, it seems to us, the significance of the first part of Anselm's approach (the one that corresponds to chapter II of the *Proslogion*).[1]

Agreed, Caterus and Kant would perhaps say. But it is not always just a question of the thought of absolute Reality. Let us admit that to think of the greatest, it should be thought of as absolute Reality, but we remain nonetheless in the order of thought: "From the fact that that which is indicated by the name God is conceived by the mind, it does not follow that God exists save only in the intellect."[2] The Thomist and Kantian objection can be presented then in the following manner: whether God does or does not exist, the understanding can always conceive of him without contradiction as the necessary being, that is to say the essence of which is being; this is the truth of its concept, but there is nothing more to say. Now this is precisely what is not at all acceptable to the semantic principle. So let us admit with Kant that the orders of being and thought are completely alien to one another. In that case the choice is: either the

1. It is in Book IV (chap. X, 437) of the *New Essays on Human Understanding* that Leibniz mentions the argument of the 'famous archbishop of Canterbury'. The formulation that he uses (God is a Being "whose greatness or perfection is supreme, containing within himself every degree of it") clearly rests on an 'infinitesimalist' interpretation of the *quo nihil majus*. But Leibniz does not lay stress on that. It seems more important to him to 'fill in the gaps' of this proof: namely, the necessary Being exists on the condition that it be possible, that is to say that it can be thought of without contradiction. The non-contradiction of the idea of God is implied by his infinity: "Nothing can prevent the possibility of that which includes no limits, no negation, and hence no contradiction" (*Monadology*, §45). A doctrine quite close to that of Leibniz is to be found in René Guénon, *The Multiple States of the Being* (trans. H. D. Fohr [Ghent, NY: Sophia Perennis, 2001]), chap. 1, in which Guénon formulates the notion of universal Possibility. Finally, we encountered later an interpretation of Anselm's argument more or less equivalent to our own in Father A. Gratry's book *La connaissance de Dieu* (Paris, 1864, tome I, 297ff. et tome II, 79–98).

2. *Summa Contra Gentiles*, bk. 1, chap. 11; likewise: *De Veritate*, q. X, a. 12, and *Commentary on the Book of Sentences*, I, Dist. I, q. III, a. 2. The thought of St. Thomas is surely not the same as Kant's. He even affirms in all the texts cited that in God essence and existence are identical, and therefore that the existence of God is evident by itself. But this is not so for us, because we truly do not conceive of the Divine Essence because of the weakness of our intellect. Not by nature then, as with Kant, can human intelligence attain to pure Being, but, in some fashion, by accident, which lets stand the possibility of a demonstration starting with the, for us, naturally knowable material being, and going back toward its Cause—while, for Kant, the collapse of the ontological argument entails that of all the other demonstrations.

Kantian proposition according to which the order of being differs radically from that of thought pertains to the order of being (and is therefore true), but then Kant's thinking is able to miraculously emerge from itself; or it still pertains to the order of thought and the being in question is a being of thought, like the whole of the *Critique of Pure Reason*, and consequently, by very virtue of this impossibility of passing from logic to the ontological, it is absolutely impossible to know whether anyone can pass from logic to the ontological, since the ontological is no more than ontological thinking, that is to say logic. By this it becomes clear then that the contradiction of Kantianism is truly Epimenidean, and this is rightly what his critique of the ontological argument shows. When he writes: "*Being* is obviously not a real predicate . . . ,"[1] he clearly forgets that, by virtue of the limits of pure reason, no proposition on *Being* is directly concerned with Being, but only the concept of being in general, which, properly speaking, is not even a concept, but an ideal never attained by pure reason. Or else, then, *Being* is not (solely) a concept (which we will agree to), but we cannot know it, and Kant can know it only with the help of a certain *intuition of being*, the possibility of which his whole system strives to deny.[2]

In the same way an objection could be raised against St. Thomas: how can we know that God's existence is obvious by itself if such a proof does not exist for us? How is it possible to think that necessary being does not exist? What do we have in mind when we think about such a proposition? On the one hand we admit that the concept of God implies his existence, on the other we affirm that that is true only *in intellectu*. What then is the thought of an existence that is only necessary in thought, *secundum rationem tantum*, if not the *thought of a nonexistence*? In any case, everyone is in the same boat, whether their names are Anselm, Aquinas, Leibniz, Descartes, or Kant. What they say (and what we say) is after all only of the order of thought, and there is a tautology here. Now, either this thought attains to being and the true—in one manner or another—and our discourse makes sense, or else not, and the refutation of Anselmian or Cartesian discourse is only another meaningless discourse. To think that a reality necessarily exists, but can be nonetheless nonexistent: this is to think that its existence is contingent and therefore

1. *Critique of Pure Reason*, 567; AK, III, 401.

2. The best Kantians (E. Weil, *Problèmes kantiens*, Vrin, 22) make a case here for the Kantian distinction between thinking and knowing: we think about necessary being, we do not know it. The semantic principle consists precisely in denying the radical or absolute validity of such a distinction—we do not deny its relative validity—quite particularly when this involves the idea of the absolute.

contradicts itself. If we think that God necessarily exists, we think that he exists necessarily, that is, that he is necessarily real. The question we pose to St. Thomas is not, moreover, simply aimed at identifying a contradiction *in terminis*, but also at pointing out a pure and simple impossibility: how is it *actually* possible to think about the divine *esse* as necessary, all the while thinking about this necessity as if it were a purely ideal and unreal necessity? How is it possible to affirm, along with Father Garrigou-Lagrange, that "This abstract idea which the mention of the word 'God' awakens in us, though it differs from all other ideas in that it implies aseity or essential existence, abstracts, like all other ideas, from actual or *de facto* existence,"[1] since precisely the whole strength of the argument consists in becoming aware that such an abstraction is, in the instance of such an idea, rightly impossible?

This objection would be valid only on the condition that the idea of God was an idea like any other, and this is what St. Anselm's adversaries presuppose implicitly: that is, that the existence of God be conceived like that of all the other sensory beings. In this case we would be required to distinguish in fact between possible existence and real (= verified) existence. And so the existence of the hundred thalers is perfectly possible: that is non-contradictory,[2] but not necessarily real. However, we ask, what exactly does this mean? Nothing else than this: the hundred thalers are not necessarily in Kant's pocket, that is, are not necessarily existing here and now. It is then the here-and-now (the possibility of experiencing the determinations of space and time) that is taken as criterion for the distinction between the possible and the real. And this is clearly also what the Thomistic notion of an "actual or *de facto* existence" seems to imply. Now the spatio-temporal determination is what specifically defines contingency, and therefore excludes necessity. Hence, and vice versa, to speak of a necessary being is to exclude spatio-temporal contingency as criterion for its reality. To say that "because Kant thinks that a hundred thalers are in his pocket, it does not follow that they are actually there" is an obvious sophism, for no one can think (except by mistake or insanity) that there are a hundred thalers in his pocket when there are none. And conversely no one can

1. *God, His Existence and His Nature*, vol. 1, trans. B. Rose, O.S.B. (New York and London: B. Herder Book Co., 1934), 68.

2. As we have seen for Leibniz, the non-contradiction of the idea of the infinite stems from the fact that no negation is involved. To which Garrigou-Lagrange objects (op. cit., 70): "there is nothing negative about the idea of the swiftest possible movement, which yet involves a contradiction." But that is not altogether exact, for motion, being a relative reality, is not capable of an absolute degree, and so the idea of the swiftest possible movement involves the negation of every movement's relativity.

think about divine Being, infinite Reality, without at the same time thinking that this 'Being' is nowhere in any place or time, and under no determination; It simply is, absolutely, and this thought, when we become aware of it, is by itself a much more certain, much more indubitable experience than all those upon which, according to Kant, we base our judgments about existence; for, after all, even if they are found to be there, a hundred thalers do not stay long in any pocket.

THE HIGHEST POSSIBLE CONCEPTION

We must try to grasp then the Anselmian argument in its true nature and ultimate significance, for clearly, from all the preceding analyses, the negation of this argument leads to more glaring contradictions. And yet its power to convince is far from recognized. The reason for this is, we think, that the importance of this argument—or more exactly its function—is not to *demonstrate* the existence of God, in the ordinary sense of this term, that is, in the sense that certainty concerning this existence would be a function of this demonstration's validity; this would be in fact to subject such an existence to the jurisdiction of demonstrative reason. But the function of the Anselmian argument is to lead the intellect to become aware of its true nature, by making it *experience* the presence of God within it when it carries out an act of the highest possible conception. This is *the highest form of the semantic principle.*

All discussions of this argument are only interested in what happens on the side of the object to be proven, and ask questions in order to know whether the proof is valid, legitimate, efficacious, and whether yes or no, one should now, for a compelling reason, agree that God exists—this seems too facile to the believers, and too inept to the atheists. But, to the contrary, we think that we should be interested in what happens on the side of the intellect that follows Anselm's approach, and this is moreover the only way to take into account the rather surprising formulation proposed to us. This is not in fact a series of premises either known or obvious by themselves offered to our sagacity, but an exercise, an intellectual exercise, in realizing a speculative experience, from which we are next asked to extract the *truth* most certainly contained there. For want of viewing it in this way the argument remains misunderstood, and the attempts at logical formulation intended to test its rigor (and to deny it) arise from an *ignoratio elenchi.*

What this is, in fact, is a subjecting of the human mind to an ordeal. This ordeal consists in asking it to produce the highest conception possible for a human intelligence. Now there is no other means for the intellect to experience its highest possibilities than to give itself the task of thinking about a being such that a greater cannot be imagined. By introducing this 'object' into human thinking, it is subjected to an effort such that it is in some manner compelled to surpass itself: the object of thought works on thinking from within, so that it is truly no longer a thought thinking about an object as if it were a product to be shaped, which would depend on it, and to which it would as a result be superior, but it is quite really the object itself that is 'thinking' within it. The 'thought' of the *nihil quo majus* should obviously not consist in a concept, that is, in a mental *term*; it can only consist in a movement to surpass and transcend *accomplished under the effect* of That which is thinking within us. Otherwise, if we were still dealing with a concept, clearly we could imagine something greater, for by every concept's being determined it admits of being surpassed by another concept, just as every number admits of a successor. To tell the truth, it is rather surprising that no one has perceived what this is really all about, and that the objectors have all treated this argument as if it were a concept, some saying: we do not have this concept; others declaring: existence cannot be deduced from a concept—both forgetting the speculative act that was *actually* proposed in a manner nevertheless quite clear and forthright. Surely the reason for this is that the argument's apparently dialectical (or logical) bearing draws the reader's mind into adopting an equally logical 'attitude', that is, to see in all this just a series of concepts. But if we keep to the indisputable Anselmian formulation of the *nihil quo majus cogitari nequeat*, then our mind becomes aware that, for this speculative act to work, it has to stop thinking by itself and allow itself to think through That which is thinking within it. Or rather, it becomes aware that this is precisely what is in the process of occurring at the very moment it strives to accomplish this task. It discovers itself to be and experiences itself as an effect and no longer as a cause. The reversal to which the mind bears witness within itself is rightly the reversal of the logical to the ontological, and this is where symbolism and metaphysics are bound together.

THE CONCEPTUAL SYMBOL AS METAPHYSICAL OPERATOR

In a certain manner St. Thomas and Kant are altogether right: we cannot conceive (in the ordinary sense of the term) of the infinite or the Divine Essence. This is precisely why, when we try to conceive of it in the only manner still accessible to us as that with respect to which nothing greater can be conceived, the mind discovers in *this very operation* a reversal of speculative intention: whereas the direction of this intention goes, as is normal, from the mind in act to its object, the mind sees itself compelled to recognize that it is itself the 'object' of a pure Act of being that is thinking within it, the concept of which is only a mental trace.

We propose to call such a concept the 'metaphysical operator', meaning to signify by this that the terms of metaphysical language should be regarded, not as points of arrival (for terms), but as points of departure: they indicate a mental operation to be effected rather than designate the thing about which they are speaking since, precisely, there is no 'thing' involved. We will characterize this operation as *anagogic*[1] (in reference to Dionysius the Areopagite), that is: an operation according to which the mind has to surpass itself by conversion to the transcendent contents about which it is thinking. In such a perspective, the concept is only the manner in which the infinite Reality reflected in it is drawn on the surface of the mental mirror. In this sense, Descartes too declared that the idea of the infinite is the mark in us of the Workman imprinted on his work, *instar archetypi*. Here, in the full truth of the term, the concept is experienced as *symbol*.

Now it is understood why it was necessary to conduct our reflections on the semantic principle as far as the Anselmian metaphysical operator, the *id quo majus nihil cogitari potest*. Within it, in fact, is revealed the *articulation of transcendence and immanence in the speculative order: the immanence to the human mind of divine infinity opens this spirit towards the transcendence of absolute Reality*[2] by obliging it to surpass and even annihilate itself. It is understood by That which it strives to understand, and That induces it to go back from the speculative reflection towards the supreme Model. In such an intellectual act it becomes aware that to conceive of the infinite is to be, in a certain manner, in Its presence—the explicit aim of our own presentation is to make such an awareness possible—at the same time that this presence, by its very infinity, manifests its symbolic nature, that is to say reveals that it can be in the soul only in the manner of an effect indicating its cause, or a reflection by participation, or an analogical correspondence between the human understanding of Reality and the light of the divine *Logos* that shines forth from Being. This is also then *the place where metaphysics and symbolism are articulated in the order of human knowledge, since this is where metaphys-*

1. In Greek *anagoge* literally means: the act of climbing above.

2. Absoluteness refers rather to transcendence, insofar as the term 'absolute' denotes that which excludes every relation, and therefore that which is not bound to anything: Supreme Aseity. Infinity refers rather to immanence, insofar as the term 'infinite' denotes that which leaves nothing outside of itself, because it is greater than any given greatness, and consequently that which includes within itself all things by overflowing and surpassing all boundaries, that which the theologians have sometimes called the presence of immensity. Obviously, these distinctions are real only as a function of the vantage point of human existence, although they possess an objective basis in the very nature of the Principle.

ical knowledge reveals its symbolic nature. If we envisage now the metaphysics of the symbol for itself, we see that the Anselmian metaphysical operator assumes its full meaning only through the doctrine of the Immaculate Conception understood in its most universal sense, with the conception of the *id quo majus nihil cogitari potest* only participating in it in speculative mode.[1] Now it is also precisely in this 'metaphysical place' that is realized in the loftiest sense the passage from Symbol to Reality, or again from Being to More-than-Being.

Conclusion: From Semantic Principle to Sacred Symbolism

The semantic principle as ontological argument should not be interpreted in the perspective of the ontologism of Gioberti or Ubaghs, at the very least as understood (and condemned) by the Catholic Church in the nineteenth century.[2] In particular, the fear of seeing in this a source of pantheism seems unfounded since, to the contrary, the *nihil majus* concept implies, as we have shown, an awareness of the finitude of the mental mirror in which it is accomplished, and this is, basically, the sole reason why it is required to pass from logic to the ontological. For how could finite being become aware of its finitude if not by this speculative 'encounter' with infinite Reality? How could it *know* its limits if it were not capable of exceeding them one way or another? Certainly, there is also an affective or existential experience of our finitude—such as evil, suffering, failure, powerlessness, death—but it is also not, by itself, a knowing of finitude, as is fairly proven by the incomprehension (denial, revolt) that it elicits from us. Thus a finite being would have no awareness of its own finitude or the finitude of the world in which it finds itself if there were not within it a knowledge, however obscure, of a beyond. In the same way, no body can find the limits of space, and yet the least thought is outside of space and constitutes an insurmountable limit for this space. Far from striking against the walls of its prison (that is, the human condition), such a being, if

1. We have dealt with this question from a theological point of view in our book *Amour et Vérité* (2011), 295–305. At the very least it is curious to note that, speaking of what would be later called the "immaculate conception," St. Anselm employs a formula altogether similar to the one designating a 'conception' of the infinite: "It is fitting that a purity such that none greater can be conceived outside God should shine in the Virgin" (*Decens erat ut ea puritate qua major sub Deo nequit Virgo illa niteret*; *De conceptu Virginis*, c. 18; *Œuvres*, tome 4, 1989, 174–175).

2. G. Dumeige, *La Foi catholique*, Decree of the Holy Office, September 18, 1861, nos. 73–76.

Kant were right, dragging its prison along with it, would have no experience of them, no more than a body can experience the trans-spatial, so that for it space is really indefinite. In others words, we repeat, if the *Critique of Pure Reason* were right, no one would have ever written the *Critique of Pure Reason*.

Now it is a fact that man is conscious of his finitude, especially since we can write this sentence. Such an awareness implies, as its necessary condition, an awareness of the infinite as such, that is, not of a particular extension of the finite, but of what is absolutely limitless. In the order of finitude, whatever it may be, such an awareness is a *symbol without referent*, for it refers to nothing we might experience in ourselves or the world. Quite precisely, we mean that the very existence of such an awareness is an 'erratic' fact, a sign, a witness. Choose one, then: either this awareness, this idea, this concept, this sign without referent (in the order of nature) by itself has no significance, or it does: the mental 'incarnation', the reflection, or the speculative appearance of the infinite Transcendent. The first alternative is the one followed by all of western thought since the advent of scientific rationalism, but not by Descartes, who, as the true metaphysician that he was, affirmed in the Third Metaphysical Meditation (§24, Veitch trans.): "In some way I possess the perception (notion) of the infinite before that of the finite, that is, the perception of God before myself." From Galileo to Derrida we have traversed major stages of this first alternative: they have led to the impossible rejection of the semantic principle.

Just as a low light at dawn makes the least form on a flat surface stand out, likewise the Sun of reason, by sinking towards the earth, accentuates first the irrationality of the most obvious religious forms. This is why criticism has begun by attacking them. Next, from the most exterior and distant religious forms and sacred symbols, it has gone on to the more interior and closer forms, such as language. But it is in the critique of the most interior form, that of thinking itself, that it has attained its most decisive power, for which Kantianism remains the model. The erratic shadow of the sacred has lengthened inordinately on the human world, so far as to plunge all reason into darkness. By wishing to integrate, that is, by wishing to eliminate them by explaining them—these transcendent appearances that are the signs of the divine outside of us and within us—through a radical immanentism, reason was progressively led to define itself as a contingent structure, completely unintelligible and indiscernible in the surfacialism of all the forms. The Sun itself was just an erratic element among other elements. Here we had the triumph of the great producers of the modern world: Marx made philosophy descend from heaven to earth, while Freud, in the midst of a finally reestablished darkness, attempted to stir up the nether regions. Philosophy was however not so earthly, the darkness was not so total, that there did not remain, perfectly inaccessible to these degradations, the properly celestial light of the semantic principle. Under the demands of this light, and ascending back to the zenith that is the idea of the infinite as metaphysical operator, we have been compelled to reject the Epimenidean contradictions of reductionism, and even its most sophisticated and noblest form, that of Kantian criticism. Starting with a critique of religious symbolism, that is, with the signs of the Transcendent, we have ended, through the semantic principle, with *the very idea of the Transcendent as a sign of Its presence*. The light of the intellect appears then as a veritable internal revelation, only possible because the intellect, the universal openness of the human being, implies its anticipated and *a priori* effacement. We are by no means contesting that the human being, such as it presently exists, is only real by its individual closure upon itself as an indivisible psycho-biological structure—quite the contrary. But there is something about knowledge which, by nature, is universal. This is an anti-paradox expressed specifically by the semantic principle. Man is an individual being in whom is manifested the possibility of the universal, but *only in intellective mode*, at the surface and at the center of the mental mirror.

This is why Thomism is right to demand, in order to demonstrate the *existence* of God, that we start from the *existence* of the world, not only because in that case we would remain *in linea entis* and not pass unduly from logic to the ontological, but because it is true that, chronologically, the knowledge that we have of existence comes first from the exterior world, other people, and things,[1] to the extent that they resist us or we come up against them. It is the existence of a natural order that gives us, according to ordinary experience, our first idea of reality. As a consequence, the ontological argument, which is not a demonstration but a

1. At least as to conscious knowledge, for we do not know what the intellectual intuition of being among the newborn or infants might be. Basically, the Thomistic way is quite close to the Heideggerian thesis. As Jean-Luc Marion recalls: "... Being, which coincides with no being (ontological difference), nevertheless gives itself to be thought only in the case of a being" (*The Idol and Distance*, trans. T.A. Carlson [New York: Fordham University Press, 2001], 13). Is this to say for all that that Thomism is reduced to onto-theology, and succumbs to the original sin (according to Heidegger) of western metaphysics, which only thinks of Being under the mode of the Supreme Being? We think we have shown, however, that this was not the case with St. Anselm. But neither does St. Thomas confuse (despite terminological appearances) God and the world: "God was not 'Lord' until he had a creature subject to Himself" (*Summa Theologiae* I, q. XIII,

coming-to-awareness, can only come in second place, when we transpose, analogically, the idea of reality to the highest degree. But it is also its completion, because, in a certain manner, it is necessarily present in all stages of our experience. We learn about existence from the exterior world, but no sensory experience, contrary to a too-widely-held conviction, is able to communicate the idea of *being* as such: it gives us only sensations, perceptual data. What apprenticeship to the world elicits is in truth the progressive awakening of the memory of infinite Reality which rests in the substance of our intellect.

And so it goes as well for the sensible signs of the Transcendent that are sacred symbols. The *semantic principle* has established that the path of reductionism, which leads to its negation, is forbidden us. This principle has received its most perfect form by means of the *metaphysical operator* that is the *concept of the infinite*, which is disclosed at the same

---

a. 7); and: "This name 'good' is the principal name of God in so far as He is a cause, but not absolutely; for *existence* considered absolutely comes before the idea of cause" (ibid., I, q. XIII, a.11). This is why Meister Eckhart, all the while accepting the theology of 'Brother Thomas', can write: "God becomes when all creatures say 'God,'" and this saying is their very being (cf. *Sermons & Treatises*, vol. 2, trans. M. O'C. Walshe, 81). Now it is not only the Christian faith that has preserved them from ontological idolatry. For it seems that neither Plato or Plotinus have committed a like error.

Besides, it is not out of the question that our conclusions agree with those of St. Thomas. Some interpreters in fact lean in this direction, rightly so we think, and see in St. Thomas an intuitive knowledge of God by immanence of the divine *esse* to the spiritual nature of the human intellect. This is especially so for F.M. Génuyt, in his book *Vérité de l'être et affirmation de Dieu* (Vrin, 1974). He first recalls some texts from St. Thomas, drawn chiefly from the *Commentary on the Book of Sentences*: "Speaking about God however in relation to us, he can thus be considered . . . in two ways. The first is according to his likeness and participation, and in this way, his existence is self-evident (*ipsum esse*). . . . The second way is according to supposit [divine subjectivity]. . . ." (*I Sent.*, d.3, q.1, a.2). And this knowledge is an intuition; in fact, "to perceive by the intellect is nothing other than an intuition, which is nothing other than the presence of the intelligible for the intellect, whatever the mode, so that the soul always has an understanding of itself and God" (*I Sent.*, d.3, q.4, a.5). And Génuyt adds, "this knowledge rests in some fashion on the ontological presence that the soul maintains with itself and with God" (op. cit., 188). For "if, as we have stated, the *lumen intellectuale* coincides with spiritual existence, it must be concluded that it is the same for the spirit to be and *to be light*, and consequently the same as participating in both the *esse subsistens* and the light of divine knowledge" (ibid., 191). "Every demonstration of the existence of God—a demonstration necessary for knowing what it means to exist and therefore based on experience of the sensible world—presupposes however this presence of absolute being in the depths of the soul, which discovers this presence within itself, immanent to every intellectual activity" (ibid., 192–194). "At origin, the soul perceives *being itself*; reasoning leads it to think little by little about being as God, and what it affirms at the end of this reasoning is *subsistent being*" (op. cit., 194). As you see, we are quite far from Kant.

time to be an *idea-symbol* of the Transcendent, causing us to rediscover in this way, as a confirmatory result, the notion of symbol with which we started. Thus the semantic principle not only opens the way to a metaphysics of symbolism: *it imposes itself.* In light of this principle, sacred symbols rise up in all their humanly irreducible reality: they command us to understand them, and nothing can excuse us from this task, but to understand them we must undergo an apprenticeship to them. The experience of religious symbolism is not then an ordinary experience; neither is it a marginal experience. Quite the contrary, it is the central experience of human existence, so that, correlatively, neither is a metaphysics of symbolism a secondary zone of speculative knowledge, but it rightfully claims the central place in philosophic thought. By this we are enjoined to renounce seeing in sacred symbolism only the sign of an economic alienation, a collective neurosis, or the functioning of a chain of signifiers: it is neither social or psychic illness, nor mechanical absurdity. We need to let symbolism itself have its say and be heard, however strange its discourse may be and however far it might lead us.

PART V

The Hermeneutic Principle or the Conversion of the Intellect to the Symbol

Introduction

The Irreducible Symbol

We have now arrived at the end of our inquiry, that is, at the threshold of the metaphysics of sacred symbolism to which it introduces us. We have traversed a long and arduous road, across nearly the entire field of European thought, because that is the place where the most radical rejection of the symbol was carried out. Besides, we have no other solution. How are we to rediscover the sense of the symbol after three hundred years of a critique of the symbol, the varied phases of which have deposited in our cultural soul a succession of sedimentary layers under the weight of which symbolism has progressively suffocated? Everything proceeds as if the major stakes of modern 'civilization' were to destroy humanity's religious soul, and to endow western reason with a well-ordered system of speculative reflexes ideal for preventing a return of the sacred.[1] Disregarding the interdict they have imposed, we had to examine then, one after the other, each component of this system, and rediscover the truth they sought to erase. Surely, it was not the spectacle of a daunted Catholicism that gave us pause—to the contrary. It was enough to observe its theologians, historians, and liturgists giving way before the pitiless hue and cry raised against anyone the modern world counts as thinkers of note. Everyone comes to the system's defense: since Kant has shown that . . . we can no longer say that...; since Freud..., since Nietzsche..., since Heidegger and Foucault...: one after another the speculative avenues of approach are closed, the space in which a religious life can be lived is narrowed more and more, every escape route blocked; the beast is ready for the kill. Never, it seems, has any era of human culture known such intellectual terrorism, never have such nar-

1. This is clearly a *fact* that should be taken into account by Christian traditionalists who wish to save western civilization and religious faith at the same time, the first being the active negation and destruction of the second. In its 80th and last article, does not the *Syllabus* declare the following proposition to be an error: "The Roman Pontiff can and should be reconciled and come to terms with progress, liberalism, and modern civilization"?

row limits been assigned to intellectual activity. And each time one of these speculative dictators announces his interdicts, sets up his barriers, and amputates a new portion from the intellectual field, then the clergy applaud, clamor for the liberation of the spirit, and eagerly submit. In truth, it is urgent to proclaim a metaphysical revolution, to reject every limitation of activities proper to the intellect, and to give our intellect the right to think all that is possible, that is to say infinite Possibility: the only object worthy of our minds is absolute Reality. The infinite is the true homeland of the intellect.

However, the more the unlikelihood of faith is brought out, the more apparent will be the irreducibility of its signs as well. Such is, in any case, the dilemma summarizing three centuries of philosophic critique: either the signs are false and faith insanity, but then there never was and never will be any reason; or else reason would have to renounce itself, by virtue of the semantic principle, and in that case symbols are truthful and testify first to an invisible Transcendence. Basically, our approach has complied with two correlative demands that grew in strength as we progressed in an examination of the critique of symbolism: it is just as impossible to renounce the radical semanticity of thought as it is to diminish the radical transcendence of the symbol. The first is in some manner irrefutable in its transparency and self-evidentness; the second is revealed in the end to be indestructible. One resembles the light, the other those megaliths that stand in silence, a sign in stone that the millennia have been unable to obliterate. These demands are inseparable: the semantic demand was revealed by means of the symbol and in connection with its impossible reduction; the symbolic demand is revealed by means of the intellect and from its contradictory negation. We have complied with the semantic demand; now we must comply with the symbolic demand, that is, to state the basic principle of the hermeneutics of sacred symbolism. For, if it is true that the intellect discovers *its own truth* by understanding itself to be *a symbol of the divine* and *an irrefutable theophany*, conversely, the truth of the symbol only becomes apparent to the intellect, since the function of the symbol is to illuminate the intellect. And therefore the irreducibility of the symbol makes sense, for hermeneutics, only with respect to symbol interpretation. In other words: we have identified the semantic demand of every symbolism; now we must identify *the symbolic demand proper to every semantic, that is to every intellection.* This correlation of *logos* and *mythos* proves on the whole that, even though every symbol requires in the final analysis the production of an intelligible understanding, intelligibility is only produced under the action of a symbol.

11

Of the Hermeneutic Principle in Its Essence

The Reversal of Hermeneutics

In *Histoire et théorie du symbole* we have explained what a hermeneutics attuned to the symbol is and the three forms it should assume: instituted, speculative, and integrative. Now we are no longer dealing with forms, but with principles, and a fundamental principle. No need to show what functions are fulfilled by hermeneutics, or what particular rules hermeneutics uses in its interpretation of symbols, but rather what principle it follows in the very essence of its act. Now we have made use of this principle from the beginning of our study, since we have in fact done nothing but seek to understand what a symbol is. And is it in exactly this that hermeneutics consists in the first place? Is it not at first a reaction to the very fact of symbolism? We began by saying that the symbolic sign was first a sign of itself, a sign that symbolism exists, just as we declare now that hermeneutics is first an awareness of symbolism as a fact, and this awareness rightly makes it a hermeneutic.

In what then does this awareness consist? In other words, since hermeneutics is a cognitive act, what distinguishes this act from the ordinary activity of our intelligence? Surely the act as such cannot differ, by itself, from other speculative acts; like every cognitive operation, it consists in the intellectual grasp of an object. Only the object itself, then, can make the hermeneutic act specific. In other words: what difference is there between an 'ordinary' reality and a symbol, a symbolic entity? We have stated this many times: the knowledge of a particular being brings along with itself an immediate ontology of reference. What is given *with* the perception of a reality is the world in which it is contained and to which it refers. This does not, we repeat, does not involve the portion of space, the empirical milieu, or the environment from which the being perceived is inseparable within the perception itself. It is quite exactly a matter of an ontology of reference, and therefore of a certain 'idea of reality' which deals, surely not with an abstraction, but to the contrary with an objective though unobservable reality. It is as objective as a language system, which, for Saussure, functions within language, and yet no one can see or hear it. Similarly, the implicit and immediate ontology of reference is *a particular system of reality that requires the existential possibility of a specific being.* To be complete, we should also speak of several systems of reality that hierarchically envelop each other according to their greater or lesser *competence*,[1] each partial system being defined by the extent of existential possibilities that it is capable of taking into account. Thus the system of absolute Reality is the one that takes into account all possible existence, and even the possibility of being as such: in such a 'system', which is no longer a system, performance is indistinguishable from competence. At an infinitely lower degree, we notice that a living being demands of its reality-system a competence the reality-system is not capable of, which implies a merely physical being; likewise for a thinking being. However, this in some fashion involves sub-systems of a more general whole that define certain conditions such as form, time, life, etc., and constitute what we will call the world of *common experience.* This label, which is equivalent to the 'sensible world' for Plato, the 'visible world' for the Nicene Creed,[2] 'formal manifestation' for Guénon, etc., is more precise than it first seems. Even though not based on the objective character of this world, as with Guénon, it is so named, as with Plato and the Creed, from the fact that *this* world is perceived by men. The 'world of common experience' signifies that which is commonly experienced by men, that is to say: that which the human community experiences as world. And so we are making human experience the determining criterion, not of the real as such, but of such an ontology of reference, of such an order of reality, which seems to conform to the nature of things, insofar as man is actually the determining center of this order of reality. This has

1. We use this term in the sense employed by Noam Chomsky to designate the sum total of the possibilities of expression a given language offers to its users. The *competence* of a language so defined is to be distinguished from performance, which designates the actual use a given language's speakers can make of its competence; even though *competence* is determined, *performance* is not. *Competence* corresponds, in a certain manner, to the rules of the game, *performance* to the game itself. In the same way we could distinguish between a system of reality (or ontology of reference) and actually existing realities, which are part of the 'cosmic game'.

2. Trans. — Creed = *Symbole* in French.

nothing to do with any idealism whatsoever, but expresses the necessary correlation that joins knowing subject to known object—and this is also Kantianism's 'share of truth'. Furthermore, in the definition of *this* world we are implying the human community, that is not only what each individual has in common with all (the generic essence), but also what everyone together experiences collectively. In other words: *this* world is not only what anyone can experience, but equally the ontology of reference that each person discovers to be *communal*, that is, discovers to be the one to which everyone else refers. In other words, there is never any world *for* a single individual. It is the *essence* of 'this world' to be one for each and all at the same time, except precisely in dreams, where each one is alone in experiencing oneiric objects. This definition of the world as communal horizon of ontological reference seems foreign to Kantianism, where the order of phenomena is purely relative individual consciousness. This is a serious gap because this characteristic is essential: the world is potentially a world for everyone, the world "is other people," while the Kantian universe could be that of a sleepwalker.

However that may be, it is precisely the absence of such an implicit ontology of reference, the absence of a reference to the world of common experience, that characterizes apprehension of the symbolic object; or, rather than absence, it would be better in this connection to speak about an impossibility of referring to the ordinary world. Clearly, a symbolic entity is presented in *this* world, but does not refer to it. Now this symbolic experience is twofold: not only does it make us aware of a common ontology of reference, which, by reason of its implicit nature, should not be posited as such, and is only revealed in its being suspended, its being put in parentheses; it also calls for 'another world', another system of reality. In short, to experience the world, *this* world, is to inseparably experience a world 'beyond', at least as a need, although we would have no 'idea' if we did not, in a certain manner, feel its connaturality within ourselves. Here we come back to the discovery of 'meaning' described in *Histoire et théorie du symbole*, by which we are given at once an awareness of the world (or worlds) and of our own thinking. It is in fact by an experience of the symbolic sign that the universe of things and beings is distinguished at a single stroke from the universe of knowledge, and that they become at once equally real. Awareness of the sacred symbol is a disruptive and dazzling experience from which springs a consciousness of reality and the intellect. We say 'of the sacred symbol', that is to say referring to another world. We could just as well say: experience of the Transcendent within the experience of its sign's presence. Such is truly the point of departure for humanity, and therefore also for language.[1] Language is apprehended in meaning and meaning is apprehended in experience of the sacred, and therefore of the symbolic, which is *the* presentification of it. That the linguistic sign is extracted next from the symbolic sign is certain. But it is first the symbolic sign that makes this possible; for, if the essence of a sign consists in making present what is absent, how would this be possible outside an awareness of an 'essential' absence, of what is essentially absent because impossible in the world of common experience and exceeding its competence? Thus the possibility of an 'essential' absence, which is none other than the experience of the religious and the divine, wholly conditions every experience of the sign and is found to be implied by it. How can we not but see here the origin of technology, since tools are created only through a semantic tearing away of an object from the world, abstractively grasped as capable of being inserted into a multitude of unactualized contexts, in short by putting it in reserve, setting it aside? Now is not the first, original and founding, *because essential* (or *radical*), setting-aside, the very one which in any fashion cannot but be separated, isolated, different, unactualized, the *sacred*?[2] And how could an object be set aside from actual relationships unless because a meaning is read into it, the presence of a reality that in any case refers to a transcendence, that is as a support for the divine, in short: a symbol? If the staff, wheel, ax, fire, thread, or flintstone were not at first invested by Transcendence, venerated as signs of the divine, and so set apart from the rest of the universe, they would not have become those instruments upon which human life relies.[3]

But as much must be said for thinking, insofar as it becomes aware of itself as a real activity and as 'a world apart', which means that, since thinking too discovers itself in the experience of the symbolic sign, the intellect is metaphysical in nature. Metaphysics—that is, knowledge of what surpasses the visible order—is therefore prior to all other knowledge. Just as religion precedes technology, so a speculative awakening to what is beyond the sensible necessarily precedes thought of the world

1. In the previous chapter we have seen that (ontologically) the first experience of the Transcendent is that of the 'concept' of absolute and infinite Reality infused into the very substance of the intellect; *it is therefore the presence of the idea of God within him that makes man a human being.*

2. We think that such is in fact the primary meaning of this term (which is found again in a kindred term: *secretum*, from *secernere*, to set apart) and which is itself derived from *sancire*: to separate.

3. The necessary correlation between the appearance of language and that of technology is recognized moreover by scholars as eminent as Leroi-Gourhan in *Gesture and Speech* (trans. A.B. Berger [Cambridge, MA & London: The MIT Press, 1993], 215ff.).

here below. As Helen Keller's experience, which we have meditated on at length[1] shows, her discovery of language and, correlatively, of the objective world of things and the subjective world of thinking has to do first and foremost with the essence of water, and not with its actual, physical, 'here-and-now' existence. This is an authentically mystical experience, immanent to all human thought, and only our attraction towards the useful and dispersive multiplicity of things has caused us to forget it. We have said enough about this now to be able to clearly formulate the hermeneutic principle, which we have constantly utilized and is the unifying principle of our entire study. Besides, as we have just shown, it is not just we who have applied it more or less explicitly. Humanity itself has practiced it constantly and, even more, this principle itself is incorporated into humanity. However, because our civilization has almost completely forgotten it—and we know very well why—we can only formulate it in a deliberately 'revolutionary' manner: until now, we will say in parody of a famous aphorism: *symbolism has only been interpreted starting with common reality, now common reality must be interpreted starting with symbolism.*

Principial Hermeneutics and Speculative Hermeneutics

This hermeneutic reversal that could be characterized as an anti-Kantian reversal, to the extent that it subordinates the knowing subject to the symbolic object, expresses clearly what we have called the conversion of the *logos* to the symbol. A conversion—or rather a reconversion—is involved since the human *logos*, forgetful of what has given birth to it, conceives of its relation to the symbol only as a fading away and reduction. *Practically* identifying the symbolic sign with the linguistic sign, it saw in the latter only a language to be deciphered, unimportant in itself, the referent of which was simply more difficult to know than that of the linguistic sign. Because, as we have stated, the intelligence is intent on objectivity, a cognitive act ordered to being (and that by definition and according to the irresistible evidence of its own nature), when it grapples with a symbolic entity it spontaneously only sees a mediation for an accidentally absent object, and it is for this reason *represented.* When it perceives that, in the end, the object represented is not detectable in the world of common experience, it continues however to retain its intent on objectivity in this respect, all the while striving to bring this undetectable object back to an object detectable in the world of common experience, with the most widely held solution consisting in reducing the

1. Cf. *Histoire et théorie du symbole*, chaps. IV and V.

existence of this transcendent object to the existence of a human belief that affirms its reality. Now human beliefs, like humans themselves, belong to a common existential order, even though it may be susceptible to various genetic or structural explanations. It can even be admitted that, in practicing this 'reductive' hermeneutic, critical reason has the feeling of simply accomplishing the task of every hermeneutic, which would be defined as determining the referent towards which a symbol points, therefore doing nothing but what the most religious hermeneutic already does, the only difference being that the latter believes in the reality of the supernatural referent denied by the former. This is moreover just how theological treatises define hermeneutics: "The art that delineates the rules for interpreting Holy Scripture is called *Hermeneutics*; and the application of these rules to the interpretation of Holy Scripture is named Exegesis."[1] But this fine unanimity is deceiving. Or rather, it can only be obtained subsequent to a profound altering of sacred hermeneutics itself. This ends up treating symbolism as another discourse and seeing in the meaning of symbols no more than a product of convention and, in addition, one of the ingenious resources of the generally so-called 'eastern' sphere of imagination.

It is certainly true that the function of hermeneutics is also, and even principally, about stating the meaning of symbols, in other words, determining their ontology of reference: the symbol 'makes us think'; still, it is necessary to know *about what*. The rolè of hermeneutics is to respond to this question. However, this function, which defines speculative hermeneutics, despite being the main one—which we do not deny—is not principial, but secondary. And that explains perhaps why there are cultures where this secondary and explicit hermeneutics seems, so to say, absorbed by primary or principial hermeneutics, unformulated in itself. We are by no means asserting by this the pure and simple absence of 'metaphysical commentary' in these cultures. To the contrary, we think such a commentary is always present, unless the tradition is definitively dead and its meaning entirely lost. A living culture not only conveys a 'formal' tradition, it also transmits a metaphysical one, however concise. Thus explicit hermeneutics is not doomed to improvisation, but the culture itself teaches how symbolic figures must be spoken of, or at least it provides the minimal indications indispensable for that. In short, sacred culture includes within itself the preliminaries of its own metalanguage, which speculative hermeneutics would only have to develop.

Furthermore, this metalanguage can be inconspicuous; this is even,

1. Abbé J. Berthier, *Abrégé de théologie dogmatique et morale*, E. Vitte, 1927, num. 214.

we think, most often the case. It constitutes an *esoteric commentary* held and orally transmitted by a few initiates. Such was the case for Christianity,[1] Hinduism, and, closer to us, the Dogons with Ogotemmêli or the Sioux with Black Elk. That an outside observer, even a famous ethnologist, is ignorant of this is by no means surprising. In any case, this metalanguage is written down and eventually delivered up to be profaned or misunderstood only so to 'save what might be saved'. And this is why, quite often, these symbolisms seem to be directly experienced by those to whom they pertain, without them thinking to give any explanation. However, prior to a 'minimal commentary' that fixes the meaning, the existence of an implicit hermeneutic must be posited, formed by the very *reality* of the symbolic sign. Before any interpretation, and as the basis for all interpretation, there is the semantic presence of the symbol which is, by itself, a hermeneutic of reality. Such is, for example, the *existence* of Christ, the Word incarnate, by Himself a 'sign of contradiction', or again, at the 'biblical' origin of humanity, the *existence* of Adam, symbol of God "in the image and likeness."

And so we perceive the radical significance of the hermeneutic principle: common reality does not interpret the symbol; it is the symbol, as such, that interprets common reality. In fact, if an explicit and secondary hermeneutic risks treating the symbolic sign as a merely transitive, linguistic sign, the 'semiotic materiality' of which would be without interest, this is by totally forgetting its specificity: in symbolism, it is not words but things that signify, even when these things are said with words. In the final analysis, as we have so often stressed, what makes a symbol is the *naturality of the signifier*. The symbol is a thing, a being, or an event endowed with a semantic property, a reality of the world of common experience in which *is presentified a transcendent intelligible*. It is that, or it is nothing. Consequently, there can be a speculative hermeneutic of the symbol only if there is first and necessarily an awareness of the symbol's true essence; otherwise someone might write thousands of books on the symbol and quite simply not know what they are talking about. This condition is absolute and it differentiates drastically between every authentic philosophy of the symbol and discourses dealing only with a pseudo-symbol. Now, to be truly aware of the symbol is to be

1. Actually, a secret teaching is known to have existed, given by Christ to certain apostles and transmitted by them down to Clement of Alexandria and Origen; cf. J. Daniélou, "Les traditions secrètes des Apôtres," *Eranos Jahrbuch*, 1962. Christians can reclaim paternity of the term 'esoteric', since its first usage is found in Clement of Alexandria (*Stromata*, V, 9; *P.G.* IX, col. 90). This question is dealt with at length in *Guénonian Esoterism and Christian Mystery*, trans. G.J. Champoux (Hillsdale, NY: Sophia Perennis, 2004), 296–393.

open from the outset to this existential hermeneutic implemented by the symbol with respect to the ordinary world, to this 'semantic transfiguration' that it continually realizes through its mere presence as symbolic sign.

On the Symbol Attuned to Hermeneutics

As just explained, the hermeneutic principle is understood to allow an escape from the pitfall of speculative reduction, as well as that of the contradiction between the symbol's singularity and the concept's universality. The former has been pointed out many times and, truth to tell, rarely avoided. If hermeneutics consists in translating a concrete symbol into abstract ideas, what good is the symbol? When great hermeneuts—Origen, Ibn 'Arabi, or René Guénon—interpret a symbol, they are obviously convinced that they are drawing this entire science from the symbol itself. But, if their commentaries happen to fall beneath the gaze of a 'lenient skeptic', at the least they are accused of being ingenious inventions. It is hard to avoid the conviction that the symbol was only a provisional intermediary and what is essential is in the commentary. At the extreme, and if that is true, the result would be that, in case of necessity, one could do without the symbol. There is surely much pleasure in seeing impenetrable enigmas so elegantly resolved, a pleasure akin to that of astutely unraveling a detective mystery, with the hermeneut in the detective's role—a pleasure too of sharing in the 'secret of the gods', and perceiving the most profound significances under—to the eyes of the ignorant—the most commonplace appearances. One enters into complicity, then, with men from the dawn of time. Down through the millennia a shared secret invisible to profane eyes is formed, and minds are joined together in the deep certainty of a silent communion. But ultimately these are pleasures consummated in a basically intellectual act. Some will even grant that, in the strict sense, it would be hard to say the same thing with words; difficult, but not impossible. And they begin to regret that the symbol, from which so much pleasure was derived in its interpretation, is not necessary with an absolute and irreducible necessity.

However, if this were truly so, we would then have to conclude that every hermeneutic rests on a contradiction, that between the singularity of a natural form, as understood moreover by a particular and contingent culture, and the universality of a concept that considers only abstract and necessary truths; so that, in reality, it would be impossible in this case for *logos* to spring directly from the symbol. These hermeneutics dream about it, it is said, talk about it, turn it over and over, try to tame it, and catch it in the snares of their reasoning, but ultimately they are adding to it from without and do not attain to it: a forever-unbridgeable gap, which surely accounts for the indefiniteness of hermeneutic discourse,[1] but also strikes it with vanity.

Everything proceeds differently, we think, if we allow for the distinction between the just-mentioned secondary and explicit hermeneutics and the principial hermeneutics which is the very origin of metaphysical intellectuality, which bursts forth under pressure from the symbol's semantic presence. Quite true, secondary hermeneutics can display an immense range of the loftiest speculations; this is its right and duty, for *only in this way* can man enter into possession of the message held by symbolic forms before being united with it in ritual activity and realizing it inwardly. It is even inevitable that this hermeneutic, having done this, forgets the disruptive experience that had given birth to it. But this forgetting cannot last long, unless the hermeneutic is corrupted and degenerates into allegorical babble. A living hermeneutics, one that feels the mysterious power of its object passing through it, continually experiences the need to return to its source, to drink of its wellspring, to stand in the contemplative presence of the symbol which invisibly nourishes it with its reality, and unceasingly gives the intellect to itself by breaking the undifferentiated unity of world and consciousness. And, besides, the beautiful thoughts that flourish with the vision of the symbol . . . it is the symbol itself that sows them in our spirit. Surely it is only in our spiritual substance that they are able to bloom and live, and this to the extent that we receive them within ourselves; but only the symbol gives them and communicates their principle to us. Thus there should be no difficulty in attuning the singularity of the symbol with the universality of the *logos*. In its proper activity and pure nature the intellect is universal, as is the intellective light in which it participates and is, after all, that of the immutable Word, the Sun of the spirit, "which enlightens every man coming into this world." And, precisely because it is universal, this light does not enter into contradiction with any singularity; it welcomes them all and is opposed to none.

However, whether explicit or principial hermeneutics is involved, two conditions are required for the cognitive act to occur. First, the intellect must be provided with an object, for it only illuminates what is given it to illuminate; it does not invent its object but receives it, being by itself a quest for being, a desire for being as for its own good. Second, it must be able to receive it, for although this receptivity is universal as to the diversity of objects that it welcomes, it is not so as to the mode of receptivity which is cognitive in nature: but only what is 'recognized' is

1. An indefinite discourse because never done with catching up to its object.

known, that is, what is *semantically assimilable.* Only what makes sense, and insofar as it makes sense, is knowable to the intellect. What, then, can 'define' what makes sense for the intellect (to the extent that such a question can be posed, for we are at the limits of the conceivable here)? How do we determine the 'sense' of the intellect in such a fashion that we can discern what makes sense for it? The answer is simple: it is the *possible,* since the intellect is the sense of *being.* In other words, 'to make sense' means 'to be possible', to be 'possibly' real. By definition, the intellect is the sole judge and norm of this possibility (and we can say nothing else). What must be understood by the 'possibility' of a being with respect to the human intellect is, purely and simply, the capacity possessed by the intellect to conceive of this being, the intellective 'matrix' that the reception of known being presupposes and actualizes in the intellect.[1] In the end, this possibility of a being is identical to its essence. In any case, we will say that possibility is the essence *seen from the intellect's point of view;* but surely the essence should equally be seen from the point of view of the being itself.

The outcome of these reflections is that the intellect requires, for the cognitive receptivity of being, that this being be presented eidetically, in the absence of which no semantic assimilation would be brought about. Now the symbol, under the general form of religious culture,[2] realizes precisely such an eidetic presentation. Considered in this way, traditional culture constitutes for the intellect, as we have explained on many occasions, a true human *cosmos noetos,* an 'intelligible world' or 'world of essences', which is in turn an image of the macrocosmic *topos noetos,* itself a projection, into the sphere of 'celestial' manifestation, of essences contained in the divine Word, the 'place of the possibles'. Through the mediation of these cultural essences (plastic or rhythmic forms, myths, rites, language, etc.), the intellect is eidetically 'informed', actualizes its own conceptive capacities and cognitively takes possession of the world. Also in this manner do the two inseparable modes of the intellectual act make their appearance: one, more informal, resides in the intuition of being; the other, more formal, resides in the capacity of the intellect to 'read' cultural forms, that is to be informed by them and make use of them as a mental toolkit. It is in the conjunction of these two modes

1. In this strict sense, infinite Possibility designates the capacity of supreme Reality to conceive of universal Being: it is identical to the Immaculate Conception metaphysically transposed and therefore 'surpasses' the ontological degree.

2. This is also the reason why, in *Histoire et théorie du symbole,* we began by collecting from various cultures what they had to say about the symbol, the essence that they had to communicate.

that the passage from principial to explicit speculative hermeneutics is realized.

At the end of these analyses two conclusions emerge:

- The hermeneutic reversal simply consists in becoming aware that, even though the symbol apparently demands only to be interpreted, even though the sign seems to be there only to refer to the reality it designates, more profoundly and by its very being the symbol imposes on our intellect an interpretation of cosmic realities which, under the effect of its disruptive (incongruous, unassimilable) presence, is transformed into visible signs of a transcendent Reality: what was sign becomes reality, what was reality becomes sign.

- Not only are we saying that the symbol imposes this reversal on the intellect, but it even makes this possible, because the symbol semantically informs the intellect, awakening within it a remembrance of the metaphysical Ideas, opening the eye of the soul to Realities never contemplated by the eyes of flesh.

12

Three Figures of the Hermeneutic Principle

Adam the Hermeneut

As indicated earlier, we find under the figure of Adam an example of what we call principial hermeneutics. It will be good to comment somewhat on this example because it unites together in itself, in a remarkable way, many themes of our reflection, and therefore brings to it a confirmation from tradition.

First of all, this encounter is not surprising. Quite the contrary: its absence would seem troublesome, given the fact that, if the hermeneutic we are discussing is indeed principial, it should be able to be verified in Adam, who, from the Judeo-Christian point of view, is at the source of every culture and hermeneutic. But primordial Adam not only verifies the hermeneutic principle as we have formulated it; he also shows its necessary connection with the semantic principle, or rather, as stated, he shows that the hermeneutic principle is only another relatively more 'exterior' way of expressing the semantic principle, since it involves the *logos* in its relationship to its object, whereas the semantic principle in some manner envisions the inverse relationship, that of the object (and ultimately the divine Object) to the *logos* thinking about it.

We have seen that the 'idea' of God, or again the act by which the intellect conceives of the Infinite, is revealed in its essence as the effect or result, within the intellective mirror, of the act by which the Infinite 'thinks about itself' within it. In other words, the 'idea' that we have of absolute and infinite Reality, when it becomes self-aware, by no means knows itself as the cause of this idea; otherwise this would no longer be precisely supreme Reality that it would be thinking about, but as if the effect of a relative reality. This is why, by itself, this 'idea' is in its very existence like an intellective act, the *proof* of the divine Reality's existence. The intellect can only *interpret* the presence of this 'idea' within itself as the image of its Model. Having done this, in this act that corresponds to its loftiest possibility, the intellect is necessarily identified with its own essence since here it reaches its existential limits. It finds itself to be a *symbol* of What is reflected and presentified within it. It is the 'image' of God by its very nature and 'according to its likeness' when it actualizes this nature in an act of the loftiest conception possible. Viewed in this way, we see that 'thought' of the Infinite by itself accomplishes the primary realization of principial hermeneutics. This is indeed the *symbol*, the intellect, which in its existential actuality reveals to the reflecting intellect the transcendence of the Reality of which it is the image.

From this we can conclude then that the intellect is by nature Adamic. It constitutes what remains, in fallen man, of the paradisal state, what has escaped the corruption of Edenic nature, because it was, in that very state, 'more than paradisal'. Having been created "in the image and likeness," Adam is in fact the first symbol of the Transcendent in the proper sense of the term, its first sign. Not that the macrocosm as a whole is not a natural manifestation of the divine, or that the innumerable multitude of forms, beings and relationships out of which it is composed are not adequate revelations of God's innumerable 'aspects' —man does not reflect a particular determinate aspect of God. By his nature *he is the very image of God in the world*, a nature which is therefore indistinguishable from his function, from his 'central' position, from the relationships he maintains with the rest of 'his' universe. From this point of view, Adam is *a microcosm within the macrocosm*, that is, a world in miniature, or again and specifically a representation of the world within the world, which is quite exactly a definition of 'symbol'. But he can be a cosmic icon, a mirror toward which the entire world converges and in which it is reflected, represented and present to itself, and by this is unified, only because he is a divine image and therefore witness to the transcendent and metacosmic. Man is himself that 'semantic surplus' about which we have spoken, that semantic presence in the world that exceeds the world, that world boundary within the world itself, that hole or emptiness at the center of the universe which *is* the center of the universe, that empty hub of the cosmic wheel thanks to which it turns, Adam himself, the lord of time and space. And this is how he is *a sign, a mediating symbol in whom God is presentified to the world and through whom is semantically effected the return of the multiple to the One* (to the extent that the image realizes its likeness) which is, as we have seen, the very function of the symbol: "Above all, when it is a question of God and creatures," explains Father Javelet, "*likeness* is a certain *ontological mediation between the Creator and his creature*, between the One and the many."[1] Primordial Adam, by his very existence, effects then a radical *diacrisis* of universal manifestation. He 'signifies' it in the proper sense of the term: he makes it a sign within himself, therefore tearing it away from itself and, setting its boundaries, positing it as a world within himself because he is himself this boundary.

Now what is the universal mirror of the world, its all-reflecting center, that which causes Adam to be a microcosm? It is first and foremost the intellect, the 'universal reader'. This is why Christian tradition[2] interprets the "in the image" of God as designating primarily the divine *Logos*, which is the first *Image* of divine Reality within God himself, the *Logos* which is, dare we say, '*Microtheos*' within the '*Macrotheos*'. Through this consideration we are led back to the semantic principle and the intellective conception of the divine as essence in act of the human intellect. The human intellect, the created *logos*, accomplishes here below what the principial Theomorph, the uncreated *Logos*, accomplishes from all eternity. The human *logos* is in the image of the essential Image, and only by participation in this essential theomorphism of the supreme *Logos* (a participation which is the work of the Spirit) does our intellect, the image and symbol of God, actualize its likeness, its creaturely theomorphism, by becoming a "conception of the Infinite."[3] Adam is principally then an intellect-symbol of the *Logos* and, as such, he accomplishes his hermeneutic task. For lo and behold: the first of all the hermeneutics, their unique and permanent prototype, is given to us in the divine Word, the eternal knowledge of the Father. It is He who 'interprets' the Essence[4] and deploys the non-quantitative multiplicity of the archetypal meanings that It inexhaustibly discloses. This truly 'principial' hermeneutic makes the Word to be the 'place of the

1. *Image et resemblence*, tome 1, 343. We can only refer in passing to this study, but it is a study magisterial in the extent of its scholarship and the mass of evidence gathered, as well as in the depth of its views and the authentic, metaphysical spirit that animates it. There is no Patristic or theological study to which we feel more akin.

2. Ibid., 72–88.

3. This 'conception of the Infinite' is obviously the 'immaculate conception' that *is* Mary (who declares to Bernadette: I *am* the Immaculate Conception—an astonishing formulation). Mary, from this point of view, corresponds to the *passive intellect* that Aristotle compares to the virgin wax of a tablet "on which nothing has been written" (*On the Soul*, III, 4, 430a). It is only insofar as it is perfectly empty—"without mixture," says Aristotle—that the intellect can 'conceive of' the *Logos* and so become "*theotokos*," mother of God.

4. Cf. John 1:18; literally: "God, no one has ever seen him, the One God—the Only Son, the One who is in the bosom of the Father, that very One has performed his exegesis."

divine possibles', the place of the intelligibles, that is, the Word by which is expressed and within which is formulated the infinity of the divine *Names*. Similarly, the Adam-symbol by his very existence as theophanic intellect is constituted *hermeneut of cosmic reality*: he names beings and things; he becomes, that is, the word of their essence. Just as a mirror, in order to reflect an image, has to do nothing but be itself without it being at all possible for an object to prevent its image being captured by it, so it is enough for the intellective mirror to be placed at the heart of the world for the macrocosm to become the microcosm, for it to assume a semantic form within this mirror: "And the Lord God having formed out of the ground all the beasts of the earth, and all the fowls of the air, brought them to Adam to see what he would call them: for whatsoever Adam called any living creature the same is its name. And Adam called all the beasts by their names" (Gen. 2:19–20). Adam is the logothete, the hermeneut of creation, just as the divine *Logos* is for the divine Essence. Is it in this way that the natural character of language is established, as say "Ramban, Judah Halevi . . . and others,"[1] or rather its conventional character, as Maimonides declares? But both views need to be upheld, for a name, by definition, should not be purely and absolutely natural. What is purely and absolutely natural is nature itself, being as such. A name is, by definition, a symbol and therefore other than that which it names; it is of the cultural or hermeneutic order. This is why this order is as if a new creation within creation; institutional by nature, it 'doubles' the macrocosm in a mirror, signifies it, and therefore presents this vertical discontinuity, a witness to Transcendence, which we have already and often stressed as essential to the sign and its *diacrisis*, to the disruption that it effects by its very existence. And in fact the text clearly states that God led the beings to Adam *to see* what he would declare, as if this were something new, 'unpredictable'. But, conversely, what is 'new', what is disruptive or diacritical with respect to the natural order, can only be the manifestation of the metaphysical or supernatural essence of things, and this is precisely what names have as their function to express. What is instituted in the bestowal of names is the revelation, in this world, of intelligibles not of this world which necessarily show, then, a discontinuity in its regard. For the microcosm, the symbol of the world, does not 'double' or mirror it *uselessly*. It reveals what is hidden beneath the veil of natural forms, a manifestation of which is necessarily impossible due to the world's limiting conditions. This is why we will never find entirely satisfactory relationships between the form of a language and

1. Elie Munk, *The Call of the Torah*, vol. 1: *Bereishis*, trans. E. S. Mazer (Brooklyn, NY: Mesorah Publications, 1994), 39.

the natural form of beings (even though certain basic laws valid for all languages can undoubtedly be established[1]). But this is not, as many linguists all too easily affirm, because of the arbitrary nature of a sign. Quite the contrary: it is because of a metaphysical (and not physical) relationship, a relationship that unites the primordial language with the essence of what it names. And this is why the biblical text can affirm: "whatsoever Adam called any living creature the same is its name."

Such is the solution brought by the hermeneutic principle to this difficulty. For, although it is indeed *the essence being expressed in the Adamic naming, this is by virtue of the hermeneutic operation accomplished by the intellect-symbol*, the semantic revealer of the macrocosm, insofar as it is a mirror participating in the light of the divine *Logos*, the prototypical Hermeneut of the Divine Essence. Just as the uncreated Names are only various aspects of the thearchical Essence in the thought of the Word, so *the created names pronounced by Adam are in reality the various intellective forms taken by creatures when they are thought of by him*. The Adamic hermeneutic is therefore the reflection, at the paradisal level, of the supreme Hermeneutic by which the infinite field of the divine Glory is deployed. *From this point of view we should consider the different traditional, symbolic and sacred cultures as prolonging and diversifying Adamic culture (or primordial tradition)*, and therefore as constituting a third hermeneutic, or rather a third degree of the same principial hermeneutic. We might designate each of these degrees by the terms 'divine' or 'metaphysical', 'Adamic' or 'primordial', and 'cultural' or 'traditional'. And it will be understood then in what more rigorous and less metaphoric sense it could be said that *culture* truly constitutes for man an '*intelligible world*'. This is because, insofar as the intelligible world is identified principially with the 'contents' of the Word, the place of the possibles, the hypostatic synthesis of the divine 'ideas', the cultural hermeneutic, which derives from the divine hermeneutic through the mediation of the Adamic hermeneutic, necessarily participates in its nature and reflects its structure, insofar as this is possible and according to the 'angle of vision' or the 'focus of perspectives' which determine it as one particular culture and not another.

Thus, as anticipated, all the themes of our study are gathered together under the figure of Adam. One after another they have been set in place and clearly ordered, and in this figure find their confirmation and coher-

1. We are thinking especially of the research that Marcel Locquin has pursued and into which he has given a glimpse in a rather astonishing article in the review *Science et vie* (special issue, *La planète des hommes*, June 1980), under the title "Le fond common des langages et des écritures," 53–63.

ence. However, although Adam the hermeneut establishes the principial hermeneutic of every culture, although it is in him that this hermeneutic is constituted as the basic act of our intellect, which is Adam within us, it is not in him that every culture is *completed*. Cultural or traditional hermeneutics, the hermeneutics of sacred symbols, is addressed to historical man and speaks of a common reality. It is quite far from primordial hermeneutics, with which it no longer has, so to say, any contact. It has undergone trials and known crises, the most substantial of which is the *Babelian failure*. This is why, in the depths of its humiliation, it requires for its completion, its perfection, another figure to be given to us "under the sign of Jonas." These are the two remaining points to be seen.

The Babelian Failure of the Symbol

The text of Genesis cited a moment ago continues in this fashion: "And Adam called all the beasts by their names, and all the fowls of the air, and all the cattle of the field: but for Adam there was not found a helper like himself" (Gen. 2:20).[1] The Adamic hermeneutic is therefore in direct relationship with the transcendence of the theophanic symbol, which, in its interpretation of the world, finds no referent of itself. Within it and by it the essence of every being becomes name, because everything is in relationship to it and finds its referent within it; but it itself, insofar as it is theomorphic essence, is in relationship with nothing in the world, and in this world finds no *name* for itself, no symbol revealing its essence (because in reality this name is within itself). If therefore (principial) man is the image of God, then (principial) woman is the image of man; she is the microcosm of the microcosm, what the man did not know (profound sleep) he was: his "luminous reflection," to cite Fabre d'Olivet's translation, his *likeness*, the name of his essence.

This transcendence of the Adamic symbol, under the action of which the world is revealed in its semantic forms, accounts for the principle of every hermeneutic of cosmic reality. It is the basis, as we have stated, for the origin of human cultures. But it does not signify their end and completion. And this is why:

1. Fabre d'Olivet gives the following 'literal' rendering of this verse (we are using the Vulgate): "And-he-assigned Adam, names to-the-whole quadruped-kind, and-to-the-fowl of-heavens, and-to-the-whole living-nature earth-born and-for-Adam (collective man) not-to-meet with-an-auxiliary-mate as-a-reflected-light-of-him" (*The Hebraic Tongue Restored*, vol. 2, trans. N.D. Redfield [New York and London: G.P. Putnam's Sons, 1921], 87).

In conformity with our proposed definition of a symbol as 'semantic operator', Adam the hermeneut, by interpreting reality and through this very interpretation, inaugurates a 'semantic alchemy' of the cosmos. However, in its very principle, this work of intelligible Transfiguration is effected by calling reality into question. What is 'suspected' here is not as before the sum total of sacred symbolism, but it is symbolism itself that is suspicious of the sum total of the created world, the semantic competence of which is challenged: nothing within it is similar to the transcendence basic to Adam the hermeneut. The microcosm denounces the illusion (a certain illusion) of the world in its pretension to be a large-scale analogue of man. The world may be reduced to man, but man is not reduced to the world. Must we see in turn, in this hermeneutic of reality, a reductive interpretation of the real that is simply the reverse of the reductive interpretations encountered throughout our critique? Must a naturalist and ultimately semiotic reductionism be opposed to an idealist and semantic reductionism in the contrary direction? Is that the principial significance of a symbol?

This is possible. But the reason is that the hermeneutic of the real by the Adamic symbol can be only realized by instituting an autonomous order of the symbolic as such. This thesis of the autonomy of the symbolic is constant in our study. It corresponds, as we have shown, to the transcendence of its presentifying function; it is its *image*, its proof. The symbol is transcendent, that is, independent of every set order of reality, in its symbolic being and to the very extent that it is a sign. This independence, this autonomy of the symbolic order (an essential autonomy that in no way excludes multiple dependencies on all those natural or cosmic elements used to create a symbolic order), this autonomy, we say, also founds *the possibility, for this order, of being taken for an end.* The symbol is itself only insofar as it offers itself as a pure mediation for the transcendent presentified in it, but it can only be in this way annihilated in its function because it is altogether itself, and in some manner enjoys a privilege of complete independence with respect to every set order of reality. This is also true for instrumentality in general, which is not at all surprising since the symbolic sign defines the essence and the possibility of *every* instrumentality, as brought out a moment ago. Without the sacred, without the ontological 'separation' instituted by the presence of the symbolic sign, there would be neither technology nor civilization. The idea of an instrument is basically the idea of a reality independent of every actual articulation, but capable of every possible articulation. This is why, by very virtue of this independence, the only end the cultural order can pursue is its own development. The characteristic of an instrument or a mediation is, in fact, to be rich with a (potentially) indefinite multiplication of mediations. To join two terms with a third is to eventually pose indefinite intermediate relationships, which basically proves the 'third man' argument. Therefore every cultural mediation, and the first of all, the mediation of the sign, is rife with a possible proliferation of itself, the indefiniteness of which *imitates* on its own level the inexhaustible richness of being, and can even lay claim to rivaling it. Obviously, for this to happen, we have to lose sight of the basic presentifying function of a symbol, the purely participative nature of its reality as a semiotic entity. In this perversion of the nature and function of the sign—but, we repeat, a perversion whose possibility is implied by the very existence of symbolic mediation—the symbolic order, that is to say the order of discourse as well, is progressively substituted for being itself, which it was charged with making known. Everything is reducible to its mirror. *Since everything can be said, the saying of it is everything.*

Such is, quite specifically, the case with the Hegelian perversion of true gnosis. Clearly it is a philosophy of mediation, but of a mediation, or even an instrumentality, which becomes for itself its own end and, in the final analysis, is identical to Hegelian discourse itself. Such is also the case with the sophist, whom Plato shows as condemned to encyclopedism and *mimesis*; for *mimesis* is the only means of being equal to everything, since one can always reflect everything.[1] In general this is so with every idealism, for which gnosticism and sophistry are on the whole equivalents. But this is also what the biblical text teaches under the notable figure of the 'Tower of Babel': "And the earth was of one tongue, and of the same speech. And in emigrating from the east [*l'Orient*],[2] men found a valley in the land of Sennaar and dwelt there. And they said to each other: Come, let us bake bricks and let us burn them well-burnt. So for them brick was like stone, and bitumen was for them like mortar. And they said: Come, let us build a city and a tower, the top of which may reach to heaven, and *let us make ourselves a name*, lest we be scattered over the face of all the earth" (Genesis 11:1–4). The tower in

1. *The Sophist*, 232–234. How do we account for the impression of universal knowledge given by the sophist? A 'unique art' capable of producing everything is needed. This is the art of *mimesis* and illusion, which is thus playing the same role as the Vedantine *Maya*. For the sophist is within us. He is one potential corruption of intellective mediation implied by its mirror-nature. It is the *other* of being, the 'potentiality' of which scandalizes the Parmenides within us, who 'works' the intellect from within and compels it to surpass the 'ontological idea'; for Non-being clearly in some manner needs to 'be'.

2. Some translations render this as "towards" the East. But the *LXX* and the Vulgate have understood: "from," and we think this meaning stands out for metaphysical reasons as well.

question is therefore also the 'name', that is to say the symbolic order in its entirety and ultimately sacred culture in all its progress, a culture Adamic in origin. It had survived the exile from Eden; it was the immemorial and primordial tradition, humanity's semantic memory, a memory forming the unity of the human species, which was "one tongue, and of the same speech."

But this memory could be preserved, whatever the estrangement humanity experienced with respect to the principle of tradition, only on the condition that it remain turned towards this principle, towards Eden; that it remain 'oriented'. Now men had specifically turned away from the east [*l'Orient*]: "their hearts forsook the One Who preceded the world."[1] Since then symbolic mediation, forsaking the transcendent goal that it should attain, is turned in upon itself and *poses as transcendent*. A 'Name' is raised toward heaven, with the ambition of assuming for itself alone the verticality out of which it is composed, a verticality which was, as we have often mentioned, altogether essential for it. What was a sign bestowed is transformed into a fabricated sign; what should 'lead to heaven', to the Transcendent and the divine, itself becomes, by taking itself to be the goal, an obstacle and unbreachable barrier to the immanence of the divine within it.

This is why God himself is then obliged to 'descend' and inaugurate the first 'supernatural' or revealed covenant with Abraham. Up until that time Adamic culture was bearer of the divine Light which was presentified within it, in some manner, through nature. It is the continuation of the earthly Paradise, that is, of the state wherein Heaven touched the earth, wherein the earth was as if naturally and immediately present to Heaven. Through the Fall there 'appeared' a distance between Heaven and earth; but, in accord with its 'orientation', tradition naturally remained a bearer of the divine immanence, of its light and power, because the intellect is in some manner naturally supernatural. However, for cultural mediation, there also appeared the temptation to 'fill in' the vertical distance that separated earth from Heaven, and to utilize its own transcendence for its own benefit: let us make ourselves a name. In response to this semantic usurpation, which not only could not by itself unite earth to Heaven, but which, additionally, at the same time and necessarily destroyed the unitive function of the Symbol, its conjunctive power—hence the confusion of tongues—God could only 'descend'. But this time He descended by an act proper to his Transcendence which reestablished communication between heaven and earth only by maintaining the distance separating them, and on the basis of

1. *Rabbah*, 35; cited by Elie Munk, *The Call of the Torah*, vol. 1, 142.

an election, a choice; that is, by restricting his immanence to the exclusive point of a man, a form, a religion,[1] the point being the trace of the vertical on a horizontal plane. In truth, the confusion of tongues was not so much a punishment as the actual realization of the desire that animated the 'builders of a name', for God only punishes by 'abandoning' us to the consequences of our acts. Now the confusion of a language's signifying entities evoked by the 'symbolic' etymology of Babel, so named "because there God *confounded* the language of the whole earth" (Gen. 11:9)—from the verb *balal*, which signifies: to muddle, mix up, confound, literally "to babble"—this confusion, we say, indeed realizes a unification of language, but through an indistinctness of forms: unity, which on the level of forms implies their distinction, is succeeded by uniformity. But for unity to be no longer realized except by uniformity, by an indistinctness of forms: this is inevitable precisely from the very moment semiotic entities are no longer 'oriented' towards the Transcendent but reduced to their formal being. And this is why what had been *Bab-el*, "doorway to God," that is to say the opening towards the One, the privileged place of the One's immanence, becomes the source of a confusing multiplicity, the privileged place of a dispersive separation.[2] Claiming autonomy for itself, that which had a unitary function necessarily becomes an obstacle and cause of division.

Under the Sign of Jonas

The Babelian failure of the symbol is basically analogous to the failure and contradiction of every reductive hermeneutic. But here, because this involves the principial hermeneutic effected by the symbol with respect to the real, it is the real itself that disappears as such and is transformed into a semiotic entity: the flesh becomes 'wordy'. Symbolism

1. This is where the *de facto* plurality of religions is born, each of which affirms itself to be unique; cf. "The Problematic of the Unity of Religions," trans. G.J. Champoux, *Sacred Web*, 17, 2006, 157–182.

2. There is no need to ask then how we could pass etymologically from *balal* to *Babel*, which appears to be 'scientifically' impossible. On this topic we repeat: the spiritual and theological indigence of the commentaries on Holy Scripture, in the entirety of the Catholic editions consulted, is truly scandalous. But this is not just a recent problem; seventeenth- and eighteenth-century editions are no less scandalous in this respect. A most external and least trustworthy kind of knowledge concerning the letter has completely killed the spirit—whereas reading the Fathers, and even the commentaries of St. Thomas Aquinas, is enough to bring us into the presence of overflowing life. From this point of view, Elie Munk's commentaries, to be found in an edition of the *Torah* addressed to the general public, reveal metaphysical and mystical depths without any comparison to what is to be read in Christian publications.

must therefore be reestablished in its true function. The symbol must prove that it is truly that which transfigures reality, the semantic operator that actualizes the intelligible nature of the real, but without 'reduction', without 'residue', without a 'corpse'. This is the world of common experience in its entirety, right down to its most opaque root, a world which should be transformed *assumptively*, or rather, the truly spiritual nature of which should be finally revealed. Yes, we should go even that far: otherwise all the rest is either a lie or a metaphor.

The first figure of the Adamic hermeneutic shows this as an inaugural, 'initiatic' proof, because it calls upon us to choose: either, along with the genetico-critical hermeneutic, 'suspect' symbolic utterances, or, along with the symbolic utterances, 'suspect' the world of common experience, forswear its illusory autonomy, its lying consistency, and be open to the Word's light made flesh within it, which sparkles in the darkness of cosmic exteriority. The alternative is clear: either be converted to the symbol so that it draws us after it towards an anagogical ascension into the glory of the world "such as it will be when eternity finally changes it into itself," or be projected indefinitely (and that is hell) into the Babelian dispersion of difference. And yet this must still involve the glory of the world "such as it is within itself." In this anagogical ascension, time, space, colors, events and the virtues must still not be volatilized into abstractions, but, to the contrary, enter by this ascension into possession of their full reality. For, in truth, the dazzling promise at the heart of the experience of the symbol is the promise of a real world at last, a world at once serene like immutable Being and bursting with joy like God dancing in the exaltation of trinal brightness. This is why principial hermeneutics demands another figure, a figure no longer inaugural but terminal, within which are held the promises of the symbol, for want of which every metaphysic of the symbol is only vanity and a chasing of the wind. *This figure is that of the sign of Jonas, that is to say of the resurrected Christ.* It is truly a terminal sign, since it is written: "An evil and adulterous generation seeketh a sign: and a sign shall not be given it, but the sign of Jonas the prophet. For as Jonas was in the whale's belly three days and three nights: so shall the Son of man be in the heart of the earth three days and three nights" (Matt. 12:39–40). This is here the first announcement of the resurrection, the figure of which we will now briefly investigate *under the sign of Jonas.*

The book of Jonas is one of the most disputed in Holy Scripture, one of those which has stirred up the liveliest discussions, as much with respect to its historicity as to its meaning. All this we will set aside. But it is so much the more important that Christ had chosen it as a 'sign' of the chief event of his life's 'story', of the event that gathered together and united all his acts and words: the resurrection. This is important because *the resurrection is thus identified as a sign*, which means not only that it points to and puts a signature on the Christic 'deed', but also and above all that it proves, guarantees, and authenticates it. And that is something that should not be lost sight of in the contemporary discussions relating to this miracle. It is almost always asked what proves the resurrection, at least in Scripture, while it is forgotten that, in reality, first for Christ, and next for St. Paul and the apostolic tradition, it is the resurrection that proves. What are, then, the indications given by this enigmatic "sign of Jonas"?

First let us stress, among the numerous aspects of the symbolism of Jonas, those that seem more directly linked to the question raised. The Matthean verse is, moreover, explicit enough, since it draws a parallel between the stay of Jonas in the belly of a sea monster and the three-day burial of Christ in the depths of the earth. Now we cannot underestimate the fact that, in Hebrew, *yona* (Jonas in Greek) means 'dove'. And the dove has been seen as a symbol of Israel, because of a verse in Hosea (7:11) where this comparison is found; and, with greater likelihood, in reference to Psalm 74:19, which sings of the sorrow of the chosen people, captive in Babylon, and which prays God to "not deliver the soul of thy dove to the wild beasts." It is then not impossible to see in Jonas eaten by the sea monster an allusion to Israel held prisoner in Babylon.[1] And, by the same stroke, the symbolic relationship that unites the sign of Jonas to the Babelian or Babylonian sign is established. Moreover, Jonas is the "son of Amittai," as the text tells us (1:1), that is to say son of the "truth" (*emet* in Hebrew), and hence he is also the sign, the mark, the effect of it. And just as at Babel men had detached the sign from its intelligible referent and stolen it to do their work, a sign of themselves and no longer of the divine (let us make ourselves a *name*), so the Babylonian monster swallowed Jonas-Israel, himself a sign of the truth, the prophet-people. However, the darkness cannot contain the light, the sign-thief cannot keep it prisoner: someday he has to give back the stolen spirit (dove). But, for such an operation to be possible, the distance that the transcendence of the Abrahamic covenant maintained between heaven and earth had to be abolished. In its abasement, its *kenosis*, the divine 'descent' has to go lower than all lowness, further than every distance, so to manifest that there is nothing outside the divine, that in the depths of the thickest darkness still resides the pure light of the *Logos*. What is more, this 'descent' has to prove that the flesh itself partakes in

1. This is also an interpretation supported by several exegetes; cf. in particular M. Delcor, *La Sainte Bible*, Letouzey et Ané, tome VIII, *Introduction à Jonas*, 1961, 274.

the dignity of the *Logos*, that the Uncreated innately resides *in the heart* of the creature, in such a way that the glory that it is called upon to know within It, as supernatural as it might be, is nevertheless ordered to the perfecting of its nature. This is why the 'Son of Man' is swallowed up in death—"in the heart of the earth," says the text; that is, at the most extreme point of separation and abandonment. The semantic Being par excellence, the Word, is not only made flesh but also corpse; It is identified with a semantic husk deserted by meaning, with the scattered fragments of the Tower of Babel; and, from the depths of the abyss, It re-arises, gathers together, reunifies, re-erects the Tower-symbol towards the Transcendent, and reopens the 'Gate of Heaven'.

For the 're-surrection' is a new 'surrection', *anastasis* in Greek, a 'righting' of what had been laid low, a raising, a restoration. But the Tower-symbol, raised again and re-situated within the axis of its Transcendence, carries everything away with it: *si exaltatus, omnia ad Me traham*: "If I am lifted up, I will draw all things to myself," Christ says in St. John (12:32), alluding to His being displayed on the cross. The cross of contradiction makes of the Son of Man the visible *sign* of reconciliation.[1] Everything is fulfilled. Everything is complete and brought to its perfection: there is no longer any corpse, *the tomb is empty*, the semiotic husks themselves are reabsorbed at last into the Supreme Referent that they presentified. This anagogic ascension leaves neither trace nor residue. Resurrection and ascension are therefore two phases of the same process,[2] to which must be added the Pentecostal descent, that is to say

1. As we have pointed out (*Histoire et théorie du symbole*, chap. I, article III, sect. 1), the Gospel of St. John identifies the cross of Christ with the brazen serpent that Moses had raised in the desert, healing all those who looked upon it, that is to say identified with what Scripture itself calls a "symbol of salvation."

2. As we know, St. Luke in the *Acts of the Apostles* indicates a lapse of forty days between Resurrection and Ascension, which is evidently considered by exegetes as purely 'symbolic', it being understood that we are dealing with a sacred number, "the one number available to denote a fairly long period of time"! Such is the explanation given without hesitation by one of the most reputable exegetes of our time, Walter Kasper (*Jesus the Christ*, trans. V. Green [New York: Paulist Press, 1976], 148). Now this is altogether inexact, for there are several other sacred periods in Jewish symbolism—that of the Jubilee, for example, which lasts 7 x 7 + 1, or that of 70, symbolizing the week and the Sabbath. In reality, 40 = 10 x 4; 10 indicates the end of a cycle, 4 is the number of earthly matters; 4 x 10 signifies the entire course of an earthly cycle, and therefore the time between a birth and a death, that is to say the 'exhausting' of one state of existence and the passage to another one. The symbolic significance is therefore altogether precise and rigorous. As to its historical reality, this is obvious for anyone who truly has a sense of the symbolic, for, far from denying this reality, the sacred value of a number confirms and guarantees it. And this is equally valid for the mention of the cloud that hid Christ from the apostles' sight at the time of the Ascension.

the outpouring of the Spirit, of *yona* the dove, by whom death has given birth and who prepares the universal resurrection, the renewal of the face of the universe.

The 'Son of Man' designates the quintessence of human reality, the manifestation of what is most interior and most real in it, that is to say its deiformity, its quality as God-symbol. And therefore the Son of Man is the symbol of symbols; this is the symbol rediscovered in its primal and saving truth. In it the truth of the symbol is realized in its fullness, as well as the efficacy of its presentifying function and therefore its function as redeemer of all 'flesh'. It is the prototypical symbol. It is also necessary that there be accomplished in it, visibly, that which will be verified for the symbolic order in its entirety (the macrocosm) only if this assumptive integration is actualized in the one who is the heart of the world (microcosm). It is from the very heart of the earth, night, darkness, and the tomb that the Newborn One should spring forth, the newly Living One, the One who appears beyond the waters of death in the 'land of the living'. This is the first phase of re-ascent from the abyss. Next He must 'prove' that the world in which He Himself, the new Adam, lives is already a new heaven and a new earth. Surely, during these forty days it is a question of witnessing to the truth of the Resurrection, but also of witnessing to the glorious, spiritual and 'pneumatic' nature of all creation—in other words, of all earthly forms. This is a taking-possession of the human world and the 'verification' of its luminous essence; tied also to this realization—but this time in *breadth*—is situated the sending of the apostles on their mission, a mission precisely by which should be realized the regenerative baptism of the entire world. Then the second phase can occur, that of ascension, by which is manifested the reality of the sense-world's assumptive integration into the spiritual being presentified in it. The Resurrection, the forty days, and the ascension trace a cruciform diagram: from the 'heart of the earth' to the arising from the tomb, this is the re-ascent along the lower portion to the center of the cross; the forty days 'actualize' the horizontal portion, which is reintegration with a renewed, glorious, and spiritual human state; lastly the ascension actualizes the upper vertical portion, which is integration with the superior states of the being—in other words, with the angelic worlds and spiritual hierarchies. This ascension also takes place in the sensory world and the space of shared experience, by virtue of the rigorous nature of analogical correspondences between the symbol and its referent. If the vertical symbolizes a going 'outside of the cosmos', this is also because it, in its own manner, actualizes it. Therefore, only by ascending bodily (but gloriously) along the vertical is there any visible and active (and not passive as in death) means of leaving the world. In truth, the very *reality* of the vertical is of the semantic order and should not be conceived of outside of its symbolic function.

As for the third phase, the descent of the *Pneuma*, it corresponds exactly to the Anti-Babel, the reuniting of tongues. As Hans-Urs von Balthasar so accurately states: "Lastly, the Resurrection of the Son is the revelation of the Spirit."[1] This is then the revival of the 'power of signs', the reawakening of their semantic dynamism, the re-actualization of their operative virtue. Symbolism, consecrated anew in the baptismal blood of the God-Man, can henceforth communicate the intelligible fire which dwells in it, the *Pneuma*, to all who consent to let themselves be informed by it, so that, in this communication, the entire world may also know the revelation of the Spirit, and so that the spiritual fire in all things may be set alight.

Such, briefly stated, is the significance of the "sign of Jonas." But how is this possible? It is here that the radically hermeneutic function of the Resurrection is manifested. Because this symbolism is presented in such a way that it is posed as a final and *decisive* proof, by no means can it be distorted. Either symbols are true, or they are false inasmuch as they are symbols: that is our choice. In the figures of Adam and Babel, symbolism prevails by far over event; at the very least we are entitled to suppose so. But, in the Resurrection of the Son of Man, it is the symbol itself that becomes the event, since, if it is true, as all theology manuals teach, that Christ in rising *reunited* his soul to his body, then by this very fact He has carried out the *restoration of the symbol.* In other words, He has proven that, among the innumerable symbolic forms, there was at least one that integrally verified its ontological relationship to the reality that it presentified. In short, there was at least one sensible form that was nothing but its transcendent content. *But then this restored symbol is just enough to make the world of common experience tip to the side of glory.*

We will not, after so many others, return to the exegetical question of the Resurrection.[2] We will make only two remarks, one having to do with scriptural texts and the other with Patristic literature.

1. *Mysterium Paschale*, trans. A. Nichols (Grand Rapids, MI: Eerdmans, 1990), 210.

2. A rather detailed account of it is found in W. Kasper's *Jesus the Christ* (p. 130 ff), and above all in Hans-Urs von Balthasar's *Mysterium Pascale.* The "International Catholic Review" *Communio* has dedicated a special issue to it: "He is Risen," tome VII, 1982, num. 1. Exegetically, we think the whole question boils down to the following items: — 1) we are limited, in any case, to the New Testament texts, whose divergences are explained in various ways; —2) but they all agree on one point: Jesus, who was dead, is alive again in his body; despite the staggering imagination of the exegetes, no doubt is allowed in this respect; in particular, it cannot be honestly maintained that these texts

As for the first point, we need to know that from the first half of the eighteenth century (chiefly in Protestant Germany) down to our own time, the calling-into-question of the Resurrection has given rise to an abundance of historical, exegetic, hermeneutic, philosophical and theological hypotheses, which by itself alone proves that the European intellectual climate feels itself directly affected by this event in what is most essential to it (its sense of the real, its ontology of reference) and finds itself unsuccessful in effectively neutralizing it. As for the second point, it gives rise to exactly the opposite observation. After a search through Patristic literature, we come away with two constant facts: although the textual differences are not ignored,[1] on the one hand the fact of the Resurrection is not disputed by anyone,[2] and on the other hand nearly all of the Fathers tie Christ's Resurrection to the resurrection of mankind and even of all creation, Christ's Resurrection being the proof and cause of theirs. The Resurrection is unanimously seen, then, by tradition as a hermeneutic of the reality that causes a transfigured universe to pervade us, as an event revealing the glorious nature of the created.[3]

If we effect, then, the conversion of the *logos* to the symbol, if we agree to let ourselves be 'worked' by this implicit hermeneutic, what are we reading into this event?

were intended to be 'mythic' or 'symbolic' but were afterwards poorly understood; —3) hence there are three possible answers to the question: "Why do these texts affirm the Resurrection?": *a.* because the Apostles saw and touched Christ; *b.* because they believed that they saw and touched Christ (perfect hallucination); and *c.* because they—either they themselves or the later Christian community—lied so to, by inventing and giving credence to this legend, justify and guarantee the *fact of the faith*, that is to say the fact that, surprisingly, they continued to believe in Jesus beyond the setback of his death. But, if the invention of the legend is late (second century), this is because it was useless for the transmission of the faith. If the invention is apostolic, it is even more unlikely, first because it would be 'bizarre' (an account of a few apparitions in subtle mode, 'in the astral', would have sufficed), next because it would be risky (how could they be certain no witness would come forward to contradict such an enormity?), and finally because the differences in scriptural settings become altogether inexplicable (the success of a lie requires a uniformity of witness). We are left then with hypotheses *a* and *b*. But what is the difference between a real perception and a perfect hallucination…?

1. For example, St. Ambrose, *In Lucam*, lib. X, num. 153, *P.L.*, t. XV, col. 1242; St. Jerome, *Letter CXX ad Hedibiam*, cap. IV, *P.L.*, t. XXII, col. 988.

2. This means that it seems easy for them to believe in it. St. Augustine declares: "It is incredible that Christ rose in the flesh (*incredibile est Christum surrexisse in carne*) and with his flesh ascended to heaven" (*City of God*, XXII, 5).

3. This indeed has to do with the *nature* of the created, but with an aspect of its essence to which it still does not have access. The proof is the Transfiguration. Christ's body is already, before the Resurrection, of a spiritual nature, as is all flesh. But this glorious nature is not visible. Recall that, for Byzantine iconography, every icon of Christ is

First of all we are impelled to totally reject Bultmann's thesis by which, in Karl Barth's words, "Jesus is risen in the kerygma,"[1] which is to say that kerygma, the proclamation of faith in Christ, is itself the only locus where Christ is truly alive for us, the only acceptable meaning that we can give to the Pascal affirmation. Basically, this thesis makes kerygma, makes faith the true miracle, the only sign, the only symbol. We should reject it because it itself begins by rejecting the text and the objective proclamation of scriptural fact. Bultmann does not deny that revealed data, what we would readily call symbolic data, exerts a veritable hermeneutic. To the contrary, it is even a constant of his doctrine that kerygma, and especially the kerygma of the Resurrection, "opens up to us the possibility of understanding ourselves."[2] However, it is not the kerygma as such that interprets for us; it is an already-interpreted kerygma, a kerygma demythologized by Bultmann's reductive hermeneutics. Thus, in reality, we do not even possess the right to let the word of God be heard in us, such as it was actually proclaimed. Undoubtedly, they will retort, but is it possible to proceed otherwise when the meaning of this word completely calls into question our ontology of reference, when it demands a "forced *sacrificium intellectus*"?[3] Yes, this is possible, provided that we understand that what is at stake is not only 'self-understanding', but also an 'understanding of the real'. Now what is said in the Paschal kerygma is simply the following: the relationship of the person to the body that presentifies it is completely transformed; we pass from a presence endured to a presence mastered.

What about the relationship that maintains the person with the body, through the mediation of which the person is present to the world of bodies and, in general, to what is called historical and common reality? This relationship is dual, if not even contradictory. On the one hand, the person is the basis for presence; every presence is that of a subject spiritual in nature, free, master of its own acts and especially of that

first an icon of the Transfiguration: "In former times, every iconographer-monk began his 'divine art' by painting the icon of the transfiguration" (Paul Evdokimov, *The Art of the Icon: A Theology of Beauty*, trans. Fr. S. Bigham [Redondo Beach, CA: Oakwood Publications, 1990], 299). And the pictorial symbol is in fact painted, as Evdokimov tells us, "not so much with colors as with the Taboric light," that is with the light of the Holy Spirit which imbues all flesh.

1. Cited by Walter Kasper, op. cit., 132. Bultmann's thinking on the Resurrection is expressed in *The New Testament and Mythology*, 36–41.

2. Ibid., 39; likewise "The Relation between Theology and Proclamation," in *Rudolf Bultmann: Interpreting Faith for the Modern Era*, 239, where he explains that faith is a "self-understanding"; to understand the *New Testament* is to understand oneself.

3. *New Testament and Mythology*, 3.

essential act, the act of presence. Its 'being-there' is only through the person; there is a *there* only if *someone* is there. On the other, however, the person can 'be there' only if his presence is mediated through his body. Now, insofar as this body is in space, it dispossesses the person of his presence, since the body, through its objective 'surface', limits spiritual subjectivity irremediably and, powerless, surrenders it to the gaze of all. To the very extent that a person is present bodily, he is inevitably and involuntarily absent. He is ex-posed, posed outside himself, inseparably attached to what he can not entirely inhabit, which, at death, necessarily becomes then a simple object, a corpse. Merleau-Ponty says that to see an object is to go around it. To be corporeally visible is to *be seen*, then, from wherever one is not, to be *surrendered* for inspection, to be affected by an *essential passivity*, to yield indefinitely to an ontological impotence. Such is corporeal life, a sensory mode that claims to secure the reality of its object. 'My' bodily presence is 'my' visibility, but 'my' visibility is precisely not 'mine'. Unbeknownst to us and without being able to do anything about it, it belongs to all possible on-lookers. No one is master of his bodily presence, that is to say of what is identifiable, able to be experienced, knowable of the person within the framework of historical and common reality. To be present in this world *is* precisely not to be the master of it.

The Resurrection effects a complete reversal of this endured relationship. Having done this it accomplishes the most profound desire of the spiritual person; far from doing violence to him, it realizes the perfection of his nature. The body, through which the entire world is corporified, and starting from which is deployed, for the person, the indefiniteness of the non-personal, the indefiniteness of objectifying and exteriorizing limitations, this body is fully infused by the spirit. Enveloped by the spirit, it becomes a mode of the spirit and entirely at the disposal of the spirit's will. This body literally becomes a *corporeal act*. Its passivity has disappeared; it is *no longer seen*, it is made to be seen. Through this total mastery of its presence, all of its relationships with the rest of the corporeal world are equally transformed. It eludes every power of the world over it. Not that it is 'above' the sensible world, unless in a figurative sense (but real in view of the imperfections of any formulation); quite simply, this spiritual body is no longer *subject* to the conditions of the ordinary world. Its corporeal presentification is entirely dependent upon the spiritual reality of the person, a presentification that it can bring about as freely as our thinking, in its ordinary state, can produce any number of concepts or none at all. We are not saying that such a spiritual body, if it corresponded to a real possibility, could no longer be precisely identifiable in the sensible world. For, if this were so, it would mean that the sensible world is entirely closed and impervious to the spiritual world, whereas it is exactly the reverse: that which can do more can do less.[1] Spiritualized, the body is by no means evanescent, nor does it disappear; for this is the realization of its true nature, because the body is an integral part of the essence of a person—otherwise, the fact that we are corporeal beings would be devoid of sufficient reason; our bodies would be second-hand cast-off clothes. What has risen is therefore clearly the very body of Christ, not another body coming inexplicably from elsewhere. It is this body, finally reestablished in its true essence, which will become not another but itself, as proven by the transfiguration on Tabor, because, we repeat, the body *is also the person* according to one of its dimensions. Nor are we saying, as is more and more the tendency today, that this body is visible only to the eyes of faith. Surely something true is expressed by this, since Christ is absolutely no longer visible as an object, for he has quite simply ceased being an object. By his death He has 'exhausted' the radical *passivity* of his corporeal objectivity (this is entirely the meaning of the *Passion*, which is essentially the bringing-about of corporeal passivity and therefore its transformation into corporeal act). But He also shows Himself to those who do not have faith, the most striking example of which is St. Thomas. It must only be said then that the risen Christ manifests Himself to whomever He wishes, and besides this is how it is expressed in Scripture: "Him God raised up the third day, and gave him to be made manifest, not to all the people, but to witnesses preordained by God" (Acts 10:41–42). And this is also why the tomb is empty, why there should no longer be any corpse; for the corpse is only the proof and ultimate realization of the essential passivity of our bodily presence, or again of the heterogeneity of the fleshly form with respect to the spirit. But this heterogeneity has been conquered, 'exceeded'; the symbol has been restored.

1. Thus, it is not because a being lives in a three-dimensional space that it cannot move in two dimensions. Quite the contrary. Yet the converse is not true. As already stated, for a bi-dimensional being a treasure is hermetically sealed in a rectangle, and it will never understand how to cross this uninterrupted and perfectly sealed enclosure. Another example: it is not because thought is spiritual and non-spatial by nature that it cannot be present in space. Quite the contrary. The words of a sentence are in space and thought can be present in each or all of them together. Transcendence is not absence, it is rather *the possibility of a non-conditioned presence*. As for the spiritual body, the more technical details have no place here. We will point out however that we are not dealing with the subtle (or psychic) body, which is not immortal in the full sense of the term, and is related rather to the phantom body, a possibility excluded by St. Luke (24:39). We have explained at some length the cosmological doctrine of the three bodies in *Amour et Vérité*, 77–88.

Conclusion: *O vere beata nox!*

We are now fully aware of the demands of the hermeneutic principle. We have set them out in their totality; they alone show forth the true nature of the symbol. That western thought refuses to accompany us this far is only too evident. And yet we believe there is no other way out. Either we recognize in the symbol this semantic operator that initiates the alchemy of the sensible world and its assumptive reintegration into its intelligible archetype, that saving sign of which Christ's Resurrection is as if its crucial and decisive figure; or else we abandon ourselves and will be more and more abandoned to the disintegration of reason and to the indefinite pulverization of the world of phenomena. Only the symbol, as unification, as dynamic conjoiner, the energy of *yoga*, that is to say of 'junction', can check the expansion of the cosmos toward its annihilating exteriority and lead it back to its subjacent and interior unity. It is on this that everything turns. Either we are converted to the symbol so that it draws us in its wake towards an anagogical ascension, into the glory of the world "as it finally is in itself," or we are projected (and this is hell) into the sheer dispersal of difference.

But it is not only against the backdrop of three centuries of scientific materialism and agnostic rationalism that we have to rediscover an understanding of the symbol; it is also against the backdrop of its rejection and forgottenness on the part of those who should have remained faithful to it. What an example Christian (Protestant or Catholic) 'lack of understanding' sets which, rejecting every symbolic vision of the cosmos, fancies itself, through total blindness, advancing towards the concrete, the substantial and the solid! They tell us we must have a sense of the Incarnation. But *this* sense of the Incarnation is only the sense of a corpse—for what is flesh without the indwelling of the Word? And so, reduced to itself, this 'concrete' universe is only graveyard decay. Do they think that they have established the reality of a sacred fact because they have enclosed it in the narrowest circle of its existential conditioning? Who does not see that time, space, and bodies are indefinitely divisible, apt to disintegrate, to vanish? Who does not see that to reinforce conditions is to increase this conditioning and therefore limitation, that is to say the nothingness that encompasses and haunts reality? Who does not see that to lead everything back to a tangible and observable order, for fear of seeing things fly away to Heaven, is to condemn them to their own finiteness, to their own decomposition? History is brandished as a definitive argument, and looks down with disdain on all those ancient peoples engulfed by what is imaginary, deprived like cripples of the magic criterion of reality. But history is strictly *nothing* without the symbol and myth presentified in it, that is, without the God who reveals himself therein and establishes its unity and meaning.

And so modern Christianity, too, is invited to this conversion of the intellect to the symbol, which basically consists of a change in our way of seeing the real. As long as hermeneutics will not question its own conception of the reality of things, as long as it does not understand that, by its very existence, the symbol upsets the world of common experience, shakes it to its so-called foundations, overthrows it, and returns it to the side of the invisible, the trans-spatial and the semantic, the 'apostasy or cultural neurosis' alternative will not be resolved.

Such is the most radical message addressed to us by the symbol. But we will understand it in all its truth only if we hear it as it is pronounced in the glorious immanence of divine creation.

...Day is done. The sun has set. Then, in the Paschal night, the joy of the *Exultet* ascends: "*O vere beata nox*..., O truly blessed night, which alone deserves to know the time and hour when Christ rose again from hell. This is the night of which it is written: And the night shall be as light as the day, and the night is my illumination in my delights... O truly blessed night... A night, in which heaven is united to earth, and God to man!"

SYMBOLISM & REALITY
The History of a Reflection

Each created being is a symbol
instituted not by the arbitrariness of men,
but by the divine will,
to render visible the invisible wisdom of God.
～ Hugh of Saint Victor

Originally published in French as
Symbolisme et réalité, Histoire d'une réflexion

Introduction

Every thought has its own history. Doubtless the *more geometrico* explanation of a philosophy (such as Spinoza's *Ethics*) lets us experience its logical coherence by disclosing its objective structure to the reader's eyes. But we all know that things did not happen in just this way. No thought is worked out in the field of principles and axioms alone: every philosophy is a dialogue, an effort to answer certain questions posed to the philosopher by the cultural climate of his time, in which the climates of previous times are tallied up and summarized. Surely, any philosopher worthy of the name always aspires to discover truth in its timeless essence and, after a fashion, tends to sum up the history of his thought with this truth. But this aspiration for the timeless is carried out by those conceptual means provided by his times and according to his own speculative preoccupations; and so this aspiration is open to the dialogue between the present and the future, or perhaps may also be condemned to oblivion by them.

This is why it seemed opportune to retrace the genesis of my reflections on sacred symbolism; not for its autobiographical interest, but so to render its discourse more intelligible, by showing to exactly which questions this reflection has tried to bring answers. As will be seen, these questions are also posed to all people today, at least insofar as they are religious and strive to cling to the truth of the Christian revelation, especially under its Catholic form. In actuality, by our scientific certainties and their accompanying mental attitudes, all of us find it extremely difficult to believe in the truth of those sacred facts related to us by the Old and New Testaments. My entire philosophical advance has sprung from the conviction which has imposed on me the duty to speculatively accept this formidable challenge.

Science and Revelation

In 1950 Pope Pius XII decided to proclaim the dogma of the Assumption, affirming that "Mary, after having completed the course of her earthly life, was raised up body and soul to heavenly glory." We can see in this proclamation the final dogmatic message of the Catholic Church, since it concerns She who was at the origin of the Christian revelation and since it celebrates the ultimate event of her earthly existence. But this message was poorly received. At the time I was studying philosophy at Nancy and, among the young Christians of my circle, incredulity reigned. "Speak of the ascension of her soul to heaven if you will," they said, "but the ascension of her body.... What does that mean? Or then is this a matter of a symbol?" Oddly enough, far from shocking me, this criticism elicited from me what seemed a self-evident response: beyond the divisions and oppositions of analytical reason stands the truth of the real, one within itself, inseparably both historical and symbolic, visible and invisible, physical and semantic. This self-evident response rested upon a kind of direct and sudden intuition in which was revealed, obscurely but without any possible doubt, the ontologically spiritual nature of the matter of bodies, without for all that casting any doubt on the reality of their corporeity. What this perception bestowed upon my understanding was truly real, but only corresponded to a mode of this reality (corporeity) which included other modes by definition invisible, although even more real. In other words, what most held as scientifically unthinkable seemed cosmologically possible to my intuition. What I disavowed, what I held to be impossible, was that the reality of Mary's Assumption was exclusively of the historical order or exclusively of the symbolic order. I dismissed both the fundamentalist literalists who, to save the truth of revelation (Scripture or Tradition), reject all symbolism, and the demythologizing 'figurists': by denying all presence of heavenly reality in the earthly, the otherworldly in the world, the divine in the human, and therefore every miracle, they exile the religious dimension outside history and human existence, and transform into mere figures, into simple 'manners of speaking', the traditional expressions of faith.

It was in fact clear that both shared the same idea of history and symbolism. What distinguished them concerned only their relationship to faith. For the first—fideists at heart—commitment to faith implied an exclusive commitment to the letter of the text; otherwise faith would risk disappearing. For the others—rationalists at heart—commitment to faith (a faith emptied of all content and reduced to its own affirmation) would imply rejection of the letter of the text and its unconsciously mythical nature; otherwise faith would risk becoming impossible. But for both the field of the created world would have to be identical to the material order, and would necessarily show itself in the form of physical beings and historical events in a space/time setting, while the symbolic would only make sense as something unreal or a fictional substitute for the real. To admit that certain events of sacred history did not happen and develop in the way that we have been told, or that they obey semantic necessities and not the physical necessity of the conditions of our sensory world alone, is, for the literalist, to deny their existence. Obviously these fideists cannot tell us how events contrary to all physical laws might have happened, and they invariably appeal only to the miraculous. To this naive fideism corresponds the learned fideism of demythologizing rationalism. According to Bultmann, it is precisely because sacred facts and miracles are physically impossible and theologically false, and we—being actually unable to believe in them—are forced, in order to save our faith, to interpret them as simple figures of religious discourse which betray, while cosmologizing it, the truth of faith: understood in this way, religious mythology appears for what it is, a collection of symbolic figures. As we see, both share in one and the same conviction: the real and the symbolic are mutually exclusive.

Exegesis and Symbolism

Obviously, I knew that figurists and literalists only represented a twin minority tendency on the margins of Christianity. At least this was the case with the Roman Church at the time. But during the following thirty years I saw, to the contrary, the winds of demythification swell immeasurably and change into a tornado threatening to carry away all of the Catholic faith before it; even today its destructive power is far from exhausted. But, in particular, if the edifice of Christian thought had been built upon a principle other than that of the mutual exclusion of the real and the symbolic, it could have resisted the Bultmano-Modernist tornado. Alas, this was not at all what happened! Since the sixteenth century, and above all from the end of the seventeenth, the most official Catholic (and Protestant) exegesis, repudiating the spiritual sense of Scripture—counter to Patristics and the Middle Ages—has been exclusively engaged in proving the historicity of what, in the Divine Word, might be reasonably detached from the mythical. Nevertheless, as could be foreseen, the biblical terrain of historicity has been, from Richard Simon and Dom Calmet down to our own day, continually reduced. The result is unmistakable: we have an immense accumulation of knowledge, but at the cost of suffocating the sacred text under the minutiae of a contentious erudition; as for the spiritual sense, a shallow moralism and pious sentimentality have been charged with taking its place. Thus one renowned theologian declared speculations on the symbolism of the Cross to be backward and superfluous: crucifixion was nothing but the "cruelest and most shameful" mode (Cicero) of being put to death according to Roman law; under another law Christ would have been hanged, burnt, drawn and quartered or beheaded. Besides, it is important to insist that this Cross is made of two pieces of wood, and that is its sole reality; every significance that can be attributed to its form is adventitious and without any connection to the physical existence and structure of this death-dealing instrument.

Now an irrefutable philosophical demand forbids me to acquiesce without examining the principles of this exegesis. How do they not see that, in their textual studies, they display a critical rigor which they quite unconsciously abandon when dealing with the results of physics? For, ultimately, is not this exegesis basically determined by one's ideas about the order and substance of the world? And what are these ideas if not the inheritance of a scientific materialism already outmoded around 1900, at the moment when first Poincaré and then Einstein announced their initial theoretical outlines of Relativity, and when the observation of radioactivity would soon lead Bohr and Heisenberg to call back into question the idea of both matter and physical reality. Of this at least I was absolutely certain—I mean of the *duty* imposed on a philosopher by the post-Galilean scientific revolution: the idea of the world inherited from Galileo and Newton was completely undone in its space/time framework (the theory of Relativity) as well as in its material contents (Quantum theory). And even if some deem it necessary to end up with maintaining a materialist vision, was it not at least legitimate to do so only at the conclusion of an equally complete critical examination? And so, I thought, is it not foolish to hear Bultmann declare that we cannot turn a radio dial and believe that Jesus has ascended into Heaven at the same time, while at that time he himself, in some respects even more than his teacher Heidegger, continued to think about science under the form of an entirely obsolete ideology? How can we ignore that he rejected "New Testament mythology" in the name of a scientific concept which is itself only a myth? This *lack of cosmological awareness* is clearly even much more accentuated among the majority of the other exegetes and theologians who are, generally, far from possessing Bultmann's intellectual breadth.

Physics and Symbolism

Thus I was led to rediscover the undeniable truth of Platonism and even of a certain Aristotelianism, the only speculative keys allowing us to think intelligibly about the reality of the real. In this I was aided not only by my Christian faith and the Platonic orientation of my early education, but also by reading René Guénon, who opened my eyes to an understanding of oriental metaphysics, and finally by the teachings of Raymond Ruyer, the greatest 'natural philosopher' of our times. And in fact, to the basic cosmological question—"What is the reality of the physically real?"—materialism is incapable of providing the least response. By wanting to reduce, as do so many so-called exegetes and theologians, the historico-physically real to its space/time conditioning and to material substantiality, far from saving it, far from assuring its consistency, far from rooting it in being and guaranteeing its objective existence, they positively destroy it, since they subject it to fragmentation, to the indefinite divisibility of space/time and matter. If the Cross of Christ is *only* made of two pieces of wood, then it vanishes in the scattering and inconstancy of an electron mist.

Should we, for all that, embrace the idealism which certain physicists do not hesitate to profess? Here again, Ruyer taught me to recognize the same materialist 'hang-up'. The realism/idealism alternative only makes sense provided that a reality other than matter does not exist: in that case, yes, since matter shows itself incapable of establishing the reality of the real, we have to resign ourselves to idealism. But what if modes other than the material one exist? What if the soul, as the dynamic and non-spatial unity, as the 'form-organizer' of bodies, of *all* bodies, is also a reality, even more real than the body since it would assure it of its consistency and permanence with the passing of time? And what if the essence, as intelligible, transpatial and transtemporal unity, were also a reality even more real than soul[1] and body, since the molecule-essence (of water or gas), lion-essence, oak-essence survive all transformations, all destructions? And is it not precisely to the extent that each human body, each gas or water molecule, each lion or oak tree participates in that essence, is 'informed' by it (in the Aristotelian sense) and actualizes it here and now in our world, that each corporeal being is real, consistent and objective?

And what then basically is a being or thing of our world, if not a 'mode of presence' of the essence or the archetype in our space/time continuum, a mode under which we can know this essence, 'see' it, become aware of it then and perhaps re-ascend to it? Obviously none of these modes of presence are apt to relinquish the archetype as such, in its own being. But still a true presence is involved: the archetype does not hover over physical beings like a cloud; it 'inhabits' the thing itself, endowing it with the being that it is, letting it subsist in the very bosom of becoming, impermanence and indefinite limitation. Nevertheless, because existence in this world implies being subject to those existential conditions which define it, this presence (or immanence of the archetype) cannot be total or plenary: the essence itself is only in the order of essence, which is to say in God. This is why corporeal reality is only a *mode* of presence, a 'modalized' or limited presence, and hence, in certain respects, also an absence.

1. In the sense of a 'natural mental form'.

Thus all beings, all realities are at once an archetypal prophecy (or revelation)—inasmuch as they actualize a mode of *presence*—and an archetypal reminiscence (or memorial)—inasmuch as every *mode* implies a certain absence of what it modalizes: this is why every created being heralds the archetype it manifests and summons us, by the remembrance which it awakens in us, to re-ascend to it.

Nature and Culture

Having arrived at this point of my reflection, and convinced of the soundness of my conclusions, I observed that, in short, by their 'prophetico-anamnestic' function, the beings of creation are identical to symbols: in fact every symbol is a sign which, on the one hand, 'presentifies' or renders present (and not only represents) the reality signified, and which, on the other, by the same token reveals its absence among us, and is thus calling out to us to rejoin it.

Having done this, having found creature and symbol to be identical, I clearly saw that I had crossed from the natural order into the cultural order, since, although creation is to be first encountered in nature, the symbol is *first* encountered only in human cultures, particularly in the discourse of religions and those rituals and artistic works which illustrate it. But precisely this passage from nature to culture was, in some way, only the verification of the opposite trajectory which had led me from culture to nature. For I began with the Christian discourse and those sacred facts disclosed in it. Meditating on the rationalist negation of these facts, I shed light on the materialist ontology upon which it was based and the scientific ideology which inspired it, both of which have been incontestably refuted by the theory of relativity and, even more, by quantum physics. They should be abandoned. Symbolism had led me to physics and physics, in its turn, led me back to symbolism. But, in this exchange, symbolic reality and physical reality were transformed: the symbol was no longer a fictitious substitute, an arbitrary sign, a simple being of reason, since it exercised a presentifying function and was, in a certain manner, identical to the reality it symbolized; conversely, physical reality was no longer only, inasmuch as it is real, a pure 'being-there', a pure object, a simple contingent datum, the *en-soi*, the *in-itself* impenetrable to my understanding and which is real, truly existent, only on the condition of repelling every attempt of the understanding to grasp it; physical reality was itself inhabited and formed in its subsistence (*subsistentia*) by an essence, a 'semantic form', whereas its sensory appearance, its mode of manifestation, was only its symbol.

The symbol revealed itself to me then as the place of passage and conversion from nature to culture, since its physical elements (forms, colors, sounds, substances, gestures, etc.) have been inserted in a significance-bestowing process on the basis of an analogical correspondence between the physical nature of the symbolizing element and the metaphysical nature of the symbolized reality. But, conversely, within this process culture also seemed to be converted to nature, since that which is a work of the mind and therefore mental in nature, a sign, assumes visible forms, is made flesh and becomes a (mysterious) thing among other things. Thus, the passage from the natural order of physics to the cultural order of the symbol is found to be clearly marked in the very essence of the symbol: by effecting this passage philosophically, by affirming the ontologically symbolic dimension of every creature, I was only making myself conform to this essence. I was justified then in defining my ontology as a *symbolic realism*, which means that it is the notion of the symbol which lets us *think* the notion of reality.

Sense and Symbol

This point, however, demanded that I pause here. The symbol, as I said, is the key to ontology—this much is certain. But, in order to play this role, it first needs to be noticed as a symbol, for a model is only a model provided that it be distinguished from that of which it is a model. It therefore needs to be acknowledged as an entity of a particular nature which is neither a being of the world—in which case it would be eliminated from the general class of existing things which we ordinarily experience—nor a simple concept, a purely mental entity, a thought—in which case it would be eliminated from the general class of psychological events of which our soul is the seat. Let us venture further. We can become aware of the existence of this particular category of entity that is the symbol (solely along the lines by which we can adequately think about what a being is) only if we have an experience of the signs which refer to nothing of the directly knowable according to common experience, without for all that referring to a simple fiction, but which to the contrary points to the only true Reality. Every sign which refers to something naturally existing, whether externally or internally observable (the outline of a tree, the words 'tree' or 'concept'), can be replaced by the reality designated: this is called the *referent* of the sign. Thus it loses, by being absorbed in the referent, its own existence as sign. Such a sign is called 'transitive'; we hardly pay any attention to it. This is the case with words in plain speech, signs used in writing, road markers: awareness passes through them with only their referent in mind.

Likewise, when the referent of a discourse is imaginary (novels, fairy tales, etc.) the sign as such is absorbed, surely not by the referent (which should not be substituted for it), but by the fabulary function, by something that is entirely deferred to the functioning of the sign itself. This is the reason why the fable is called 'fable' (etymologically: 'what can be said,' 'the sayable'), and is simply nothing but a playing with the powers of language—here the signifying function absorbs the sign.

What is left then is religious discourse, a discourse in which the pure symbol can be 'discovered': it alone speaks of something that the speaker and the spoken-to holds to be perfectly real, but which cannot be shown or imagined, or in any way substituted for the sign. It might be objected that religious discourse does not present us with pure symbols alone. Indeed, it often speaks of sacred facts, of saving events, which certainly have a symbolic sense, but which also have—and this according to the clear intention of the discourse itself—a historical sense, as in the crossing of the Red Sea, the Burning Bush and Christ's Ascension. The text which speaks of them first tells us *what has happened*, and it is this what-has-happened that possesses (or is possessed by) a symbolic sense. When I read a particular text in faith, I am not yet experiencing the symbol, and therefore I have not yet awakened to an awareness of symbolism. Clinging to the reality of the event spoken of by the text, I am led to this reality by the text itself, the signs which compose it being forgotten as soon as they are understood; the deciphered signs are obliterated during the accomplishment of the signifying function.

But when the text proclaims: "In the beginning was the Word," or: "who is seated at the right hand of God"; or when it speaks of "the tree of the knowledge of good and evil," then we find ourselves confronted by signs in some way *intransitive*. They clearly have an altogether real referent (for a Christian), but one which we cannot observe in any way here below, not even in the imagination. The only *place* where we can touch on these referents is the sign itself, and it is by dwelling in it, inhabiting it and listening to it, not by passing through it and forgetting it, that we might have some chance of knowing this referent, insofar as this is possible. Such an intransitive sign, which refers to nothing in the world, is a pure symbol, not in the sense of a fictional substitute, since nothing can be appropriately substituted for it, but in the sense that it is offered to us in this very world as a residual entity, an entity reducible to nothing whatsoever of this world. To experience such an entity is truly to gain access to the con-sciousness, the 'shared knowledge' of symbolism, and therefore to gain access to the intelligibility of being.

Clearly, the possibility of such an experience is only offered by religious discourse. This discourse is in fact the only one to give an instance of a knowledge, the knowledge of faith, which consists in adherence to a symbol, that is to say to 'knowing in the symbol' what we cannot know in any other way (according to the usual order of our knowledge), and yet for which the symbol is only a mode of presentification: "*who has seen me has seen the Father*," Christ says in St. John (13:9). This means: I myself, the pre-eminent Symbol, I alone am the visibility of the Father, the only mode by which the Father can be visibly present; but also: who sees *Me*—it is not I myself that is seen, it is the Father! Here, in the mystery of the sacred sign, is accomplished the marvelous alchemy of the visible and the Invisible. Far from being negated or obliterated, the visible is established in its loftiest dignity, corporeal matter has been affirmed and honored in its original glory. But this is only so on the condition that it is not reduced to its cosmic accidentality, to its contingency, to its vaporous limitations. It is the gaze of faith looking out on what is seen. What it does see is not seen by passing through and beyond its appearance, but by contemplating it, entering into it, being united with it.

The Key of Knowledge

Far from being opposed to philosophical discourse, religious discourse offers it that model by which it can think about cognitive activity. For, as Plato teaches, to know is to know what is—this at least is the truth of knowing; the knowledge of what is not is ignorance. And yet our senses only present us with changing realities, which belong then not to being but becoming, so that we can neither impugn the reality of the cosmos, nor find within it the wherewithal to satisfy our need for being. This is why we have to *learn* to grasp the identity of being within the alterity of becoming, to perceive essence in existence as that which accounts for the reality of the real, that which bestows on it its subsistence (*subsistentia*, *hyparxis*), and, however generously it gives itself to be known by phenomena, nevertheless remains secret and absent in its very presence. This apprenticeship to true knowledge is first made at the school of religion, in the adherence of faith to those symbols which it proposes to us.

But this is not all. Not only does religious discourse initiate us into the knowledge of essences through the experience of sacred symbols; not only does it teach us to 'descend' back into the sensory world, or, rather, to reverse perspectives, exchanging the within for the without and the without for the within; it leads us into a discovery of the sensory *within* the intelligible itself, as reposing within it, existence finally reconciled and reintegrated into its essence. Now metaphysical knowledge demands such a reversal. For ultimately, if the intelligible is truly the essence of the sensory, if the spirit is truly the essence of the body, this may be actualized to a certain extent and experienced in a certain manner. The metaphysical hope for a reality that will finally be itself, for an existence that will finally be what it is, for a tree that will finally be a true tree, for a rose that will finally be a true rose, "such that within itself eternity finally transforms it," the hope for a world that will finally be good, just as God saw it on the day of its creation: this hope cannot be completely and definitively deceived. What the philosopher wants is everything, the one and the other, and the one in the other and the other in the one, total and perfect knowledge; what he indefectibly expects is the finally accomplished realization of that promise inscribed in the very substance of his intellect, the hope of a being which is pure light, an obscurity which is transparency, since only in this way can I be united with what I know.

Actually it is not enough to know that which is; we also need to be that which we know, in default of which the so-called known being, remaining foreign and external, would remain unknown. But how are we to be what we know, if what we know, the that-which-is, is found to be absent from our own presence here below, incompatible with our bodily reality, and if it can only be attained in 'spirit', indirectly and by reflection in the mental mirror (*through a glass in a dark manner*, says St. Paul [1 Cor. 13:12])? Do we need to be either abstracted by ecstasy or made absent through the death of our own existence to unite intelligible with intelligible? This would spell the ultimate dashing of the philosophic hope. If the corporeal, the limited, the finite, the transitory, the contingent and the historical are definitively excluded from the wedding-feast of knowledge, then it would be completely unintelligible, completely unknowable, and all philosophy vain, which would mean that human intelligence—philosophy being simply its highest activity—is vain. What is the intellect in its most profound nature if not the 'instinct for the Real', the hope for total Being? An intellect which knows itself should have no doubts about this. If, therefore, the perfection of knowledge requires the unity of knowing and the known (it is, says Aristotle, their common act), it also and necessarily requires the unity of knowing as such with the known as such. If that unity which knowledge should realize always entailed the separation of the knowing being from the being known in a sensory and intelligible way, in short, if knowledge were only *abstract* and *abstractive*, then this would truly be a mutilation and a tearing-apart.

It is therefore necessary that the philosophic hope be realized to a certain extent, in default of which either an angelist or a materialist dualism (it is all the same) would triumph and, with it, the empire of indefinite death. We need to have bestowed on us the grace of a corporeal reality which, in its very flesh, is entirely *assumed* by the Spirit which dwells within it, informs it and incorporates it in being. An existence transparent to the glory of the essence is needed, a matter so totally ransomed, so totally integrated with the spiritual Act which gives it being, that it ceases to remain outside of It, to ex-ist (*ex-sistere*), that it finds itself wholly united to It, a flesh altogether and without remnant united to its Word: in short, a body without a corpse.

Jesus Christ: Symbol and Savior

The revelation of Christ, of His incarnation, redemption and ascension into heaven, is then the realization of the philosophic hope. It is so within the distinctiveness of a saving manifestation which, in the midst of time and space, unifies being and triumphs over all contingency, reveals itself as the transfiguring nucleus of creation. Undoubtedly, the universe of people and things, seen to its full extent, still awaits its assumptive glorification, in an end of times and spaces which will also be their full realization. But at least it is enough for the philosophic hope that creation is glorified in the "first-born of every creature," so that it is in each and all virtually, to the extent of their good will, and so that every intellect is at least able to know that the flower of knowledge has ripened and that "its fruit is good to see and to eat." For not only does religious discourse affirm the event/advent of the body's glory in a point of human history and on the earthly globe, it also proposes its saving diffusion to the entire world, through the mediation of the Church, under the form of the consecrated Bread and Wine become the Body of Christ, by which our own bodies can be fed and transfigured. And since chewing is the very symbol of knowledge, by eating of the Bread/Body I know through my body and *realize* the perfection of knowledge.

I therefore saw understanding and faith, philosophy and religion, reason and revelation reconciled to each other. They were not reconciled by reducing one to the other or vice-versa, but because religious faith accorded the philosophic intellect the realization of its hopes and needs, while the intellect accorded to faith the light and meaning by which it might be illuminated and understood. In Mary, ascending into heaven in her glorious Assumption, I contemplated existence "become one single flesh" with its essence; I contemplated human nature crowned with its eternal archetype.

The Asphyxiation of the Intellect

There is almost nothing about what has just been said in the two works which I have already devoted to the symbol: *Histoire et théorie du symbole* and *The Crisis of Religious Symbolism*. The reason is simple. The reflection which I have explained above is rightly concerned with the metaphysics of sacred symbolism under its triple—ontological, noetic and ritualic—aspect. However, for it to be accepted by the Western *intelligentsia*, this metaphysic demanded something besides the testimony of a youthful intuition. It demanded prolegomena intended, not to prove its truth—which should not be the object of a demonstration in the mathematical sense of this term—but to establish its *legitimacy* with respect to critical reasoning, to establish the *right* which this metaphysic has to be accepted as such.

To appeal to modern man in the realm of thought is, in fact, an appeal to critical reasoning; it is this that governs philosophical activity. And to legitimize a doctrine before the tribunal of reason is to show its (relative) necessity. But such a necessity can only be established negatively: I know that such and such a function is necessary to a being's life only when its disappearance entails the death of this being; and the function itself is identified only when the cutting-away of an essential organ entails the being's demise. For example, I can be unaware that (and this is so for the modern world) the metaphysics of religious symbolism is necessary for the life of our intellect; but if, having proceeded with the cutting-away of this organ that the sacred symbol is, I observe that this cutting away entails the death of reason itself by "semantic asphyxiation" (that is, by the inability to inspire *Meaning*, which is the life of our mind), then critical reason will be constrained to recognize that such a metaphysics cannot be disqualified a priori (as philosophically unacceptable), and that it ought to accept the possibility of its truth, although it cannot (and never will be able to) take into consideration the particular symbolism of such or such a religion. To rationally explain any given symbolism would be equivalent to reducing *mythos* to *logos*, thereby rendering it useless: what is completely explained disappears in this very explanation. And even though I might clearly understand the need for symbolism, I can only be apprised of a symbolic system (the sum of a religion's symbols).

Clearly, my thesis is that the symbol not only 'gives food for thought', but even endows thought with its very self. The irreducible duality, with respect to the human mind, of symbol and understanding, revelation and intellection, is the very source of the life of the mind and the unsurpassable horizon of its activity. Although no symbol completely actualizes the intellect, which therefore, in itself, in its pure superconscious essence, surpasses every horizon. And the intellect's recognition of this duality as the source of its life is precisely the only way it has here below of integrating and unifying this very duality: it is the ritual life which brings the unity of being and knowing.

And yet, before showing that the emptying-out of the *mythos* and sacred symbols entails the asphyxiation of the *logos*, I first had to say what I understood by 'symbol'. I would have preferred to avoid this methodological obligation, but it was imposed on me by the sudden appearance of a major cultural event, unnoticed by me until then, even though it had totally permeated some of the human sciences: Structuralism.

The Crisis of Religious Symbolism

Hence *Histoire et théorie du symbole*, my initial book on the symbol, a work in which, after having rejected the analyses and theses of Structuralism, I elaborated the theory of the symbolic sign by reflecting on the word 'symbol' (*symbolon*) examined in its historical development, and by the philosophical shaping of its concept according to the demands of sacred symbolism. I was then ready, after this technical elaboration, to subject my prolegomena to the metaphysics of the symbol. My self-appointed task was the following: to induce modern reason to consent to the need for this metaphysic by showing it the list of dilemmas into which the rejection of sacred signs had led it. This was my second book, *The Crisis of Religious Symbolism*, about which I will say a few things. This book tells of a passion, the passion suffered by the western soul when critical reason undertook to purify humanity's cultural consciousness by ridding it of its religious representations. I resolved to induce a cultural 'kenosis' in the believing intellect, a kenosis modeled on Christ's Passion, He who was "made sin for us" (2 Cor. 5:21). I decided to accompany the major atheists of the modern world, Feuerbach, Marx, Engels, Freud, Lévi-Strauss, etc., in their negations of the symbolic forms of the divine, so to cancel out these negations by revealing the unperceived contradiction in their make-up.

For to deny that sacred forms are messages of the Transcendent necessarily makes of them simple unconscious products of human consciousness. True, the origin of these products varies from author to author; and yet it always comes down to a process of alienation. Now, that consciousness can hide within itself an illusory power in the very make-up of its cognitive act, a consciousness which is therefore only one with this very act, and which totally falsifies it without being able to recognize it—this is a completely contradictory thesis. On the one hand it affirms that alienation of consciousness is universal and necessary, since the hidden principle of its origin resides in the structural situation of this consciousness. On the other hand it presupposes, in order to simply formulate a thesis that makes sense, that the one who upholds it is not affected by this vice, that his own consciousness has not been falsified, and therefore that this illusion is not structural and universal, in other words that it is not an *illusion* but a simple accidental error, which cannot but occur and which, henceforth, should explain this ineluctable religious alienation. Thus the revealers of alienated consciousness, inexplicably and miraculously, exempt themselves from the universality of alienation. Such a pretense demands, for its justification, the privilege of a divine revelation transcending all structural conditioning, and this exalts its supporters as veritable but impossible prophets, since their revelation consists precisely in declaring all revelation an illusion, like someone who shouts: "Speech does not exist!"

Modern criticism reposes then on a scandalous imposture. Surely, I am not disputing an atheist's right to undertake a demonstration of the non-existence of God, a demonstration to which a believer could respond with a refutation. But both should be made on the common basis of a necessary reliance on the *logos*, on its ability to attain to what is true, and therefore on its *unalterable rectitude*. This is what I have called the "semantic principle." The intelligence, or intellective consciousness, can err, be mistaken and even momentarily and accidentally deceived, but it cannot be completely falsified in its essential cognitive act, and no one should suppose, in order to account for some error, that this is necessarily so. Certainly this is a psychological possibility: one can, using various means, *persuade* men that their consciousness is totally upside-down—this is called "brainwashing"—but this is logically and ontologically impossible. There always comes a time when some people realize that the very act by which someone wants to persuade them that they were or are 'fools' contradictorily presupposes that there is in them a sane intelligence, capable of understanding this discourse. One cannot brainwash a chair, a pumpkin or the extremely feeble.

Demented Cosmology

Such then was the end-result of the 'critique of symbolic reason' undertaken by modern thought. For the Structuralists and their successors it inferred the dismissal of reason and the end of 'logocentrism'. But this critique was, in fact, only the philosophical result of the unprecedented scientific and cosmological crisis unleashed by Galilean physics. And this is where I began my study, since it is this which, destroying the mythocosm of antiquity and the Middle Ages, has rendered sacred symbols illusory by making the cosmic presentification of the celestial realities which they are meant to disclose impossible.

Having done this, I seemed to have 'exhausted' the analysis of the cultural roots of contemporary atheism; and I thought that I had arrived at the end of my philosophical prolegomena, however imperfect the execution. But it was then that, retracing my philosophical route from its starting point, there suddenly appeared before me the key to all of my reflections on the symbol. Yes, I should have perceived it from the start. And yet I had to reach the conclusion of this critique of sacred symbolism to become aware of it—how clear it is that the simplest truths sometimes take the longest time to master. Hence the last part of *The Crisis of Religious Symbolism.*

Parodying Marx's famous formula in his eleventh thesis on Feuerbach ("the philosophers have only [*seulement*] interpreted the world in different ways, what matters is to *change* it"), I declared: until now the hermeneuts have sole-ly [*seule-ment*] interpreted the symbol in different ways in terms of the reality it designates (or gives the lie to); what matters now is the understanding that, more radically, it is not (common) reality which interprets the symbol, but the symbol which compels us to interpret this reality, to see it otherwise than under the reductionist appearance which it has assumed in our eyes, and to go beyond it.

If in fact, as I have shown, the existence of sacred symbols is refractory to all rational explanation, what can be the attitude of a philosophical understanding confronted by these monsters, monsters whose disquieting strangeness is admitted to the extent that reason expels them from itself and purifies itself of them? The only attitude is one of a 'conversion to symbols'. But what is to be said about this?

To be converted to symbols is first of all to become aware of how the symbol is monstrous only with respect to our ordinary ideas about physical reality, according to the laws of which the symbol should not be ontologically real, although it too may be allowed to *exist* in the order of cultural representations: inexhaustible and unimpeachable, that is what a symbol is. As a consequence, to be converted to the symbol is also to observe that the symbol is not a simple passive support for an interpretive act, but that, through its unimpeachable and inexhaustible cultural essence, it puts commonplace reality, the world of ordinary experience, in doubt, threatens it in its reassuring banality. To be converted to the symbol is, in this case and ultimately, to accept to follow it into putting the real in doubt, to accept to enter with it into the metaphysical conversion of the real to which it welcomes us, to be open to the transfiguration of the world's flesh of which it is the prophetic witness and the salvific allure. For it is said: *heaven and earth will pass away, but my words will not pass away*; and also: *behold I make all things new.*

Being and Entities

Such is the history of a meditation pursued by one Christian philosopher for nearly fifty years. What remains now is to write this metaphysics of the symbol, a project which took shape in me so long ago and some themes of which I have outlined. It could be placed under the banner of symbolic realism, a term by which I designate all that forms the depths of my philosophical conviction: there is, in the intellect, an innate sense of being or of the real as such (and therefore also of what is not); but all we ever have is just the experience of such or such a reality. Without this experience, the sense of the real would not awaken me to self-awareness, and the intellect would be ignorant of its own ontological nature. And yet no experience fully gratifies the being's desire, a desire inherent to its intellective goal. Being only a limited (determined) presentation of being, a single reality (one existing among others) can give no more than an ontological promise, a being-symbol of true Being, of the truly Real. By this is recognized and justified what there is of the incontestable in Heidegger's analysis: to be, true Being should not be identical to single being, to an entity. But far from being the locus of Its forgetting, beingness is the occasion of Its revelation. And this revelation is dual. On the one hand it awakens the intellect to its ontotropic essence, which means that the intellect discovers in the ontic experience (or the experience of entities) the transcendent nature of its own ontological goal. In other words, the intellect discovers itself as the sense of and the desire for Being as such, and not only a grasp of such or such an entity: it perceives within, in its own life, an intention that surpasses the natural order of entities, an order to which it is not exclusively ordained. On the other hand, ontic experience is less the grasp of the entity itself than the discovery of its own ungraspability. Every objectively real being is an objection. What is is what resists me. Therefore I experience the entity's being after the manner of that which, within it, evades me: for an entity, that which is in-itself is that which is-not-for-me. This is the experience-limit of a 'beyond' of my intellective goal, the paradoxical encounter with what impedes my insight and exhausts it. Hence every judgment of existence is implicitly based on an obscure intuition of being which is also, and necessarily, a kind of act of faith. For nothing can prove that such or such a thing exists: God alone is demonstrable.

Conclusion

Now, among all of the world's realities, there is only one, the Word made flesh, in which Being as such, the supremely Real, has been made visible. Within It the ontological symbolic-ness of earthly realities from being implicit become explicit (*who has seen me has seen the Father*), and something more than explicit: saving. In and through this reality, not only is the prayer of the intellect granted beyond all measure and gratified beyond all desire; it is also the human being who has been delivered from his or her own contingency, not because it has been annihilated, but because it has been justified. The relativity of created being has been converted and transfigured into a relationship of love: we are, in the dereliction of our creaturely state, those whom the Son of God has loved and for whom He has given His life. The blood of the Cross is the justification of the world.

True, this unique reality of the Word made flesh is known and recognized only by faith. In its presence, the natural intellect enters into a kind of night, a kind of passion. The Being that it seeks in all things, the inaccessible transcendence of which it experiences in each one of these things, bestows Itself on it only in a renunciation mortifying to its own light, its own insight, although it knows, in its own depths, that this passion is with a view to its resurrection in the light of glory. And this is why entry into faith can only be conveyed and sustained through love, at least for that which belongs to the very depths of its lived reality. But as for the conscious form of its supernatural commitment to the recognition of Christ, the intellect should further understand that, having done this, it by no means negates what was the rule of its natural progress, since, to the contrary, it is accomplishing it in truth: it crosses in fact from the implicit faith of judging existence to the explicit and voluntary faith in the human revelation of the One who is. The act of faith is no longer lived then as the awareness of the ontotropic mystery which the intellect bears within it, or as the obscure intuition of 'what is not for me', but as the mode of *knowledge* in which the intellective light will accede to its ultimate blazing-forth. Thus can philosopher and Christian commune in unique and dual faith: in the unalterable rectitude of human intelligence and in the truth of the sign of God incarnate in Jesus Christ.

Nancy, 7 December 1990
and 24 *February* 1997.

INDEX OF NAMES

INDEX OF NAMES

CPSIA information can be obtained at www.ICGtesting.com
Printed in the USA
LVOW08*2339150716

496555LV00006B/29/P

9 781621 381921